Promoting Sustainable Fisheries

Legal Aspects of Sustainable Development

General Editor

David Freestone

This series will publish work on all aspects of the international legal dimensions of the concept of sustainable development. Its aim is to publish important works of scholarship on a range of relevant issues including conservation of natural resources, climate change, biodiversity loss and the role of international agreements, international organizations and state practice.

VOLUME 6

Promoting Sustainable Fisheries

The International Legal and Policy Framework to Combat Illegal, Unreported and Unregulated Fishing

By

Mary Ann Palma
Martin Tsamenyi
William Edeson

MARTINUS
NIJHOFF
PUBLISHERS

LEIDEN • BOSTON
2010

This book is printed on acid-free paper.

Library of Congress Cataloging-in-Publication Data

Palma, Mary Ann.
Promoting sustainable fisheries : the international legal and policy framework to combat illegal, unreported and unregulated fishing / by Mary Ann Palma, Martin Tsamenyi, William Edeson.
p. cm. -- (Legal aspects of sustainable development, 1875-0923 ; v. 6)
Includes bibliographical references and index.
ISBN 978-90-04-17575-4 (hardback : alk. paper)
1. Fishery law and legislation. 2. Fishery management, International. 3. Sustainable fisheries. I. Tsamenyi, M. (Martin) II. Edeson, W. R. (William R.) III. Title.
K3895.P35 2010
343'.07692--dc22
2010006370

ISSN 1875-0923
ISBN 978 90 04 17575 4

PRINTED IN THE NETHERLANDS

Table of Contents

Series Editor's Preface

I am pleased to welcome this volume by the team from the Australian National Centre for Ocean Resources and Security (ANCORS) at the University of Wollongong. It is the sixth volume in the Martinus Nijhoff monograph series on *Legal Aspects of Sustainable Development* published under my general editorship. The aim of this series is to publish works at the cutting edge of legal scholarship that address both the practical and the theoretical aspects of this important concept.

As we move into the second decade of the twenty-first century it is no exaggeration to say that global fish stocks continue to be in crisis. The crisis is precipitated not simply by the further depletion of stocks but also by major – maybe fundamental – defects in fisheries governance. For a considerable number of years the Food and Agriculture Organization of the UN (FAO) has been warning of the state of the world fisheries. Its biennial publication, *The State of the World Fisheries and Aquaculture* tells us that only 25% of global fish stocks are underexploited or moderately exploited, the remainder are fully exploited, overexploited or depleted.

The international community has at last begun to take this seriously. The UN General Assembly has become engaged; new legal and policy instruments have been developed, and new initiatives to assess and reform existing management bodies have been launched. But it is all taking a long time. In the meantime however the most pernicious defects of the fisheries management system persist. Among these is the huge incidence of 'illegal, unregulated and unreported' fishing. Each of these three practices is of course a distinct and separate problem. Taken as whole however, the significance of IUU fishing is that perhaps as much as one third of all the fish caught globally are caught without regard to the fisheries management systems, however flawed they may be, that we have in place. It is a fundamental challenge to sustainable fisheries.

This important volume documents both the measures that states have taken individually and together to address this systemic problem. These include legal and policy measure to close *lacunae* in the system but also practical measures to increase the effectiveness of enforcement. The vessels involved in IUU activities are owned by someone, are registered somewhere and fly a national flag.

I am confident that this volume will make a major contribution to a wider informed understanding of the legal responsibilities that such activities involve and the measures being taken to enforce these responsibilities.

David Freestone
Washington DC

Foreword

The fight against IUU fishing was launched by Australia in 1999 at the FAO Committee on Fisheries, in response to increased illegal activity in the Southern Ocean targeting Patagonian toothfish. In developing an international campaign against this activity we quickly realised that there were three core elements to the problem: illegal activity, unregulated activity and the under reporting of catch and effort. These elements I, U and U impacted differently on countries. While Australia and other developed countries had the commitment, assets and the financial and legal capacity to deal with the issue in their waters, developing countries who lacked the assets, were at the mercy of unscrupulous operators and large international companies who cared little for the social disruption that IUU fishing caused and were focused solely on individual profits.

As the debate matured the international community began to develop and implement a toolkit of legal and voluntary instruments to mitigate IUU fishing. These tools have included the FAO IUU Plan of Action, improved port state controls, boarding and inspection regimes in Regional Fisheries Management Organisations (RFMOs) and catch documentation schemes (CDS). What has also emerged and is somewhat unique in international fisheries, has been the regional and international collaboration between countries and in particularl within law enforcement agencies to bring some of the major illegal operators to court. Some good examples of this co-operation can be seen in the agreement between Australia and France on Southern oceans surveillance and in the development of the Regional Plan of Action to combat IUU fishing developed by ten South East Asian countries with a strong interest to mitigate a common problem.

Individually, countries passed rules and regulations to restrict the trade in IUU product without appropriate documentation. The US Lacey Act has proven to be one of the most effective domestic tools to deal with this issue.

This book makes a valuable contribution to continue this long term fight against IUU fishing. The book covers not only the extent and impact of IUU fishing but puts into context the development of the international legal and policy framework that has been developed to deal with this ongoing issue. The book will provide valuable assistance to governments and policy makers who are forced to deal with this issue and will be equally useful for RFMOs who have the responsibility to ensure that the world's migratory and shared fish stocks are sustainable.

The authors of this book are associated with the Australian National Centre for Ocean Resources and Security (ANCORS) at the University of Wollongong in Australia. This centre under the guidance of its Director Professor Martin Tsamenyi has rapidly emerged as a key centre for the development of oceans policy and maritime law particularly as it relates to fisheries and fisheries management through RFMOs.

Congratulations on a fine publication that will no doubt find its rightful place with policy makers and law enforcement officials engaged to combat the ongoing and destructive practice of IUU fishing.

Professor Glenn Hurry
Chief Executive officer
Australian Fisheries Management Authority
Canberra

Authors

Mary Ann Palma

Dr Mary Ann Palma is a Research Fellow at the Australian National Centre for Ocean Resources and Security (ANCORS), University of Wollongong, Australia. She holds a Bachelor of Arts in Public Administration at the University of the Philippines, Master of Marine Management at Dalhousie University in Canada, and a Doctor of Philosophy at the University of Wollongong. She also obtained a Diploma on the Law of the Sea from the Rhodes Academy of Oceans Law and Policy, Greece. Dr Palma specialises in international fisheries law and policy, particularly in addressing illegal, unreported and unregulated fishing. She undertakes researches on IUU fishing in Southeast Asia, the Pacific, and Australia and has a number of publications on the subject. Dr Palma also participates in bilateral fisheries cooperation between the Philippines and Indonesia and the meetings of the Western and Central Pacific Fisheries Commission as a member of the Philippine delegation.

Martin Tsamenyi

Professor Martin Tsamenyi is the Director of the Australian National Centre for Ocean Resources and Security, University of Wollongong, Australia. Professor Tsamenyi holds a Bachelor of Laws degree from the University of Ghana and Master of International Law and Doctor of Philosophy degrees from the Australian National University. Professor Tsamenyi is a world recognised expert in international fisheries law and policy and provides advice to a number of governments, international organisations, non-governmental organisations, and the fishing industry on the implementation of international fisheries laws and regulations. Professor Tsamenyi previously served as Fisheries Law Adviser to the Pacific Islands Forum Fisheries Agency and is currently the Legal Adviser to the Secretariat of the Western and Central Pacific Fisheries Commission.

William Edeson

Professor William Edeson is a Professorial Fellow at the Australian National Centre for Ocean Resources and Security, University of Wollongong, and a Professorial Visiting Fellow at the University of New South Wales Law School. Professor Edeson was Senior Legal Officer at the Food and Agriculture Organization (FAO) of the United Nations where he was involved in the preparation of the FAO Compliance Agreement, the Code of Conduct for Responsible

Fisheries, the International Plan of Action to Prevent, Deter, and Eliminate Illegal, Unreported, and Unregulated Fishing (IPOA-IUU), and the Reykjavik Declaration on Responsible Fisheries in the Marine Ecosystem. He was also an adviser to numerous governments in Southeast Asia, Caribbean, Western Pacific, Indian ocean, and southern and western Africa on law of the sea and fisheries matters. He has also advised several regional bodies on international legal aspects of fisheries, including the Ministerial Conference on West African Fisheries, CARICOM, Organization of Eastern Caribbean States, Forum Fisheries Agency, Western and Central Pacific Fisheries Commission, South West Indian Ocean Fisheries Commission, Western Central Atlantic Fisheries Commission, General Fisheries Commission for the Mediterranean, Commission for the Conservation of Southern Bluefin Tuna, and the Indian Ocean Tuna Commission. Professor Edeson is a Member of the editorial board of the International Journal of Marine and Coastal Law and is author of numerous articles on the law of the sea and international law. He also co-authored a book *Legislating for Sustainable Fisheries: A Guide to Implementing the 1993 FAO Compliance Agreement and 1995 UN Fish Stocks Agreement*, The World Bank, 2001 (with David Freestone and Elly Gudmundsdottir).

List of International Instruments

Agenda 21, Adopted at the United Nations Conference on Environment and Development, Rio de Janeiro, Brazil, 03–14 June 1992

Agreement Establishing the World Trade Organization, Marrakesh, concluded on 15 April 1994, in force 01 January 1995, 1867 UNTS 154; 33 ILM 1144 (1994)

Convention on Biological Diversity, concluded on 5 June 1992, in force 29 December 1993, 1760 UNTS 79; 31 ILM 818 (1992)

Convention on Fishing and Conservation of Living Resources of the High Seas, 1958, Geneva, Switzerland, concluded on 29 April 1958, in force 20 March 1966, 17 UST 138; 559 UNTS 285

Convention on International Trade in Endangered Species of Wild Fauna and Flora, Washington D.C., USA, concluded on 3 March 1973, in force 1 July 1975, 27 UST 1087; TIAS 8249; 993 UNTS 243, amended at Bonn, on 22 June 1979 and Gaborone on 30 April 1983

Convention on the Conservation of Migratory Species of Wild Animals, Bonn, concluded on 23 June 1979, in force 1 November 1979, 19 ILM 15 (1980); ATS 1991/32; BTS 87 (1990), Cm. 1332

Convention on the High Seas, Geneva, Switzerland, concluded on 29 April 1958, in force 30 September 1962, 13 UST 2312; 450 UNTS 11

Convention on Wetlands of International Importance especially as Waterfowl Habitat, Ramsar, Iran, concluded on 2 February 1971, in force 21 December 1975, 996 UNTS 245; TIAS 11084; 11 ILM 963 (1972), as amended by the Paris Protocol, 3 December 1982 and Regina Amendments, 28 May 1987

Food and Agriculture Organization (FAO), *Agreement on Port State Measures to Prevent, Deter and Eliminate Illegal, Unreported and Unregulated Fishing*, Adopted in November 2009, Appendix V of the FAO Council, Hundred and Thirty-seventh Session, Rome, 28 September-02 October 2009, Report of the 88th Session of the Committee on Constitutional and Legal Matters (CCLM), 23–25 September 2009, *CL 137/5,* September 2009

FAO, *Agreement to Promote Compliance with International Conservation and Management Measures by Fishing Vessels on the High Seas*, Rome, Italy, concluded on 24 November 1993, in force 24 April 2003, 2221 UNTS 91

FAO, *Code of Conduct for Responsible Fisheries,* Adopted at the 28th Session of the FAO Conference, Rome, Italy, 31 October 1995

FAO, *International Plan of Action for Reducing Incidental Catch of Seabirds in Longline Fisheries* (IPOA-Seabirds), Rome, Italy, 1999

FAO, *International Plan of Action for the Conservation and Management of Sharks* (IPOA-Sharks), Rome, Italy, 1999

FAO, *International Plan of Action for the Management of Fishing Capacity* (IPOA-Capacity), Rome, Italy, 1999

FAO, *International Plan of Action to Prevent, Deter and Eliminate Illegal, Unreported and Unregulated Fishing* (IPOA-IUU), Rome, Italy, adopted on 23 June 2001 at the 120th Session of the FAO Council

International Convention on the Harmonised Commodity Description and Coding System, Brussels, Belgium, concluded on 14 June 1983, in force 1 January 1988, 1503 UNTS 167

International Labour Organization (ILO), *Accommodation of Crews (Fishermen) Convention, 1966 (C126)*, concluded on 21 June 1966, in force 06 November 1968

ILO, *Convention Concerning Work in the Fishing Sector (C188)*, Geneva, Switzerland, adopted on 14 June 2007, www.ilo.org

ILO, *Fishermen's Competence Certificates Convention, 1966 (C125)*, concluded on 21 June 1966, in force 15 July 1969, www.ilo.org

ILO, *Maritime Labour Convention,* 23 February 2006, 45 I.L.M. 792 (2006); ILO, Report I(1B), International Labour Conference, 94th (Maritime) Sess. (2006)

ILO, *Medical Examination (Fishermen) Convention 1959 (C113)*, Croatia, concluded on 19 June 1959, in force 07 November 1961, revised by C118, www.ilo.org

ILO, *Minimum Age (Fishermen) Convention, 1959 (C112)*, concluded on 19 June 1959, in force 15 July 1969, www.ilo.org

International Maritime Organization (IMO), *International Convention on Standards of Training, Certification and Watchkeeping for Fishing Vessel Personnel (STCW-F)*, London, concluded on 7 July 1995, Select Documents on International Affairs No. 43 (1995) 12, p. 148. www.austlii.org.

IMO, *Torremolinos Protocol of 1993 relating to the Torremolinos International Convention for the Safety of Fishing Vessels 1977*, Torremolinos, Spain, concluded on 2 April 1993, www.imo.org

United Nations Agreement for the Implementation of the Provision of the United Nations Convention on the Law of the Sea of 10 December 1982 Relating to the Conservation and Management of Straddling Fish Stocks and Highly Migratory Fish Stocks, New York, concluded on 04 August 1995, in force 11 December 2001, 34 ILM 1542 (1995); 2167 UNTS 88

United Nations Convention Against Transnational Organised Crime, concluded on 15 November 2000, in force 29 September 2009, 40 ILM 335 (2001); UN Doc. A/55/383 at 25 (2000); UN Doc. A/RES/55/25 at 4 (2001)

United Nations Convention on the Conditions for the Registration of Ships, Geneva, Switzerland, concluded on 07 February 1986, UN Doc. TD/RS/CONF/19/Add.1; 26 I.L.M. 1229 (1987)

United Nations Convention on the Law of the Sea, Montego Bay, Jamaica, concluded on 10 December 1982, in force 16 November 1994, 1833 UNTS 3; 21 ILM 1261 (1982)

Vienna Convention on the Law of Treaties, Vienna, Austria, concluded 23 May 1969, in force 27 January 1980, UN Doc. A/Conf.39/27; 1155 UNTS 331; 8 ILM 679 (1969); 63 AJIL 875 (1969)

World Trade Organization (WTO), *Agreement on Import Licensing Procedures,* Geneva, concluded on 12 April 1979, in force 1 January 1980, 1186 UNTS 372

WTO, *Agreement on Pre-shipment Inspection,* http://www.wto.org/english/docs_e/legal_e/21-psi_e.htm

WTO, *Agreement on Rules of Origin,* http://www.wto.org/english/docs_e/legal_e/22-roo.pdf

WTO, *Agreement on Subsidies and Countervailing Measures,* concluded on 15 April 1994, 1867 UNTS 14, GATT Doc. MTN/FA Ii-A1A-4, 33 ILM XXX

WTO, *Agreement on Technical Barriers to Trade,* Geneva, concluded on 12 April 1979, in force 1 January 1980, 1186 UNTS 276

WTO, *Agreement on the Application of Sanitary and Phytosanitary Measures*, concluded on 15 December 1993, in force 01 January 1995, GATT Doc. MTN/FA Ii-A1A-4, 33 ILM XXX

WTO, *General Agreement on Tariffs and Trade 1994*, concluded on 15 April 1994, in force 1 January 1995, 1867 UNTS 187; 33 ILM 1153 (1994)

List of Cases

Case Concerning Delimitation of Maritime Areas between Canada and the French Republic (St Pierre and Miquelon), 31 ILM 1149 (1992); 95 ILR 645

European Community, Judgment of the Court (Fifth Chamber) of 25 April 2002, *Commission of the European Communities v French Republic*, Joined cases C-418/00 and C-419/00, *European Court Reports 2002*, at *I-03969*

International Tribunal for the Law of the Sea (ITLOS), *Case Concerning the Conservation and Sustainable Exploitation of Swordfish Stocks in the South-eastern Pacific Ocean (Chile v. European Community)*, Case No. 7, Constitution Chamber Order of 20 December 2000

ITLOS, *The "Camuoco" Case (Panama v. France) (Prompt Release),* Case No. 5, Judgment of 07 February 2000

ITLOS, *The "Monte Confurco" Case (Seychelles* v. *France), Prompt Release,* Case No. 6, Judgment of 18 December 2000

ITLOS, *The M/V "SAIGA" Case (Saint Vincent and the Grenadines* v. *Guinea), Prompt Release,* Case No. 1, Judgment of 4 December 1997

ITLOS, *The M/V "SAIGA" Case (Saint Vincent and the Grenadines* v. *Guinea),* Case No. 2, Judgment of 1 July 1999

ITLOS, *The "Volga" Case (Russian Federation v. Australia) (Prompt Release)*, Case No. 11, Judgment of 23 December 2002

New Zealand, District Court at Wellington, *Ministry of Fisheries v Tukunga,* CRI-2005-085-6992 & 6989, Judgment of 30 March 2007

New Zealand, High Court of New Zealand, Wellington Registry, *Omunkete Fishing (Pty) Ltd v Minister of Fisheries*, CIV 2008-485-1310, Judgment of 01 July 2008

World Trade Organization (WTO), Environment: Disputes 4, *Mexico, etc. v. U.S.: 'Tuna-Dolphin'*, (1991)

WTO, *United States– Import Prohibition of Certain Shrimp and Shrimp Products*, AB-1998-4, Report of the Appellate Body, *WT/DS58/AB/R*, 12 October 1998

List of Abbreviations

AALC	Asian-African Legal Consultative Committee
ACAP	Agreement on the Conservation of Albatrosses and Petrels
ACCOBAMS	Agreement for the Conservation of Cetaceans of the Black Sea, Mediterranean Sea and Contiguous Atlantic Area
ACFR	Advisory Committee on Fishery Research
ACP	Africa Caribbean Pacific Group of States
AEWA	African-Eurasian Waterbird Agreement
AFZ	Australian Fisheries Zone
AIC	Australian Institute of Criminology
AIDCP	Agreement on the International Dolphin Conservation Program
ALC	automatic location communicators
APEC	Asia Pacific Economic Cooperation
APFIC	Asia Pacific Fisheries Commission
ASCOBANS	Agreement on the Conservation of Small Cetaceans of the Baltic and North Seas
ASEAN	Association of Southeast Asian Nations
BOBP-IGO	Bay of Bengal Programme Inter-governmental Organisation
CARPAS	Regional Fisheries Advisory Committee for the Southwest Atlantic
CBD	Convention on the Conservation of Biological Diversity
CCAMLR	Commission for the Conservation of Antarctic Marine Living Resources
CCBSP	Convention on the Conservation and Management of the Pollock Resources in the Central Bering Sea
CCLM	FAO Committee on Constitutional and Legal Matters
CCSBT	Commission for the Conservation of Southern Bluefin Tuna
CECAF	Fishery Committee for the Eastern Central Atlantic
CFFA	Coalition for Fair Fisheries Arrangements
CIFAA	Committee for Inland Fisheries for Africa
CITES	Convention on International Trade in Endangered Species of Wild Flora and Fauna
CMS	Convention on the Conservation of Migratory Species of Wild Animals
COFI	FAO Committee on Fisheries
COMHAFAT	Ministerial Conference on Fisheries Cooperation among African States Bordering the Atlantic Ocean
COPESCAL	Commission for Inland Fisheries of Latin America

COREP	Regional Fisheries Committee for the Gulf of Guinea
CPPS	Permanent Commission for the South Pacific
CTMFM	Joint Technical Commission for the Argentina/Uruguay Maritime Front
CWP	Coordinating Working Party on Fisheries Statistics
DCD	*Dissostichus* Catch Document
DFQF	duty free quota free
EC	European Community
EEZ	exclusive economic zone
EIFAC	European Inland Fisheries Advisory Commission
EJF	Environmental Justice Foundation
EPA	Economic Partnership Agreements
EU	European Union
FAD	fish aggregating device
FAO	Food and Agriculture Organization
FCWC	Fishery Committee of the West Central Gulf of Guinea
FFA	Forum Fisheries Agency
FSI	IMO Sub-committee on Flag State Implementation
GATT	General Agreement on Tariffs and Trade
GFCM	General Fisheries Commission for the Mediterranean
GHP	Codes of Good Hygiene Practices
GMP	good manufacturing practices
GPS	global positioning system
GRT	gross register ton
GSP	Generalised System of Preferences
HACCP	Hazard Analysis Critical Control Point
HSVAR	High Seas Vessels Authorization Record
IATTC	Inter-American Tropical Tuna Commission
ICCAT	International Commission for the Conservation of Atlantic Tunas
ICES	International Council for the Exploration of the Sea
ILO	International Labour Organization
IMCO	Inter-Governmental Maritime Consultative Organization
IMO	International Maritime Organization
IOTC	Indian Ocean Tuna Commission
IPHC	International Pacific Halibut Commission
IPOA-Capacity	International Plan of Action for the Management of Fishing Capacity
IPOA-IUU	International Plan of Action to Prevent, Deter and Eliminate Illegal, Unreported and Unregulated Fishing
IPOA-Seabirds	International Plan of Action for Reducing Incidental Catch of Seabirds in Longline Fisheries

IPOA-Sharks	International Plan of Action for the Conservation and Management of Sharks
IRCS	International Telecommunication Union Radio Call Signs
ISO	International Organization for Standardization
ITLOS	International Tribunal for the Law for the Sea
ITU	International Telecommunication Union
IUU	illegal, unreported and unregulated fishing
IWC	International Whaling Commission
kW	kilowatt
LOSC	United Nations Convention on the Law of the Sea
LVFO	Lake Victoria Fisheries Organization
MCS	monitoring, control, and surveillance
MEPC	IMO Marine Environment Protection Committee
MOU	Memorandum of Understanding
MRAG	Marine Resources Assessment Group Ltd.
MRC	Mekong River Commission
MSC	Marine Stewardship Council
MSC	IMO Maritime Safety Committee
MSY	maximum sustainable yield
MTU	mobile transceiver unit
NACA	Network of Aquaculture Centres in Asia-Pacific
NAFO	Northwest Atlantic Fisheries Organization
NAMMCO	North Atlantic Marine Mammal Commission
NASCO	North Atlantic Salmon Conservation Organization
NEAFC	Northeast Atlantic Fisheries Commission
NMFS	U.S. National Marine Fisheries Service
NOAA	U.S. National Oceanic and Atmospheric Administration
NORAD	Norwegian Agency for International Development
NPAFC	North Pacific Anadromous Fish Commission
NPOA	National Plan of Action
NPOA-IUU	National Plan of Action to Prevent, Deter, and Eliminate Illegal, Unreported and Unregulated Fishing
OECD	Organisation for Economic Co-operation and Development
OIE	World Organisation for Animal Health
OLDEPESCA	Latin American Organization for Fisheries Development
PICES	North Pacific Marine Science Organization
PNA	Parties to the Nauru Agreement
PSC	Pacific Salmon Commission
QA	quality assurance
QC	quality control
QM	quality management

RECOFI	Regional Commission for Fisheries
RFMO	regional fisheries management organisation
RPOA	Regional Plan of Action
SADC	Southern African Development Community
SCM	subsidies and countervailing measures
SCOI	CCAMLR Standing Committee on Observation and Inspection
SEAFDEC	Southeast Asian Fisheries Development Centre
SEAFO	South-East Atlantic Fisheries Organisation
SIDS	small island developing States
SIOFA	Southern Indian Ocean Fisheries Agreement
SOLAS	International Convention for the Safety of Life at Sea
SPC	Secretariat of the Pacific Community
SPRFMO	South Pacific Regional Fisheries Management Organisation
SPS	sanitary and phytosanitary
SRFC	Sub-regional Fisheries Commission
STCW-F	International Convention on Standards of Training, Certification and Watchkeeping for Fishing Vessel Personnel
SWIOFC	Southwest Indian Ocean Fisheries Commission
TBT	technical barrier to trade
UN	United Nations
UNAFEI	United Nations Asia and Far East Institute of Crime Prevention and Treatment of Offenders
UNCED	United Nations Conference on Environment and Development
UNCLOS I	First United Nations Conference on the Law of the Sea
UNCTAD	United Nations Conference on Trade and Development
UNEP	United Nations Environment Programme
UNGA	United Nations General Assembly
UNICPOLOS	United Nations Open-ended Informal Consultative Process on Oceans and the Law of the Sea
UNICRI	United Nations Interregional Crime and Justice Research Institute
U.S.	United States of America
USD	United States dollars
VMS	vessel monitoring system
WCO	World Customs Organization
WCPFC	Western and Central Pacific Fisheries Commission
WCPO	Western and Central Pacific ocean
WECAFC	Western Central Atlantic Fishery Commission
WHO	World Health Organization
WSSD	World Summit on Sustainable Development
WTO	World Trade Organization
WWF	World Wildlife Fund

Chapter One

Introduction

In most parts of the world marine fisheries have always been the desired source of protein and play a crucial role in maintaining food and economic security. The abundance of marine fish and decline of freshwater fish, rapid population growth, increase in demand for fish, development of urban markets, long distance trade, and introduction of new technologies have all contributed to the expansion of fishing operations from the 11th century.[1] Today, the development of fisheries continues and the sector has become more globalised than ever—from vessel ownership and control, to the different elements of the fisheries chain such as fish production, processing, transportation, trade and consumption. Increased refrigeration and processing technology enable fisheries products to reach remote destinations. Fishing vessels are no longer tied to fishing grounds around a port.[2]

The rapid growth and globalisation of the fisheries sector have transformed not only global fishing patterns and operations but also the existing framework for fisheries management in a number of ways. First, the perspective on fisheries management has changed from promoting the exploitation of existing and new fisheries to achieving the objective of optimum utilisation and long term sustainable use of fisheries resources.[3] Second, the development of fisheries has necessitated an international approach to management which considers broader

[1] J.H. Barrett, A.M. Locker, and C.M. Roberts, 'The Origins of Intensive Marine Fishing in Medieval Europe: the English Evidence', 271 *Proceedings of the Royal Society B* (2004), at 2417–2421; J.H. Barrett, A.M. Locker, and C.M. Roberts, 'Dark Age Economic Revisited: The English Fish Bone Evidence AD 600–1600', 78 *Antiquity* (2004), at 618–636; J.H. Barrett, 'Archaeo-ichthyological Evidence for Long-term Socio-economic Trends in Northern Scotland: 3500 BC to AD 1500', 26 *Journal of Archaeological Science* (1999), at 353–388, *as cited in* Callum Roberts, *The Unnatural History of the Sea* (Washington: Island Press, 2007), at 18 and 21.

[2] John Rood and Michael G. Schechter, 'Globalisation and Fisheries: A Necessarily Interdisciplinary Inquiry', in William W. Taylor, Michael G. Schechter, and Lois G. Wolfson, (eds.), *Globalisation: Effects on Fisheries Resources* (Cambridge: Cambridge University Press, 2007), at 6.

[3] United Nations Convention on the Law of the Sea, Montego Bay, Jamaica, concluded on 10 December 1982, in force 16 November 1994, 1833 UNTS 3; 21 ILM 1261 (1982). *Hereinafter referred to as* LOSC, Preamble; Food and Agriculture Organization (FAO), *Code of Conduct for Responsible Fisheries,* adopted at the 28th Session of the FAO Conference, Rome, Italy, 31 October 1995. *Hereinafter referred to as* FAO Code of Conduct, Art. 7.1.1; *Declaration of Cancun*, Adopted at the International Conference on Responsible Fishing, Cancun, Mexico, May 1992. *See also* Annex, Chapter 2 of the Report of the World Commission on Environment and Development: Our Common Future, in UN Secretary General, Forty-second Session, Development and International Co-operation: Environment, Report of the World Commission on Environment and Development, *A/42/427*, 04 August 1987.

conservation objectives that take into account marine habitats and ecosystems.[4] Third, sustainable fisheries management has been embraced by the entire fisheries sector; creating a responsibility for all relevant State and non-State actors to ensure that benefits accrue from sustainable fishing activities.[5] However, despite the adoption of a conservation-orientated approach to the management of fisheries resources, the deterioration of global fish stocks persists.

The current trend in the production of global marine fisheries resources presents an alarming concern for food security and sustainable development. Some of the fisheries resources which were previously regarded as inexhaustible are now either seriously depleted or overexploited.[6] World capture fisheries production decreased from 87 million tonnes in 2000 to 84 million tonnes in 2005.[7] According to the Food and Agricultural Organization (FAO), of the major marine fish stocks or species groups, 52 per cent are fully exploited, 17 per cent are overexploited, 25 per cent are underexploited or moderately exploited, and the remaining 6 per cent of the stocks are becoming significantly depleted.[8] In areas where stocks are considered depleted, the Northeast Atlantic, the Mediterranean and Black Seas, Northwest Atlantic, the Southeast Atlantic, the Southeast Pacific and the Southern Oceans have the greatest need for recovery.[9] In addition, it is predicted that the demand for fish may reach 180 million tonnes in 2030 and neither aquaculture nor any terrestrial food production could supplement the protein production provided by marine fisheries.[10]

---

[4] Reykjavik Declaration on Responsible Fisheries in the Marine Ecosystem, Reykjavik, Iceland, 13 November 2001; *See also* FAO Fisheries Department, The Ecosystem Approach to Fisheries, *FAO Technical Guidelines for Responsible Fisheries No 4, Suppl 2* (Rome: FAO, 2003); United Nations General Assembly (UNGA), Fifty-eighth Session, Item 53 of the Preliminary List, Oceans and Law of the Sea, Report of the Secretary General, *A/58/65*, 03 March 2003, para. 117; *See also* J.F. Caddy, and K.L. Cochrane, 'A Review of Fisheries Management Past and Present and Some Future Perspectives for the Third Millennium', 44 *Ocean and Coastal Management* (2001), at 653–682.

[5] *See* Organisation for Economic Co-operation and Development and Food and Agriculture Organization (OECD/FAO), *Globalisation and Fisheries: Proceedings of an OECD-FAO Workshop* (OECD/FAO, 2007), at 13 and 114.

[6] *See* Richard Ellis, *The Empty Ocean: Plundering the World's Marine Life* (Washington: Island Press, 2003), *cf* Thomax Huxley, 'The Fisheries Exhibition', 28 *Nature* (1883), at 176–177; Daniel Hawthorne and Francis Minot, *The Inexhaustible Sea* (New York: Collier Books, 1961). *See also* Helen M. Rozwadowski, *The Sea Knows No Boundaries: A Century of Marine Science Under ICES* (Seattle: ICES and the University of Washington Press, 2002).

[7] FAO, *The State of World Fisheries and Aquaculture 2006* (Rome: FAO, 2007), at 3.

[8] *Ibid.*, at 29–31. Examples of fully exploited or overexploited stocks are anchoveta (*Engraulis ringens*) in the Southeast Pacific, Alaska Pollock (*Theragra chalcogramma*) in the North Pacific, blue whiting (*Micromesistius poutassou*) in the Northeast Atlantic, Atlantic herring (*Clupea harengus*) in the North Atlantic, Japanese anchovy (*Engraulis japonicus*) in the Northeast Pacific, the Chilean jack mackerel (*Trachurus murphyi*) in the Southeast Pacific, and the yellowfin tuna (Thunnus albacares) in the Atlantic and Pacific oceans.

[9] FAO, *State of World Fisheries and Aquaculture 2004* (Rome: FAO, 2004), at 86.

[10] *Ibid.*

The decline in global fisheries resources has been attributed to a number of interrelated factors, such as industrialisation,[11] overfishing;[12] environmental factors affecting stock productivity;[13] open access nature of many fisheries;[14] overcapacity in the world fishing fleet;[15] provision of subsidies;[16] unreliable fisheries information;[17] and unsustainable fishing practices which include the use of non-selective fishing gear that adversely affect juvenile fish, dependent, and associated species.[18] The depletion of fisheries resources is also associated with the high levels of non-compliance by fishing vessels, often reflagging to open register States to avoid fisheries regulations;[19] failure of fishing authorities to set sustainable limits on the basis of scientific advice; and the reluctance of flag States to ensure compliance by their vessels with fishing regulations and fisheries management measures.[20]

Globalisation also presents challenges to the sustainability of fisheries.[21] Economic integration, the rise of new markets and expansion of fishing opportunities have increased pressure on fish stocks and affected the way States manage

---

[11] The first industrialisation is characterised by the mechanisation of capture, particularly through the use of vessels powered by steam and diesel oil. The second industrialisation started with the invention of stern factory trawler, power block purse seiner, radar, and acoustic fish finder. *See* D.H. Cushing, *The Provident Sea* (Cambridge: Cambridge University Press, 1988).

[12] FAO, *State of World Fisheries and Aquaculture 2004*, *supra* note 9, at 28; UNGA, Fifty-ninth Session, Item 51(a) of the Preliminary List, Oceans and the Law of the Sea, Report of the Secretary-General, *A/59/62*, 4 March 2004, para. 206; *See also* Lauretta Burke, Yumiko Kura, et al., *Pilot Analysis of Global Ecosystems: Coastal Ecosystems* (Washington, D.C.: World Resources Institute, 2001), at 53.

[13] FAO, *State of World Fisheries and Aquaculture 2004*, *supra* note 9, at 28.

[14] Daniel Pauly, Villy Christensen, Sylvie Guénette, Tony J Pitcher, U. Rashid Sumaila, Carl J. Walters, R. Watson and Dirk Zeller, 'Towards Sustainability in World Fisheries,' 418 *Nature* (2002), at 692.

[15] Toward the 2002 World Summit on Sustainable Development, Johannesburg, Ensuring the Sustainable Development of Oceans and Coasts, *Co-Chairs' Report from the Global Conference on Oceans and Coasts*, 03–07 December 2001, UNESCO, Paris, at 33; UNGA, Fifty-ninth Session, Item 50(b) of the Provisional Agenda, Oceans and the Law of the Sea, Sustainable Fisheries, including through the Agreement for the Implementation of the Provisions of the United Nations Convention on the Law of the Sea of 10 December 1982 relating to the Conservation and Management of Straddling Fish Stocks and Highly Migratory Fish Stocks, and Related Instruments, Report of the Secretary-General, *A/59/298*, 26 August 2004, paras. 17, 118, and 163.

[16] Subsidy leaders in the world are Japan at an estimated USD750 million in yearly subsidies and the European Union at USD500 million. *See* Douglas C. Wilson, 'Fisheries Management as a Social Problem', in Charles R.C. Sheppard, (ed.), *Seas at the Millennium: An Environmental Evaluation, Vol. III: Global Issues and Processes* (Amsterdam: Pergamon Elsevier Science, 2000), at 155.

[17] *UNGA A/59/62*, *supra* note 12, paras. 215–216.

[18] *Ibid.*, para. 209.

[19] *Toward the 2002 World Summit on Sustainable Development*, *supra* note 15, at 33.

[20] *UNGA A/59/62*, *supra* note 12, para. 206; *UNGA A/59/298*, *supra* note 15, para. 7; *See also* Dominique Gréboval, Report and Documentation of the International Workshop on Factors Contributing to Unsustainability and Overexploitation in Fisheries, Bangkok, Thailand, 4–8 February 2002, *FAO Fisheries Report No. 672* (Rome: FAO, 2002).

[21] The impact of globalisation on fisheries is a complex issue. There are both positive and negative impacts of globalisation. Positive impacts or benefits that may be derived from globalisation include

their resources. The inability of developing States to harvest resources in their exclusive economic zones (EEZ) compels them to enter into access agreements with distant water fishing nations for fees that generate comparatively insignificant amount of revenue compared to the value of the resources harvested.[22] Furthermore, the composite linkages within the fisheries sector and the interaction of the different stakeholders involved from fish production to consumption hinder States from ensuring responsible fishing practices at every stage of the fisheries chain. Other factors such as the general inaccessibility of international markets to small scale fisheries, continuous tariff escalation and imposition of non-tariff barriers to trade and the evolving standard-setting by consumers have also placed additional burdens on producers to meet the increasing demand for sustainably caught fish.[23]

1.1. Concept of IUU Fishing

Illegal, unreported, and unregulated fishing or "IUU fishing" is a term popularised by the FAO International Plan of Action to Prevent, Deter, and Eliminate Illegal, Unreported, and Unregulated Fishing (IPOA-IUU) adopted in 2001.[24] The term encompasses most of the issues identified above as contributing to the decline of global fisheries resources. According to the Report of the United Nations (UN) Secretary-General, IUU fishing is considered as "one of the most severe problems affecting world fisheries"[25] and the "main obstacle in achieving sustainable fisheries in both areas under national jurisdiction and the high seas."[26] IUU fishing is also regarded as one of the factors that can lead to the collapse of fisheries resources or that which can seriously affect efforts to rebuild fish stocks which have already been depleted.[27]

---

access to and diversification of markets, increase in intra-regional trade, access to technological improvements, increase in the contribution to foreign currency earnings, and increased competitiveness in international markets. On the other hand, the negative impacts of globalisation include increased production costs, high prices for tradable fish products in domestic markets, and difficulties to meet scales of production. *See* FAO, *Impacts of Globalisation*, www.fao.org. Accessed on 02 February 2009.

[22] Vlad M. Kaczynski and David L. Fluharty, 'European Policies in West Africa: Who Benefits from Fisheries Agreements?', 26 *Marine Policy* (2002), at 88.

[23] OECD/FAO, *Globalisation and Fisheries*, *supra* note 5, at 16.

[24] FAO, *International Plan of Action to Prevent, Deter, and Eliminate Illegal, Unreported, and Unregulated Fishing*, adopted at the Twenty-fourth Session of COFI, Rome, Italy, adopted on 23 June 2001 at the 120th Session of the FAO Council. *Hereinafter referred to as* IPOA-IUU.

[25] UNGA, Fifty-fourth Session, Agenda Items 40(a) and (c), Oceans and the Law of the Sea; Law of the Sea; Results of the Review by the Commission on Sustainable Development of the Sectoral Theme of "Oceans and Seas", Oceans and the Law of the Sea, Report of the Secretary-General, *A/54/429,* 30 September 1999, para. 249.

[26] *UNGA A/59/298*, *supra* note 15, para. 36.

[27] UNGA, United Nations Open-ended Informal Consultative Process on Ocean Affairs (UNICPOLOS), First Meeting, 30 May–02 June 2000, Discussion Panel A Responsible Fisheries and Illegal, Unregulated and Unreported Fisheries, *A/AC.259/1*, 15 May 2001, para. 1.

The concept of IUU fishing is new; however its components, *i.e.* "illegal fishing", "unreported fishing", and "unregulated fishing" are not. What the IUU terminology did was to galvanise international efforts to address existing fisheries management concerns and problems in a more coherent and globalised way. The IUU fishing concept also confirms how fisheries problems have become more complex and global in scope. The references in the IPOA-IUU to activities of domestic and foreign fishing vessels, as well as vessels without nationality, vessels flying the flags of non-parties to regional fisheries agreements and fishing entities in areas under the sovereignty and sovereign rights of States and on the high seas recognise the extensive scope of activities which may be considered IUU fishing. Chapter 2 provides a detailed analysis of the scope and content of each of the components of IUU fishing. As will be evident in later chapters, IUU fishing does not only diminish the sustainable capture of fish, but also impacts on other aspects of the globalised fisheries sector such as trade and post-harvest activities. This necessitates the adoption of comprehensive measures that will effectively address IUU-related activities.

1.1.1. *Extent of IUU Fishing*

The first global review of IUU fishing, conducted in 2000, identified major concerns within coastal State fisheries jurisdictions, management areas of regional fisheries organisations and the high seas. These concerns included poaching, illegal fishing by vessels flying the flags of non-members of regional fisheries management organisations (RFMOs) and misreported and under-reported fishing.[28] Since then, a number of studies examining the extent of the IUU problem, including quantifying the economic loss of illegal and unreported fishing and impact of such activities on the health of ecosystems, have been undertaken.

It is estimated that IUU fishing accounts for almost one third of the total catch in some important fisheries and may represent an overall cost to developing countries of between USD2 billion to USD15 billion a year.[29] The Sea Around Us project of the University of British Columbia mapped out the number of vessels fishing illegally between 1980 and 2003 and showed that IUU fishing is a global

[28] *See* Kevin Bray, 'A Global Review of Illegal, Unreported, and Unregulated (IUU) Fishing,' in FAO, Report of and Papers Presented at the Expert Consultation on Illegal, Unreported and Unregulated Fishing, Sydney, Australia, 15–19 May 2000, *FAO Fisheries Report No. 666* (Rome: FAO, 2001), at 88–134.

[29] Environmental Justice Foundation (EJF), *Pirates and Profiteers: How Pirate Fishing Fleets are Robbing People and Oceans, London*, UK, 2005, www.ejfoundation.org, Accessed on 15 February 2006, at 2; *See also* footnote 39 of this chapter, quoting Marine Resources Assessment Group Ltd (MRAG), *Review of Impacts of Illegal, Unreported and Unregulated Fishing on Developing Countries: Final Report*, London, UK, July 2005. www.high-seas.org. Accessed on 10 February 2006, at 44.

phenomenon, occurring mainly in Central and Pacific South America, East Africa, Southeast Asia, and the Northwest Pacific.[30]

RFMOs have confirmed the wide extent of IUU fishing within their areas of competence. IUU catches of toothfish in the Indian Ocean reached 100,000 tonnes in 1996.[31] About 16.5 per cent of the total catch of toothfish in the management area of the Commission for the Conservation for Antarctic Marine Living Resources (CCAMLR) in 2003–2004 came from IUU fishing.[32] The cumulative financial losses arising from IUU fishing in the CCAMLR area was USD518 million in 2003.[33] During the 2005–2006 season, the total estimated IUU catch of toothfish in the CCAMLR area was 3,080 tonnes.[34] Similarly, the Northeast Atlantic Fisheries Commission (NEAFC) estimated that up to 27 per cent of redfish landed in 2002 was caught by IUU vessels.[35] IUU fishing in the NEAFC regulatory area includes activities of vessels flying the flags of non-members.[36] The Fishery Committee for the Eastern Central Atlantic (CECAF) identified the lack of selectivity of fishing gears employed in some demersal fisheries, particularly bottom trawling and shrimp fisheries, as one of the major IUU fishing problems in the area.[37] The Indian Ocean Tuna Commission (IOTC) reported that nearly 100,000 tonnes or 10 per cent of reported landings from its management area are derived from IUU fishing.[38]

In general, it is estimated that about 25 to 30 per cent of the global fish catch is unreported.[39] In areas where there are bilateral or regional fisheries access agreements, the degree of under-reporting can be up to 75 per cent, while on the high

---

[30] U.R. Sumaila, J. Alder, and H. Keith, 'Global Scope and Economics of Illegal Fishing', 30 *Marine Policy* (2006), at 3.

[31] John P. Croxall and Phil N. Trathan, 'The Southern Ocean: A Model System for Conserving Resources', in Linda K. Glover and Sylvia A. Earle, (eds.), *Defying Ocean's End: An Agenda for Action* (Washington: Island Press, 2004), at 83.

[32] Commission for the Conservation of Antarctic Marine Living Resources (CCAMLR), *Report of the Twenty-third Meeting of the Commission*, Hobart, Australia, 25 October–05 November 2004, CCAMLR-XXIII, Tasmania, Australia, 2004, at 31.

[33] Denzil G.M. Miller, 'Patagonian Toothfish-The Storm Gathers', in OECD, *Fish Piracy: Combating Illegal, Unreported and Unregulated Fishing* (Paris: OECD, 2004), at 111.

[34] CCAMLR, *Report of the Twenty-fifth Meeting of the Commission*, Hobart, Australia, 23 October-03 November 2004, CCAMLR-XXIII, Tasmania, Australia, 2006.

[35] Kjartan Hoydal, 'IUU Fishing in NEAFC: How Big is the Problem and What Have We Done?' in OECD, *Fish Piracy: Combating Illegal, Unreported and Unregulated Fishing* (Paris: OECD, 2004), at 161.

[36] *Ibid.*

[37] *Statement by the Coalition for Fair Fisheries Arrangements (CFFA) and Greenpeace*, The 17th Session of the Fishery Committee for the East Central Atlantic (CECAF), Dakar, Senegal, 24–27 May 2000. www.globefish.org. Accessed on 14 February 2006.

[38] Bray, *Global Review of IUU Fishing, supra* note 28, at 23.

[39] Daniel Pauly and Jay Mclean, 'In a Perfect Ocean: The State of Fisheries and Ecosystems in the North Atlantic Ocean', cited in MRAG, *Review of Impacts of Illegal, Unreported and Unregulated Fishing on Developing Countries: Final Report, supra* note 29, at 16.

seas it may be 100 per cent.[40] According to the Northwest Atlantic Fisheries Organization (NAFO), unreported fishing, particularly inaccurate recording of catch, is the most common IUU activity in its regulatory area.[41] There are various estimates of unreported fishing in regional fisheries management areas. The International Commission for the Conservation of Atlantic Tunas (ICCAT) reported that about one to five per cent of Atlantic bluefin catches were unreported from 1994 to 2002[42] while CCAMLR estimated that unreported catches in its management area increased from 254 tonnes in 1983–1984 to 28,291 tonnes in 1998–1999.[43] The North Atlantic Salmon Conservation Organization (NASCO) also estimated that about 27 to 41 per cent of the salmon catch has been unreported from 1997 to 2003.[44]

The number of vessels engaging in IUU fishing in RFMO areas has been on the rise. Vessels presumed to be conducting IUU fishing in the ICCAT area increased from five in 2005 to 22 in 2008.[45] The number of vessels on the NEAFC IUU-B List increased from 18 to 21 in the same period.[46] CCAMLR listed a total of 23 IUU vessels of both members and non-members on its IUU list from 2003 to 2006[47] and 18 fishing vessels in 2008.[48] The Inter-American Tropical Tuna Commission (IATTC) placed a total of 22 vessels on its IUU list in 2008.[49] Similarly, in 2008 the IOTC placed three vessels on its IUU list;[50] while NAFO listed 23 vessels on its IUU list.[51]

---

[40] David Evans, 'The Consequences of Illegal, Unreported and Unregulated Fishing for Fishery Data and Management', in FAO, Report of and Papers Presented at the Expert Consultation on Illegal, Unreported and Unregulated Fishing, Sydney, Australia, 15–19 May 2000, *FAO Fisheries Report No. 666* (Rome: FAO, 2001), at 222.

[41] Northwest Atlantic Fisheries Organization (NAFO), 'Annex 19, Annual Compliance Review 2008', in *Report of the Fisheries Commission*, 22–26 September 2008.

[42] Victor Restrepo, 'Estimation of Unreported Catches by ICCAT', in OECD, *Fish Piracy: Combating Illegal, Unreported and Unregulated Fishing* (Paris: OECD, 2004), at 158.

[43] D.J. Agnew, 'The Illegal and Unregulated Fishery for Toothfish in the Southern Ocean, and the CCAMLR Catch Documentation Scheme', 24 *Marine Policy* (2000), at 362.

[44] North Atlantic Salmon Conservation Organization (NASCO), 'Unreported Catches-Returns by the Parties,' *Report of the 21st Annual Meeting of the Council*, Reykjavik, Iceland, 07–11 June 2004, at 252.

[45] *See* International Commission for the Conservation of Atlantic Tunas (ICCAT), *List of Vessels Presumed to Have Carried Out IUU Fishing Activities in the ICCAT Convention Area*, 11 December 2008, www.iccat.es. Accessed on 15 January 2009.

[46] Northeast Atlantic Fisheries Commission (NEAFC), *Illegal, Unregulated and Unreported Activity B List*, 05 October 2007, www.neafc.org. Accessed on 15 January 2009.

[47] CCAMLR, Combined IUU Vessels Lists Adopted in 2003, 2004, 2005, and 2006, www.ccamlr.org. Accessed on 07 October 2007.

[48] CCAMLR, *IUU Vessel List*, www.ccamlr.org. Accessed on 15 January 2009.

[49] Inter-American Tropical Tuna Commission (IATTC), *IUU Vessels List*, 13 January 2009, www.iattc.org. Accessed on 15 January 2009.

[50] Indian Ocean Tuna Commission (IOTC), *IOTC IUU Vessels List*, www.iotc.org. Accessed on 15 January 2009.

[51] NAFO, *Illegal, Unregulated and Unreported Fishing*, www.nafo.int. Accessed on 15 January 2009.

Many States have also reported significant increases in IUU fishing activities in their EEZs.[52] An estimated 700 foreign-owned vessels were engaged in illegal fishing in Somalia's EEZ in 2005.[53] In 2005 Australian authorities apprehended or intercepted a record of 607 vessels fishing illegally in the northern waters of the Australian Fishing Zone. This represented a 100 per cent increase over the number of illegal fishing vessels apprehended in 2004.[54]

The Southern African Development Community (SADC) region reported over 600 port visits by unlicensed small longline foreign vessels in 2007.[55] IUU fishing is also a persistent problem in the Arctic region, particularly in the Barents Sea.[56] High levels of IUU fishing in the salmon, crab, and Alaskan pollock fisheries occur in the Western Bering Sea and Sea of Okhotsk.[57]

Unreported fishing is a major issue for many States. For example, under-reporting of fish catch can be as high as 50 per cent in Kenya and 75 per cent in the shrimp fishery in Mozambique.[58] From the 1950s to the 1980s, it was estimated that about 10,000 to 20,000 tonnes of catch went unrecorded every year in British Columbia, Canada for salmon and groundfish fisheries.[59] From 2003 to 2004, the estimated unreported catch of toothfish landed at Southern Africa and Mauritius ports was estimated to be 74,000 to 82,200 tonnes.[60]

There have also been increasing reports of IUU fishing activities in different areas of the world which target high value, endangered, or by-catch species. In 2007, authorities in Canadian and the United States of America (U.S.) reported smuggling of internationally protected endangered species such as queen conch meat, which are transported to South America as a seafood delicacy.[61] In Africa, there was more than a ten-fold increase in confiscated abalone between 1996 and

---

[52] Meryl J Williams, 'Enmeshed: Australia and Southeast Asia's Fisheries', *Lowy Institute Paper 20* (Double Bay: Lowy Institute for International Policy, 2007), at 57.

[53] Reuters Foundation, *Somalia: Fishermen Appeal for Help Over Foreign Fishing Ships*, www.alertnet.org. Accessed on 04 March 2006.

[54] Senator The Hon Ian Macdonald, 'Record Arrests for Northern Illegal Fishing', 03 January 2006, Ministry for Fisheries, Forestry, and Conservation Website. http://www.mffc.gov.au/releases/2006/06001m.html. Accessed on 16 February 2006.

[55] Stop Illegal Fishing Programme, *Stop Illegal Fishing in Southern Africa*, May 2008. www.illegal-fishing.info Accessed on 10 February 2009, at 6.

[56] Mark Burnett, Natalie Dronova, Maren Esmark, Steve Nelson, Asle Rønning, and Vassily Spiridonov, *Illegal Fishing in Arctic Waters: Catch of Today-Gone Tomorrow?* (Oslo: WWF International Arctic, 2008), at 10.

[57] *Ibid.*, at 21.

[58] MRAG, *Review of Impacts of Illegal, Unreported and Unregulated Fishing on Developing Countries: Final Report, supra* note 29, at 55.

[59] C.H. Ainsworth and T.J. Pitcher, 'Estimating Illegal, Unreported and Unregulated Catch in British Columbia's Marine Fisheries', 75 *Fisheries Research* (2005), at 54.

[60] Agnew, *The Illegal and Unregulated Fishery for Toothfish in the Southern Ocean, supra* note 43, at 362.

[61] 'Canadian and U.S. Authorities break up seafood smuggling ring', CBC News, 26 September 2007. www.cbc.ca. Accessed on 02 October 2007.

2006, which were traded illegally in dried, frozen, or canned form.[62] These accounts of IUU fishing are merely examples and do not fully capture the global picture of IUU fishing.

1.1.2. *Impacts and Causes of IUU Fishing*

IUU fishing is known to have negative economic, environmental, ecological, and social impacts. The Marine Resource Assessment Group Ltd (MRAG) estimates that the total loss as a result of IUU fishing in Guinea, Liberia, Sierra Leone, Angola, Namibia, Mozambique, Kenya, and Somalia, as well as in Seychelles and Papua New Guinea amount to USD372 million or 19 per cent of the total value of catch or 23 per cent of the declared value of catch.[63] More specifically, it is estimated that Guinea is losing in excess of 57,997 tonnes of fish (equivalent to USD105 million) every year due to illegal fishing; Liberia, USD12 million; Sierra Leone, USD29 million; Angola, USD49 million; Mozambique, USD38 million; Kenya, USD3.8 million; Somalia, USD94 million; Seychelles, USD7.5 million; and Papua New Guinea, USD34.2 million.[64] Non-governmental environment organisations estimate that trade in IUU fishing deprives some developing States of USD9 billion per year, of which USD1 billion is lost by African States.[65] A follow up study in April 2008 by MRAG and the Fisheries Centre at the University of British Columbia estimated that the total value of losses from illegal fishing is between USD10 billion and USD23 billion annually, representing 11.06 million to 25.91 million tonnes of fish.[66] These estimates are based on information gathered from a number of different species in 60 States and 17 FAO statistical areas, representing only 46 per cent of global catches based on FAO statistics.

In the Asia Pacific region, IUU fishing has been estimated to cost between USD4.5 billion and USD5.8 billion a year.[67] The quantity of fish taken by IUU fishing activities in the region is between 3.45 million tonnes to 8.12 million tonnes which account for about eight to 16 per cent of the total reported catch per year.[68] In the Sulawesi Sea of the Asia Pacific region for example, the financial

---

[62] Stop Illegal Fishing Programme, *Stop Illegal Fishing in Southern Africa*, *supra* note 55, at 78.

[63] MRAG, *Review of Impacts of IUU Fishing on Developing Countries*, *supra* note 29, at 44.

[64] *Ibid.*, at 37.

[65] Richard Black, 'Action Urged on Illegal Fishing', *BBC News* www.bbc.co.uk. Accessed on 28 September 2007.

[66] MRAG and Fisheries Ecosystems Restoration Research, Fisheries Centre, University of British Columbia, *The Global Extent of IUU Fishing, Final Report*, www.illegal-fishing.info/uploads/MRAGExtentGlobalIllegalFishing.pdf. Accessed on 13 June 2008, at 17.

[67] R. Lungren, D. Staples, S. Funge-Smith, and J. Clausen, *Status and Potential of Fisheries and Aquaculture in Asia and the Pacific 2006*, FAO Regional Office for Asia and the Pacific, RAP Publication 2006/22 (Bangkok: FAO-RAP, 2006), at 46.

[68] MRAG, *The Global Extent of IUU Fishing*, *supra* note 66, at 15. *See* also Frank Meere and Mary Lack, *Assessment of Impacts of Illegal, Unreported and Unregulated (IUU) Fishing in the Asia Pacific*, (Singapore: APEC Secretariat, 2008), at 27. This report also summarises available estimates of IUU fishing for different areas and member economies of the Asia Pacific region.

loss from IUU fishing has been conservatively estimated at about a third of the total annual value of the marine fisheries in the area.[69] This estimate for the Sulawesi Sea does not include the social and environmental costs associated with loss of future access to the fisheries resources. It is estimated that Indonesia loses USD4 billion a year due to illegal fishing.[70] In the Philippines, the estimated economic loss due to illegal fishing is USD894 million.[71] Another estimate from the World Resource Institute suggests that the total net loss from blastfishing in the Philippines is USD1.2 billion.[72] IUU fishing also affects the export markets of some countries.

IUU fishing activities cause damage to fisheries habitats and the marine environment, particularly coral reefs. The productive capacity of coral reefs is reduced to a fifth of their original capacity as a result of dynamite fishing.[73] An explosive of the size of a coca cola bottle can shatter to pieces all stony corals within a radius of three metres.[74] Studies suggest that fish diversity and coral areas damaged by moderate blast fishing and poison fishing may take 25 years to recover.[75] If 50 per cent of live coral is destroyed, recovery to the initial state is predicted to take about 60 years.[76] In the case of the Philippines, destructive fishing methods caused the degradation of about 70 per cent of the coral reefs and reduced the annual fisheries production by about 177,500 metric tonnes in the 1990s.[77]

IUU fishing results in high levels of by-catch of both juvenile fish and non-target species. Migratory oceanic sharks, seabirds, marine mammals are not

---

69 Mary Ann Palma and Martin Tsamenyi, *Case Study on the Impacts of Illegal, Unreported and Unregulated Fishing in the Sulawesi Sea,* APEC Fisheries Working Group, FWG 02/2007, (Singapore: APEC Secretariat, April 2008), at 24.

70 Centre for Maritime Policy and Oceans and Coastal Research Centre, *The Identification of Researchable Options for the Development of Policy and Management Frameworks to Combat Illegal, Unreported and Unregulated (IUU) Fishing Activities in Indonesian and Philippine Waters*, Final Report on ACIAR Project No. FIS/2000/163 (Wollongong: University of Wollongong, 2002), at 21.

71 Porfirio Aliño, 'Fisheries Resources of the Philippines,' *Presentation during the Australian Consultation with the Philippines and Indonesia on the Identification of Researchable Options for the Development of Policy and Management Frameworks to Combat Illegal, Unreported and Unregulated (IUU) Fishing Activities in Indonesian and Philippine Waters*. In Centre for Maritime Policy and Oceans and Coastal Research Centre, *ibid.*

72 Laureta Burke, Elizabeth Selig, and Mark Spalding, *Reefs at Risk in Southeast Asia* (U.S.: World Resource Institute, 2002), at 56.

73 Alan T. White, Helge P. Vogt, and Tijen Arin, 'The Philippine Coral Reefs Under Threat: The Economic Losses Caused by Reef Destruction,' 40 *Marine Pollution Bulletin* (2000), at 600.

74 *Ibid.*

75 Raquel Goñi, 'Fisheries Effects on Ecosystems', in Charles R.C. Sheppard, (ed.), *Seas at the Millennium: An Environmental Evaluation, Vol. III: Global Issues and Processes* (Amsterdam: Pergamon Elsevier Science, 2000), at 123.

76 *Ibid.*

77 A.C. Alcala and G.R. Russ, 'Status of Philippine Coral Reef Fisheries,' 15 *Asian Fisheries Science* (2002), at 177.

directly targeted by commercial fisheries but become a large component of the by-catch of longline, purse seine, and driftnet fisheries.[78] It is estimated that about 100,000 seabirds, including tens of thousands of endangered albatrosses, are killed each year by illegal longline vessels in the Southern Ocean.[79] In some tuna fisheries, turtle mortality is observed to be in excess of 40 per cent.[80] The use of destructive fishing methods such as trawling also exacerbates the problem of by-catch and discards.[81]

A number of social impacts of IUU fishing have been identified.[82] Some IUU fishing vessels are known to recruit their crew from States where there is a lack of alternative employment opportunities[83] and who may be unaware of the vessels' illegal operations.[84] IUU fishing has not only been equated to "stealing food from some of the poorest of the world"[85] but is also known to cause the displacement of legitimate fishers. IUU fishing can further lead to lower employment and reduction in household incomes, both of which exacerbate poverty, particularly among coastal and artisanal fishers.[86] Social impact studies show that IUU fishing not only affects industrial fishing, but is also a concern in small scale fisheries.[87]

Numerous social and economic factors have been identified as "drivers" of IUU fishing. These factors include the profitability of conducting IUU fishing activities as a result of lower administrative and management costs,[88] lesser crew costs that offset the possibilities of high maintenance and other operating costs,[89] overcapacity in the world fishing fleet, insufficient monitoring, control, and surveillance (MCS) operations, inadequate levels of penalties, inappropriate

---

[78] *See* Stephen J. Hall, *The Effects of Fishing on Marine Ecosystems and Communities* (Oxford: Blackwell Science Ltd., 1999), at 16–47; *See also* Michael Berrill, *The Plundered Seas: Can the World's Fish Be Saved?* (San Francisco: Sierra Club Books, 1997), at 65.

[79] MRAG, *IUU Fishing on the High Seas: Impacts on Ecosystems and Future Science Needs Final Report*, London, August 2005. www.high-seas.org. Accessed on 10 February 2006, at 42.

[80] Jeanne A. Mortimer, Marydele Donnelly and Pamela T. Plotkin, 'Sea Turtles,' in Charles R.C. Sheppard, (ed.), *Seas at the Millennium: An Environmental Evaluation, Vol. III: Global Issues and Processes* (Amsterdam: Pergamon Elsevier Science, 2000), at 66.

[81] EJF, *What's the Catch? Reducing By-catch in EU Distant Water Fisheries* (EJF: London, 2005), at 6.

[82] *See* Jon Whitlow, 'The Social Dimension of IUU Fishing', in OECD, *Fish Piracy: Combating Illegal, Unreported and Unregulated Fishing*, (Paris: OECD, 2004), at 231–238.

[83] *Ibid.*, at 244.

[84] *Ibid.*

[85] EJF, *Pirates and Profiteers*, *supra* note 29, at 3.

[86] David J. Agnew and Colin T. Barnes, 'Economic Aspects and Drivers of IUU Fishing: Building a Framework', in OECD, *Fish Piracy: Combating Illegal, Unreported and Unregulated Fishing* (Paris: OECD, 2004), at 169–200.

[87] FAO, Increasing the Contribution of Small-scale Fisheries to Poverty Alleviation and Food Security, *FAO Technical Guidelines for Responsible Fisheries No. 10* (Rome: FAO, 2005), at 46.

[88] Aaron Hatcher, 'Incentives for Investments in IUU Fishing Capacity', in OECD, *Fish Piracy: Combating Illegal, Unreported and Unregulated Fishing* (Paris: OECD, 2004), at 244.

[89] *Ibid.*

management frameworks, governance failure, corruption, subsidies, and poor economic and social conditions.[90]

For developing States, a major challenge in addressing IUU fishing is the limited operational capacity to manage vast expanses of ocean spaces under their jurisdiction. For example, the enforcement of fisheries laws in the EEZ of Ghana is considered weak, making it difficult to assess the level of illegal fishing and catch by foreign vessels.[91] The lack of enforcement capabilities in other African States also hinders their ability to monitor fishing operations in their waters.[92] IUU fishing has been associated with corruption of fisheries officials in some States.[93] Some IUU fishing activities are also associated with the operations of transnational criminal groups[94] and other illegal activities such as fuel smuggling, fish smuggling, and trafficking of fishing crew.[95] Political instability and governance failure exacerbate the IUU fishing problem in some parts of the world. Increased IUU fishing in the EEZ of Somalia is a good case in point.[96]

Because of the enormity of the IUU fishing in many parts of the world, addressing it has become a global concern. IUU fishing issues have been raised not only by the FAO[97] but also by the United Nations General Assembly (UNGA),[98] the

---

[90] Carl-Christian Schmidt, 'Addressing Illegal, Unreported and Unregulated (IUU) Fishing,' *Paper Prepared for the International Fisheries Compliance Conference*, Brussels, Belgium, 29–30 September 2004, at 9; United Nations Open-ended Informal Consultative Process on Ocean Affairs (UNICPOLOS), First meeting, 30 May-2 June 2000, Discussion panel A, Responsible fisheries and illegal, unregulated and unreported fisheries, Moving from principles to implementation, Illegal, unreported and unregulated fishing, Submission by the Food and Agriculture Organization of the United Nations, *A/AC.259/1*, 15 May 2000, para. 3.

[91] John Atta-Mills, Jackie Alder, and Ussif Rashid Sumaila, 'The Decline of a Regional Fishing Nation: The Case of Ghana and West Africa', 28 *Natural Resources Forum* (2004), at 16.

[92] WWF International Endangered Seas Campaign, *The Footprint of Distant Water Fleets on World Fisheries* (Surrey: WWF, 1998), at 23.

[93] Institute for Security Studies Website. *10 April 2008: Corruption in Africa's Marine Fisheries*, www.issafrica.org. Accessed on 11 April 2008.

[94] *See* 'Interior ministry reveals huge seafood contraband on Sakhalin', 03 November 2006, www.illegal-fishing.info. Accessed on 02 October 2007.

[95] Sea Resources Management Sdn Bhd, *Case Study on Illegal, Unreported and Unregulated (IUU) Fishing off the East Coast of Peninsular Malaysia: Final Report* (Singapore: APEC Secretariat, 2008), at 129–143.

[96] Scott Coffen-Smout, *Pirates, Warlords and Rogue Fishing Vessels in Somalia's Unruly Seas*, http://www.chebucto.ns.ca/~ar120/somalia.html. Accessed on 17 February 2006. *See also* M.G. Hassan, IUU Fishing and Insecurity Impacts on Somali Fisheries and Marine Resources, *Presented at the 4th Chatham House IUU Fishing Update and Stakeholder Consultation Meeting*, Chatham House, 31 March-01 April 2008, www.illegal-fishing.info. Accessed on 12 February 2009.

[97] FAO continuously undertakes activities designed to heighten international awareness on the scope and impact of IUU fishing, as well as the management efforts of regional fisheries management organisations to combat IUU fishing. *See* David J. Doulman, '2001 FAO International Plan of Action to Prevent, Deter and Eliminate Illegal, Unreported and Unregulated Fishing: Background and Progress Towards Implementation', *Presented at the FAO Regional Workshop on the Elaboration of National Plans of Action to Prevent, Deter and Eliminate Illegal, Unreported and Unregulated Fishing: Southeast Asia Subregion*, Penang, Malaysia, 10–14 October 2004.

[98] *See* UNGA A/59/298, *supra* note 15, Part IV; *UNGA A/58/65*, *supra* note 4, Part VI(A)(2).

United Nations Open-ended Informal Consultative Process (UNICPOLOS)[99] and the 2002 World Summit on Sustainable Development (WSSD).[100] The need to address IUU fishing was reiterated in the 2005 Rome Declaration on Illegal, Unreported, and Unregulated Fishing.[101] Furthermore, IUU fishing was considered a major concern in the informal consultations of the States Parties to the UN Fish Stocks Agreement.[102] During the UN Fish Stocks Agreement Review Conference, IUU fishing was identified as a problem that undermines the provisions of the Agreement which needs to be addressed collectively by States.[103]

The call to strengthen global efforts to combat IUU fishing has encouraged several RFMOs and other regional organisations to adopt measures to combat IUU fishing.[104] The European Union (EU)[105] and the Lake Victoria Fisheries

---

[99] Report of the Work of the United Nations Open-ended Informal Consultative Process on Oceans and the Law of the Sea at its Sixth Meeting, Sixtieth Session, item 76(a) of the preliminary list, *A/60/99,* 07 July 2005.

[100] *Plan of Implementation of the World Summit on Sustainable Development,* Johannesburg, South Africa, August 2002, para. 31(d).

[101] *The 2005 Rome Declaration on Illegal, Unreported, and Unregulated Fishing,* Adopted by the FAO Ministerial Meeting on Fisheries, Rome, Italy, 12 March 2005.

[102] UN, *Agreement for the Implementation of the Provision of the United Nations Convention on the Law of the Sea of 10 December 1982 Relating to the Conservation and Management of Straddling Fish Stocks and Highly Migratory Fish Stocks,* New York, concluded on 4 August 1995, in force 11 December 2001, 34 ILM 1542 (1995); 2167 UNTS 88. *Hereinafter referred to as UN Fish Stocks Agreement.*

[103] UN, *Report of the Second Informal Consultations of the States Parties to the Agreement for the Implementation of the Provisions of the United Nations Convention on the Law of the Sea 10 December 1982 relating to the Conservation and Management of Straddling Fish Stocks and Highly Migratory Fish Stocks,* New York, USA, 23–25 July 2003, ICSP2/UNFSA/REP/INF.1, August 2003, para. 14; *See also* reports of the Third, Fourth, and Fifth Informal Consultations of the States Parties to the UN Fish Stocks Agreement; UNGA, *Report of the Review Conference on the Agreement for the Implementation of the provisions of the United Nations Convention on the Law of the Sea of 10 December 1982 Relating to the Conservation and Management of Straddling Fish Stocks and Highly Migratory Fish Stocks,* New York, USA, 22–26 May 2006, Prepared by the President of the Conference with the Assistance of the Secretariat, *A/CONF.210/2006/15,* June 2006.

[104] For example Commission for the Conservation of Southern Bluefin Tuna (CCSBT), *Resolution on Amendment of the Resolution on Illegal, Unregulated and Unreported Fishing (IUU) and Establishment of a CCSBT Record of Vessels over 24 Meters Authorised to Fish for Southern Bluefin Tuna,* Adopted at the Fifteenth Annual Meeting, 14–17 October 2008; ICCAT, *Recommendation by ICCAT* Amending the *Recommendation by ICCAT to Establish a List of Vessels Presumed to have carried out Illegal, Unreported and Unregulated Fishing Activities in the ICCAT Convention Area,* 06–12 GEN, 2006; IOTC, *Resolution 09/03* On Establishing a List of Vessels Presumed to have carried out Illegal, Unregulated and Unreported Fishing in the IOTC Area, 2009; IATTC, *Resolution C-05-07,* Resolution to Establish a List of Vessels presumed to have carried out Illegal, Unreported and Unregulated Fishing Activities in the Eastern Pacific Ocean, 24 June 2005; CCAMLR, *Conservation Measure 10–06 (2008),* Scheme to Promote Compliance by Contracting Party Vessels with CCAMLR Conservation Measures; CCAMLR, *Conservation Measure 10–07 (2009),* Scheme to Promote Compliance by Non-Contracting Party Vessels with CCAMLR Conservation Measures; Western and Central Pacific Fisheries Commission (WCPFC), Conservation and Management Measure to Establish a List of Vessels Presumed to have carried out Illegal, Unreported and Unregulated Fishing Activities in the Western and Central Pacific ocean (WCPO), *Conservation and Management Measure 2007–03,* 07 December 2007.

[105] European Commission, *Community Action Plan for the Eradication of Illegal, Unreported and Unregulated Fishing,* Brussels, 28.5.2002, COM(2002) 180 final.

Organization (LVFO) have adopted their respective regional plans of action to prevent, deter, and eliminate IUU fishing.[106] The EU has adopted a comprehensive and far-reaching regulation establishing a Community system to combat IUU fishing.[107] In the Asia-Pacific region, Southeast Asian countries have adopted a regional plan of action to promote responsible fisheries and combat IUU fishing[108] while SADC has adopted a Statement of Commitment to eradicate IUU fishing.[109] Similarly, a number of States have developed national plans of action to address the IUU fishing problem.[110] The IUU related measures adopted by RFMOs and States under resolutions and national plans of actions are analysed in Chapters 9 and 10.

1.2. Legal and Policy Framework to Address IUU Fishing

Needless to say, the extent and impacts of IUU fishing described above require the adoption and implementation of adequate legal and policy measures by States and regional organisations. This is particularly important if the global sustainability of fisheries resources is to be promoted or maintained.

The legal and policy framework for the sustainability of fisheries, including addressing IUU fishing comprises a number of fisheries and non-fisheries specific

[106] Lake Victoria Fisheries Organization, *Regional Plan of Action to Prevent, Deter, and Eliminate Illegal, Unreported and Unregulated (IUU) Fishing on Lake Victoria and Its Basin*, Bagamoyo, Tanzania, 27 May 2004.

[107] Council Regulation (EC) No 1005/2008 of 29 September 2008 establishing a Community system to prevent, deter and eliminate illegal, unreported and unregulated fishing, amending Regulations (EEC) No 2847/93, (EC) No 1936/2001 and (EC) No 601/2004 and repealing Regulations (EC) No 1093/94 and (EC) No 1447/1999, *OJ L 286/1 29.10.2008*. For analysis, see M. Tsamenyi, M.A. Palma, B. Milligan & K. Mfodwo, *Fairer Fishing? Trade and Fisheries Policy Implications for Developing Countries of the European Community Regulation on Illegal, Unreported and Unregulated Fisheries* (London: Commonwealth Secretariat, 2009).

[108] *Regional Plan of Action (RPOA) to Promote Responsible Fishing Practices including Combating IUU Fishing in the Region*, Bali, Indonesia, 04 May 2007.

[109] The Southern African Development Community (SADC) IUU Statement of Commitment will become a SADC Declaration when endorsed by its Council of Ministers. SADC aims to finalise a regional action plan by June 2009 and a review of progress on implementation of the Statement of Commitment is set for the end of 2011. See TRAFFIC News, 'Southern African states move to eradicate "pirate" fishing,' http://www.traffic.org/home/2008/7/11/southern-african-states-move-to-eradicate-pirate-fishing.html. At this early stage of implementation, Mozambique has already signified its commitment under the SADC IUU Statement by undertaking immediate enforcement actions against a Namibian-flagged vessel Antilles Reefer and taking further investigation on its two sister vessels Paloma V and Aoster believed to have conducted IUU fishing in Mozambique waters. See Agencia de Informacao de Mocambique (Maputo), Mozambique: Country Seizes Namibian Pirate Fishing Ship "Antilles Reefer", 18 July 2008. Available at http://allafrica.com/stories/200807180938.html. Accessed on 10 September 2008.

[110] Examples of States that have developed their NPOA-IUU published in the FAO website are Argentina, Australia, Canada, Chile, Japan, New Zealand, U.S., and South Korea. *See* FAO Website, *IPOA-IUU National Plans of Action*. www.fao.org. Accessed on 04 June 2009. A number of other States in the Pacific, Asia, and Africa have also developed draft NPOA-IUU.

binding and non-binding instruments. The IPOA-IUU is the main international instrument which addresses IUU fishing. Aside from being elaborated within the framework of the FAO Code of Conduct for Responsible Fisheries,[111] the IPOA-IUU is also based on relevant rules of international law, particularly the United Nations Convention on the Law of the Sea (LOSC),[112] the UN Fish Stocks Agreement,[113] and the FAO Compliance Agreement,[114] as well as World Trade Organization (WTO) and International Maritime Organization (IMO) agreements, rules and principles.[115] The trade related agreements include the General Agreement on Tariff and Trade (GATT) and other WTO agreements such as the Agreements on Technical Barriers to Trade, Pre-shipment Inspection, Rules of Origin, Import Licensing Procedures, and Subsidies and Countervailing Measures. Maritime safety and labour related agreements for fishing vessels, which are part of the international legal framework for fisheries, include the 1993 Torremolinos Protocol for the Safety of Fishing Vessels,[116] the 1995 International Convention on Training, Certification and Watchkeeping for Fishing Vessels,[117] and the Work in Fishing Convention in 2007 (C188),[118] supplemented by the Work in Fishing Recommendation 2007 (R199).[119] There are also environment related instruments of general relevance to combating IUU fishing. These include the Convention on Wetlands of International Importance Especially as Waterfowl Habitat (Ramsar Convention),[120] Convention on International Trade in Endangered Species of

---

[111] FAO, *Code of Conduct for Responsible Fisheries,* Adopted at the 28th Session of the FAO Conference, Rome, Italy, 31 October 1995. *Hereinafter referred to as* FAO Code of Conduct.

[112] *LOSC, supra* note 3.

[113] *UN Fish Stocks Agreement, supra* note 102.

[114] FAO, *Agreement to Promote Compliance with International Conservation and Management Measures by Fishing Vessels on the High Seas*, Rome, Italy, concluded on 24 November 1993, entered into force 24 April 2003, 33 ILM 968. *Hereinafter referred to as FAO Compliance Agreement.*

[115] *IPOA-IUU*, paras. 5 and 66; *See also* FAO Code of Conduct, Art. 1.1.

[116] International Maritime Organization (IMO), *Torremolinos Protocol of 1993 relating to the Torremolinos International Convention for the Safety of Fishing Vessels 1977,* Torremolinos, Spain, 2 April 1993. *Hereinafter referred to as* 1993 Torremolinos Protocol.

[117] IMO, *International Convention on Standards of Training, Certification and Watchkeeping for Fishing Vessel Personnel,* adopted on 7 July 1995, in force 1 February 1997. *Hereinafter referred to as* STCW-F.

[118] International Labour Organization (ILO), Convention Concerning Work in the Fishing Sector (C188), Geneva, Switzerland, adopted on 14 June 2007. *Hereinafter referred to as* Work in Fishing Convention.

[119] ILO, Recommendation Concerning the Work in the Fishing Sector (R199), Geneva, Switzerland, adopted on 14 June 2007. *Hereinafter referred to as* Work in Fishing Recommendation.

[120] *Convention on Wetlands of International Importance Especially as Waterfowl Habitat*, Ramsar, Iran, concluded on 2 February 1971, in force 21 December 1975, 996 UNTS 245; TIAS 11084; 11 ILM 963 (1972), as amended by the Paris Protocol, 3 December 1982 and Regina Amendments, 28 May 1987. *Hereinafter referred to as* Ramsar Convention.

Wild Flora and Fauna (CITES),[121] Convention on the Conservation of Migratory Species of Wild Animals (CMS),[122] Agenda 21, particularly Chapter 17,[123] and the Convention on the Conservation of Biological Diversity (CBD).[124] Chapter 3 discusses these international instruments.

1.2.1. *The IPOA-IUU*

The objective of the IPOA-IUU is "to prevent, deter, and eliminate IUU fishing by providing States with comprehensive, effective, and transparent measures by which to act, including through appropriate regional fisheries management organisations, established in accordance with international law."[125] Significantly, the IPOA-IUU applies to members and non-members of the FAO, fishing entities, subregional, regional and global organisations, whether governmental and non-governmental, and all persons concerned with the conservation of fisheries resources and management and development of fisheries, such as fishers, those engaged in the processing and marketing of fish and fishery products and other users of the aquatic environment in relation to fisheries.[126] This covers the whole spectrum of actors involved in fisheries management; thus giving the IPOA-IUU a wider scope of application than other international instruments.

The IPOA-IUU is considered a comprehensive "toolbox", which contains a range of measures that can be used by flag States, port States, coastal States, and "market States" or States which engage in the international trade in fish, to address various manifestations of IUU fishing.[127] Measures that cut across the responsibilities of flag, coastal, port, and market States are categorised under "All State Responsibilities". These measures include the implementation of relevant

[121] *Convention on International Trade in Endangered Species of Wild Flora and Fauna,* Washington D.C., USA, concluded on 3 March 1973, in force 1 July 1975, 27 UST 1087; TIAS 8249; 993 UNTS 243. Amended at Bonn, on 22 June 1979 and Gaborone on 30 April 1983. *Hereinafter referred to as CITES.*

[122] *Convention on the Conservation of Migratory Species of Wild Animals*, Bonn, concluded on 23 June 1979, in force 01 November 1979, 19 ILM 15 (1980); ATS 1991/32; BTS 87 (1990), Cm. 1332.

[123] United Nations Conference on Environment and Development (UNCED), Agenda 21, Chapter 17, *Protection of the Oceans, All Kinds of Seas, Including Enclosed and Semi-enclosed Seas, and Coastal Areas and the Protection, Rational Use and Development of their Living Resources*, Rio de Janeiro, Brazil, 03–14 June 1992. *Hereinafter referred to as* Agenda 21.

[124] *Convention on Biological Diversity*, Rio de Janeiro, Brazil, concluded on 5 June 1992, in force 29 December 1993, 1760 UNTS 79; 31 ILM 818 (1992).

[125] *IPOA-IUU*, para. 8.

[126] Paragraph 5 of the IPOA-IUU states that Articles 1.1, 1.2, 3.1, and 3.2 of the FAO Code of Conduct for Responsible Fisheries apply to the interpretation and application of the IPOA-IUU.

[127] FAO, Fisheries Department, Implementation of the International Plan of Action to Prevent, Deter, and Eliminate Illegal, Unreported, and Unregulated Fishing, *FAO Technical Guidelines for Responsible Fisheries No. 9* (Rome: FAO, 2002), para. 16.

international instruments, adoption of national plans of action to combat IUU fishing, application of sanctions of sufficient severity, control of nationals, elimination of economic incentives for vessels engaged in IUU fishing, effective MCS measures, cooperation among States, measures in respect of vessels without nationality and vessels flying the flags of non-cooperating States to RFMOs, and publicity on IUU fishing.[128]

As a "toolbox," the IPOA-IUU attempts to embrace all existing measures which States, acting alone or in cooperation with other States or through RFMOs, may adopt to combat IUU fishing.[129] States should be able to find an appropriate tool or a combination of tools in the IPOA-IUU, to address any incident of IUU fishing.[130] Consequently, the IPOA-IUU does not necessarily require States to adopt all of the measures outlined in the instrument, but to select the measures most appropriate and applicable to particular circumstances. There are some overlaps in the application of these measures, although no contradictory measures can be found within the IPOA-IUU. Most of the measures are also addressed in other fisheries related international instruments, which will be discussed in succeeding chapters.

A closer examination of the text of the IPOA-IUU reveals the lack of connection between Part II on the nature and scope of IUU fishing and Part IV on the implementation of measures to combat IUU fishing. While Part II of the IPOA-IUU discusses the scope of each component of IUU fishing, reference is only made to the general term "IUU fishing" in Part IV. The IPOA-IUU does not specify which measures address illegal fishing, unreported fishing, or unregulated fishing. This has resulted in some States adopting flag, coastal, port, market, and all State measures as part of their compliance with the requirements under Part IV, without examining how the IUU fishing concept applies within a national context, and how such measures can address specific IUU fishing issues. This practice is reflected in the national plans of action formulated by many States. A number of States have incorporated specific measures against IUU fishing in their national plans of action, but simply adopted the IUU fishing concept under paragraph 3 of the IPOA-IUU without directly specifying how it relates to fishing activities within their EEZ or to vessels flying their flag which are conducting fishing operations on the high seas and RFMO areas.[131]

---

[128] *IPOA-IUU*, paras. 10–33.

[129] *FAO Technical Guidelines for Responsible Fisheries No. 9*, *supra* note 127, para. 16.

[130] *Ibid.*

[131] *E.g.* Ghana, *National Plan of Action to Prevent, Deter and Eliminate Illegal, Unreported and Unregulated Fishing,* June 2004; Republic of Korea, Ministry of Maritime Affairs and Fisheries, *Republic of Korea National Plan of Action to Prevent, Deter and Eliminate Illegal, Unregulated and Unreported Fishing,* no date; New Zealand, Ministry of Fisheries, *New Zealand Plan of Action to Prevent, Deter and Eliminate Illegal, Unregulated and Unreported Fishing,* May 2004.

The IPOA-IUU establishes a clear link with other international instruments.[132] Paragraphs 10 to 15 of the IPOA-IUU recognise the significance of full and effective implementation of other international instruments in addressing IUU fishing. Paragraph 13 in particular, provides that no clause in the IPOA-IUU affects or should be interpreted as affecting the rights and obligations of States under international law, with particular mention of the UN Fish Stocks Agreement and the FAO Compliance Agreement. There are also general references and repetitive clauses in the IPOA-IUU on the need to interpret the provisions of the IPOA-IUU in accordance with these international instruments.[133]

One of the problems in the IPOA-IUU is the overkill of the references, "in accordance with international law" and "in conformity with international law."[134] This redundancy may give rise to a mischievous argument that, where such a clause is not affixed to a particular provision, it would imply that international law is not a constraint.[135] Indeed, such repetitive clauses could have been replaced by a few clear paragraphs that would have elaborated on the relationship between the IPOA-IUU and international instruments and international law in general.

The relationship between the IPOA-IUU and other international instruments can be better appreciated by discussing the legal nature of the instrument. Similar to the FAO Code of Conduct, the IPOA-IUU is a voluntary instrument.[136] The definition of a voluntary instrument has not been explained any further in the text of the IPOA-IUU. However, with the non-mandatory language used, the absence of final clauses found in most multilateral treaties, and from the general context in which all the negotiations took place, "voluntary" means that the IPOA-IUU does not give rise to any legally binding obligations.[137] However, it does not mean that the IPOA-IUU is devoid of any legal effect.[138] This is because it contains provisions that may be or have already been given binding effects by or through other international legal instruments.[139] In this case, the IPOA-IUU as a voluntary instrument allows States to assume binding obligations in an indirect way.

---

[132] *IPOA-IUU*, paras. 4 and 5.

[133] *See* IPOA-IUU, paras. 20, 21, 24.10, 42, 44, 52, 53, 54, 59, 65, 66, 67, 68, 78, 79, 80.2, 83, and 84.

[134] William Edeson, 'The International Plan of Action on Illegal, Unreported and Unregulated Fishing: The Legal Context of a Non-legally Binding Instrument', 16 *International Journal of Marine and Coastal Law* (2001), at 615.

[135] *Ibid.*

[136] *IPOA-IUU*, para. 4.

[137] William Edeson, *International Plan of Action on Illegal, Unreported and Unregulated Fishing, supra* note 134, at 608.

[138] *See* William Edeson, 'Closing the Gap: The Role of "Soft" International Instruments to Control Fishing', 20 *Australian Yearbook of International Law* (1999), at 21, citing R. Wolfrum, 'The Role of the International Tribunal for the Law of the Sea,' Paper Presented in the Conference on Current Fisheries Issues and the Food and Agriculture Organization of the United Nations, Organised by the Centre for Oceans Law and Policy, University of Virginia and FAO, Rome, 16–17 March 2000.

[139] *FAO Code of Conduct*, Art. 1.1.

The point needs to be made that instruments which have been accepted voluntarily can, nevertheless, be developed into a system of well defined uniform legal consequences.[140] Although initially non-binding, once expressly accepted by a State, these instruments are applied by States, creating a legal effect.[141] This would illustrate one of the strongest possible impacts that a non-binding instrument such as the IPOA-IUU could have, not only in respect of binding agreements, but also in the development of customary international law.[142] Achievements in harmonising national, sub-regional, and regional laws have been made in the field of fisheries conservation and management through these voluntary instruments, particularly through their adoption in national legislation and regional agreements.[143]

The formulation of the IPOA-IUU as a voluntary instrument has its advantages and disadvantages. One of the advantages is that the IPOA-IUU can gain a wider acceptance among States compared to instruments such as the UN Fish Stocks Agreement and the FAO Compliance Agreement which are binding only on their respective parties. The IPOA-IUU also provides guidance to States by establishing minimum standards to address IUU fishing which have been agreed upon at the international level. The IPOA-IUU reiterates fisheries management obligations already found in binding agreements and further includes other obligations or useful fisheries management measures which may be difficult to incorporate in binding instruments.

The disadvantage of having the IPOA-IUU as a voluntary instrument is that it might not have a direct and binding effect at the national level if it is not adopted

---

[140] Peter H. Sand, *Transnational Environmental Law: Lessons in Global Change* (Hague: Kluwer Law International, 1999), at 23. The author provided the example of the international food standards of the Codex Alimentarius which was jointly drawn up by FAO and the World Health Organization (WHO).

[141] *Ibid.*, at 23.

[142] Gail L. Lugten, 'Soft Law with Hidden Teeth: The Case for a FAO International Plan of Action on Sea Turtles,' 9 *Journal of International Wildlife Law and Policy* (2006), at 167. Similar arguments have been made on the significant impact of instruments such as the 1992 Rio Declaration on Environment and Development that can have in setting out new norms of international environmental law. *See* Alan Boyle and David Freestone (eds.), *International Law and Sustainable Development* (Oxford: Oxford University Press, 1999), at 5–8; *See also* Gerhard Hafner, 'General Principles of Sustainable Development: From Soft Law to Hard Law,' in Malgosia Fitzmaurice and Milena Szuniewicz (eds.), *Exploitation of Natural Resources in the 21st Century* (Hague: Kluwer Law International, 2003), at 63.

[143] Examples include the U.S. Implementation Plan for the Code of Conduct for Responsible Fisheries, developed by the U.S. National Marine Fisheries Service, National Oceanic and Atmospheric Administration and the U.S. Department of Commerce, available at www.nmfs.noaa.gov; Canadian Code of Conduct for Responsible Fisheries Operations, developed by Fisheries and Oceans Canada, available at www.fisheriescouncil.ca; Code of Conduct for a Responsible Seafood Industry, published by the Australian Seafood Industry Council, available at www.asic.org.au; and the Regional Guidelines for Responsible Fishing Operations in Southeast Asia, formulated by the Southeast Asian Fisheries Development Centre, available at www.seafdec.org.

by States. However, the increasing interest and attention being paid by international and regional organisations and individual States to the issue of IUU fishing suggests that the IPOA-IUU is becoming one of the most accepted voluntary instruments since the FAO Code of Conduct. It can also be noted that these two outwardly similar instruments have had quite different impacts. While the FAO Code of Conduct is often cited as the principal voluntary text, the much more tightly drafted IPOA-IUU has arguably had much greater impact in practical terms than has the FAO Code of Conduct.

1.2.2. *Measures to Address IUU Fishing*

The relevant provisions of binding and non-binding international instruments that would need to be implemented by States are analysed in Chapters 4 to 8. In general, a comprehensive framework to address IUU fishing comprises the application of flag, coastal, port, market, and all State measures.

According to the IPOA-IUU and relevant international instruments, flag State measures involve the effective exercise of jurisdiction over vessels by flag States through, among other things, the implementation of a fishing vessel registration and authorisation to fish or licensing system.[144] An effective fishing vessel registration and licensing system takes into account the history of flagging and the IUU fishing activities of a vessel.[145] This allows a flag State to regulate the operations of its vessels to ensure that only vessels which comply with international, regional, and national fisheries regulations and conservation and management measures are registered and granted authorisations to fish.[146] A flag State is also required to maintain a record of fishing vessels which includes not only the basic information on a vessel, but also details of actual fishing operations.[147] Ideally, the information contained in a national record of fishing vessels would need to be easily cross-referenced with other fisheries related databases on quota allocations, catch and landing statistics and fisheries violations in order to ascertain the compliance of fishing vessels with national fisheries laws and regulations. Flag State enforcement is necessary in order to ensure compliance not only with internationally agreed fisheries conservation and management measures, but also those adopted under bilateral, subregional, or regional agreements or arrangements. Enforcement actions that may be taken by the flag State include investigation of alleged violations, cooperation with other States, and institution of proceedings

---

[144] *LOSC*, Arts. 62, 91 and 94(1) and (2); *UN Fish Stocks Agreement*, Art. 18(3); *FAO Compliance Agreement*, Art. III(2); *FAO Code of Conduct*, Art. 8.2.2; *IPOA-IUU,* paras. 34–41 and 44–47.

[145] *IPOA-IUU*, paras. 36 and 38.

[146] *FAO Compliance Agreement*, Art. III(1); *UN Fish Stocks Agreement*, Art. 18(1); *IPOA-IUU*, para. 34.

[147] *UN Fish Stocks Agreement*, Art. 18(3)(c); *FAO Compliance Agreement*, Arts. III(6), IV, VI(1), and VI(2); *FAO Code of Conduct*, Art. 8.2.1; *IPOA-IUU*, para. 42.

against vessels entitled to fly their flag which are believed to have contravened fisheries laws and regulations.[148]

Coastal State measures to combat IUU fishing focus on the implementation of effective MCS measures in the EEZ.[149] Such measures include licensing of foreign fishing vessels,[150] and the implementation of a vessel monitoring system (VMS)[151] and observer programmes.[152] Coastal State measures also include enforcement related actions such as boarding, inspection, and arrest of vessels,[153] hot pursuit,[154] and application of sanctions of sufficient severity.[155] Effective MCS measures require the establishment of procedures for the confidentiality and security of commercially sensitive fisheries information[156] and the admissibility of electronic evidence and technologies in courts.[157] Ensuring compliance with fisheries management and enforcement measures is not limited to the application of sanctions, but also incentives that will encourage obedience and self compliance among fishing vessels and fishers.

A number of port State and trade related measures have been identified by the IPOA-IUU and other relevant instruments. Port State measures include the requirement for an advanced notice of port entry,[158] fishing vessel inspections in port[159] and port enforcement actions such as denying the entry of a vessel into port and refusing the landing or transshipment of fish.[160] Trade related measures include the implementation of eco-labelling,[161] catch certification, and trade documentation schemes.[162] Trade restrictions such as prohibiting the importation or exportation of fish may be applied by States to combat IUU fishing.[163] These port State and trade related measures must be adopted and implemented in a fair, transparent and non-discriminatory manner and consistent with international trade law rules.[164]

---

148 *UN Fish Stocks Agreement*, Part VI; *FAO Compliance Agreement*, Art. III(8); *FAO Code of Conduct*, Art. 8.2.7.

149 *UN Fish Stocks Agreement*, Art. 18(3)(g); *IPOA-IUU*, para. 24.

150 *LOSC*, Art. 62(4).

151 *LOSC*, Art. 62(4)(e); *UN Fish Stocks Agreement*, Arts. 5(j) and 18(3)(g)(iii); *FAO Code of Conduct*, Art. 7.7.3; *IPOA-IUU*, para. 24.3.

152 *LOSC*, Art. 62(4)(g); *UN Fish Stocks Agreement*, Art. 18(3)(g)(ii); *FAO Code of Conduct*, para. 8.4.3; *IPOA-IUU*, para. 24.4.

153 *LOSC*, Art. 73(1); *UN Fish Stocks Agreement*, Art. 21; *IPOA-IUU*, para. 24.10.

154 *LOSC*, Art. 111.

155 *IPOA-IUU*, para. 21; *UN Fish Stocks Agreement*, Art. 19(2); *FAO Compliance Agreement*, Art. III(8); *FAO Code of Conduct*, par. 8.2.7.

156 *UN Fish Stocks Agreement*, Annex I, Arts. 1(1) and 7(1); *IPOA-IUU*, para. 24.9.

157 *IPOA-IUU*, para. 17.

158 *IPOA-IUU*, para. 55.

159 *UN Fish Stocks Agreement*, Art. 23(2); *IPOA-IUU*, para. 58.

160 *UN Fish Stocks Agreement*, Art. 23(3); *IPOA-IUU*, para. 56.

161 Agenda 21, Chapter 4, *Changing Consumption Patterns*, para. 4.21.

162 *IPOA-IUU*, paras. 69 and 71.

163 *IPOA-IUU*, para. 69.

164 *IPOA-IUU*, paras. 52 and 66.

Lastly, the IPOA-IUU provides for the application of all State measures. The specific measures discussed in this book that fall under this category are the ratification and implementation of international fisheries specific and non-fisheries specific instruments,[165] adoption of applicable legislation[166] and national plan of action,[167] State control over nationals apart from fishing vessels,[168] and elimination of economic incentives against IUU vessels.[169]

1.3. Purpose and Outline of the Book

This book analyses the concept of IUU fishing and examines the legal and policy measures adopted under international fisheries related instruments to address IUU fishing. The book also examines some of the international developments, measures and practices of RFMOs and some States to combat IUU fishing. The book is divided into 11 chapters. This introductory chapter highlighted the impact of IUU fishing on the sustainability of fisheries resources and the need for establishing and implementing an effective framework to address the problem. In the succeeding chapters, a background to IUU fishing is provided; the international legal and policy framework is reviewed; and regional and State practices in dealing with the IUU fishing problem are presented and analysed.

More specifically, Chapter 2 reviews the history of IUU fishing and examines the scope of the different components of the IUU terminology based on the IPOA-IUU, as well as other international instruments and developments. Chapter 3 examines the existing international legal and policy framework to address IUU fishing within the context of sustainable fisheries. This framework comprises both binding and non-binding fisheries and non-fisheries specific instruments, including instruments related to environment, trade, and safety. Chapters 4 to 8 provide analysis of all State, flag State, coastal State, port State, and market and trade related measures adopted in the IPOA-IUU and other relevant international instruments to combat IUU fishing. Examples of relevant practices of RFMOs, regional economic integration organisations and individual States in implementing these measures are elaborated in Chapters 9 and 10.

The concluding chapter identifies a number of emerging IUU related issues which require further consideration by States and RFMOs to strengthen international cooperative efforts to combat IUU fishing. The issues identified include the

---

[165] *IPOA-IUU,* paras. 10–15.
[166] *IPOA-IUU*, paras. 16–17.
[167] *IPOA-IUU*, paras. 25–27.
[168] *IPOA-IUU*, paras. 18–19.
[169] *IPOA-IUU*, para. 23.

responsibility, as opposed to duty, and liability of flag States for the IUU fishing activities of their vessels, the importance of port, market, and trade related measures in addressing IUU fishing, IUU fishing within the context of biodiversity conservation and environmental protection, and the possible link between IUU fishing and transnational organised crime.

Chapter Two

History and Scope of IUU Fishing

2.1. History of IUU Fishing Terminology

A chronology of events led to the emergence of the term "illegal, unreported, and unregulated fishing" and its formal inclusion in the IPOA-IUU in 2001.[1] The IUU fishing concept evolved from several discussions in regional and international fora from the early 1990s. Much of the earlier discussions on IUU fishing took place in the United Nations Conference on Environment and Development (UNCED), the Commission for the Conservation of Antarctic Marine Living Resources (CCAMLR), the United Nations General Assembly (UNGA), the Food and Agriculture Organization (FAO), the UN Commission on Sustainable Development, the International Maritime Organization (IMO), the Sydney FAO Experts Consultations and the Joint FAO/IMO Ad hoc Working Group on IUU Fishing.

2.1.1. *UNCED and Agenda 21*

The first post-LOSC instrument to identify elements of what is now termed IUU fishing was Agenda 21, one of the products of UNCED in 1992.[2] Chapter 17 of Agenda 21 identified several factors hindering the sustainable management of high seas fisheries, including "unregulated fishing, overcapitalization, excessive fleet size, vessel reflagging to escape controls, insufficiently selective gear, unreliable databases and lack of sufficient cooperation between States."[3] Chapter 17 of Agenda 21 then called on the international community to take urgent action at the bilateral, subregional, and regional levels to address these issues in the

---

[1] Food and Agriculture Organization (FAO), *International Plan of Action to Prevent, Deter and Eliminate Illegal, Unreported and Unregulated Fishing,* adopted on 23 June 2001 at the 120th Session of the FAO Council. *Hereinafter referred to as* IPOA-IUU.

[2] Other issues enumerated in United Nations Conference on Environment and Development (UNCED) as concerns for the management of high seas fisheries are "vessels flagging to escape controls, fishing practices such as large scale pelagic driftnetting and overfishing, lack of surveillance, control, and enforcement and lack of required cooperation with other States." *See* Preparatory Committee for the United Nations Conference on Environment and Development, Third Session, Geneva, 12 August–4 September 1991, Working Group II, Agenda Item 2, Annex, *A/CONF.151/PC/WG.II/L.18*, 23 August 1991, para. 59.

[3] UNCED, *Agenda 21*, Chapter 17, Protection of the oceans, all kinds of seas, including enclosed and semi-enclosed seas, and coastal areas and the protection, rational use and development of their living resources, Rio de Janeiro, Brazil, 03–14 June 1992, para. 17.45. *Hereinafter referred to as* Agenda 21.

management of fisheries resources, particularly highly migratory species and straddling stocks. In taking such cooperative action, States were to address concerns such as "inadequacies in fishing practices, as well as in biological knowledge, fisheries statistics and improvement of systems for handling data."[4] Although the term IUU fishing was not specifically mentioned in Chapter 17 of Agenda 21, the various issues identified, in broad terms, cover some of the elements of IUU fishing in the IPOA-IUU.

2.1.2. *CCAMLR Sessions*

Specific concerns about illegal and unreported fishing were first raised during the Fifteenth Session of CCAMLR in 1996.[5] A number of CCAMLR members discussed the extent of illegal fishing in the Convention area by its members[6] as well as by non-members of the Commission.[7] Vessels of non-members had been observed to be fishing with total disregard for CCAMLR conservation measures and provided no reports of their catches from the Convention area.[8] Some vessels which were fishing illegally in CCAMLR waters were also found to be reflagged to members of the Commission.[9] Illegal fishing, particularly in the longline fishery for toothfish and its expansion in the Southern Ocean was identified as a particular challenge for CCAMLR.[10] The report of the Fifteenth meeting of CCAMLR also highlighted the continued high incidence of daylight setting of longlines in contravention of established conservation measures.[11]

At the Fifteenth Session, the Scientific Committee of CCAMLR noted the high level of unreported catches in certain sub-areas covered by the Convention.[12] Although the term "unreported fishing" was not elaborated on during the Fifteenth Session, there were references to what fishing related activities should be reported by members of the Commission. It was agreed that reports to CCAMLR should not only provide a list of member vessels intending to fish in a particular season,[13] but also information on the fishing vessels of members operating in the Convention

[4] *Agenda 21*, para. 17.45.

[5] During the Fifteenth Meeting of CCAMLR, the delegation of Norway identified illegal and unreported fishing as the "greatest threat to the Commission." The UK also identified illegal fishing as a "very real and worrying problem which the Commission needs to address if the credibility of CCAMLR is to be maintained." *See* Commission for the Conservation of Antarctic Marine Living Resources (CCAMLR), *Report of the Fifteenth Meeting of the Commission*, Hobart, Australia, 21 October–01 November 1996, CCAMLR-XV, Tasmania, Australia, 1996, para. 12.13 and 13.24.

[6] *Ibid.*, para. 7.12 and Annex 5, paras. 1.38–1.46.

[7] *Ibid.*, Annex 5, paras. 1.53 to 1.58.

[8] *Ibid.*, para. 7.13.

[9] *Ibid.*, para. 7.15.

[10] *Ibid.*, para. 11.61.

[11] *Ibid.*

[12] *Ibid.*, paras. 2.9 and 4.5.

[13] *Ibid.*, para. 7.21 and Annex 5, para. 1.50.

area.[14] It was further agreed that CCAMLR members provide, in as close to real time as possible, available information on vessels that fished or intended to fish in the Convention area which were either on their registers and vessels which had been renamed, or which had been reflagged to other States.[15]

Discussions on "illegal" and "unreported" fishing during the Sixteenth Session of the CCAMLR in 1997 led to the first formal mention of IUU fishing. The term "unregulated fishing" was also used although not defined. The terms appeared as "Agenda Item 1, Illegal, Unregulated, and Unreported Fishing in the Convention area" at the Seventh Session of the Standing Committee on Observation and Inspection and as "Agenda Item 5 at the Sixteenth Session of the Commission."[16] In its Sixteenth Session, CCAMLR identified the high level of illegal, unregulated and unreported fishing as a major concern.[17]

Particular emphasis was placed on resolving problems of unregulated fishing in the Sixteenth Session. The Scientific Committee recommended to the Commission to take the strongest possible action to eliminate unregulated fishing,[18] which was estimated to be several times greater than regulated fishing.[19] It was also noted that unregulated fishing resulted in the depletion of several non-targeted species.[20] This problem was compounded by the non-reporting of data by vessels conducting unregulated fishing.[21]

Unlike in the Fifteenth Session, discussions at the Sixteenth Session of CCAMLR clearly distinguished the term "unregulated fishing" from "illegal fishing."[22] "Unregulated fishing" was used to describe the fishing activities of vessels flying the flags of non-CCAMLR members within the jurisdiction of members and on the high seas within the CCAMLR area. "Illegal fishing" was referred to as an activity that severely compromised the management of the Patagonian toothfish, resulting in an unsustainable level of fishing.[23] Mention was also made of "illegal catches" as capture that exceeded the legal catch limit set by CCAMLR.[24]

---

[14] *Ibid.*, para. 7.21.

[15] *Ibid.*, Annex 5, para. 1.52.

[16] David Doulman, 'Illegal, Unreported and Unregulated Fishing: Mandate for an International Plan of Action,' in FAO, Report of and Papers Presented at the Expert Consultation on Illegal, Unreported and Unregulated Fishing, Sydney, Australia, 15–19 May 2000, *FAO Fisheries Report No. 666* (Rome: FAO, 2001), at 53.

[17] CCAMLR, 'Annex 6: Communication Policy with Non-Contracting Parties Relating to Illegal, Unreported and Unregulated Fishing in the CCAMLR Convention Area,' *Report of the Sixteenth Meeting of the Scientific Committee*, Hobart Australia, 27–31 October 1997, SC-CCAMLR-XVI, Tasmania, Australia, 1997.

[18] *Ibid.*, para. 4.55 and Annex 5, para. 7.96.

[19] *Ibid.*, para. 2.13.

[20] *Ibid.*, para. 4.55.

[21] *Ibid.*, paras. 2.13 and 4.55.

[22] *Ibid.*, Annex 6, SCOI Report, para. 1.20.

[23] *Ibid.*, para. 5.100.

[24] *Ibid.*, paras. 5.109 and 5.130.

In the Sixteenth Session, "unreported fishing" was described as unreported catch which had deleterious impacts on the management and conservation of fisheries resources.[25] The Scientific Committee drew attention to the substantial amount of unreported catches of Patagonian toothfish, particularly in the Southern Ocean but within the jurisdiction of coastal States.[26] It was noted that the fish stock had been heavily fished around the sub-Antarctic region within the CCAMLR area and had been the most significant challenge that the Commission was facing in the protection of Antarctic marine living resources.[27] A strong relationship was established between unregulated and illegal fishing.[28] It was noted that unreported fishing, when combined with illegal and unregulated fishing, resulted in unsustainable levels of fishing.[29]

From 1997, the term IUU fishing was used regularly at CCAMLR meetings. In 1999, the term found its way into meeting reports of international organisations, including the FAO, the IMO and the UN Commission on Sustainable Development, and other regional fisheries organisations.[30]

2.1.3. *UN General Assembly*

The first reference made to "unregulated fishing" by the UNGA was in 1994 in the context of "unauthorised fishing in zones under national jurisdiction."[31] In the Forty-ninth Session of UNGA, States were called upon "to take measures to ensure that no fishing vessels entitled to fly their flags fish in zones under the national jurisdiction of other States unless duly authorised by the competent authorities of the States concerned."[32] In subsequent years, the UN General Assembly discussed "unauthorised fishing" in the context of other international fisheries concerns.[33]

---

[25] *Ibid.*, paras. 7.9, 5.124 and 5.126.

[26] *Ibid.*, Annex 6, SCOI Report, paras. 1.5 and 1.29.

[27] Stuart Kaye, 'CCAMLR and Southern Ocean Fisheries', in Sam Bateman and Donald Rothwell, (eds.), *Southern Ocean Fishing: Policy Challenges for Australia*, Wollongong Paper on Maritime Policy No. 7 (Wollongong: Centre for Maritime Policy, 1998) at 75–76.

[28] CCAMLR, *Sixteenth Meeting Report of the Scientific Committee*, *supra* note 17, para. 5.108.

[29] *Ibid.*, paras. 5.100 and 5.108.

[30] Doulman, *IUU Fishing: Mandate for an International Plan of Action, supra* note 16, at para. 40.

[31] United Nations General Assembly (UNGA), 92nd Plenary Meeting, Unauthorised Fishing in Zones of National Jurisdiction and Its Impact on the Living Marine Resources of the World's Oceans and Seas, *A/RES/49/116*, 19 December 1994.

[32] *Ibid.*, para. 1.

[33] *See* UNGA, Fiftieth Session, Agenda item 96(c), Large-scale pelagic drift-net fishing and its impact on the living marine resources of the world's oceans and seas; unauthorised fishing in zones of national jurisdiction and its impact on the living marine resources of the world's oceans and seas; and fisheries by-catch and discards and their impact on the sustainable use of the world's living marine resources, *A/RES/50/25,* 5 December 1995; UNGA, Fifty-first session, Agenda item 24(c), Large-scale pelagic drift-net fishing and its impact on the living marine resources of the world's oceans and seas; unauthorised fishing in zones of national jurisdiction and its impact on the living

In 1999, IUU fishing was formally mentioned in the UNGA when the Secretary-General of the UN noted that IUU fishing is

> …one of the most severe problems currently affecting world fisheries. IUU fishing is often undertaken by fishing vessels of States or entities that are not members of fisheries organisations or arrangements and do not consider themselves bound by the restrictions imposed by those management organisations and arrangements. IUU fishing is also undertaken by vessels that were formerly registered in a State member of regional fisheries organisations or arrangements but were subsequently registered in a non-member State (reflagging to a flag of convenience) to avoid compliance with conservation and management measures…[34]

Similar to CCAMLR, the UNGA identified the activities of fishing vessels flying the flags of non-members of RFMOs as a major IUU fishing concern. Other IUU-related issues identified by the UNGA include illegal catch; catching beyond allowable quota; non-reporting of catches; unsustainable fishing practices and use of non-selective fishing gears; by-catch, discards and incidental catch of non-target species; deliberate loss of gear to evade sighting and inspection; and over-capacity in the fishing industry.[35]

2.1.4. *FAO Sessions*

Discussion of IUU fishing at the Twenty-third Session of the Committee on Fisheries (COFI) in February 1999 was largely based on the paper prepared by Australia.[36] The paper urged the FAO to develop an international plan of action to prevent IUU fishing. Concerns about IUU fishing were discussed in the Twenty-third Session, together with the problems of fishing vessels under "flags of convenience".[37] Shortly after the Twenty-third Session in March 1999, the Rome Declaration on Responsible Fisheries was adopted by the FAO Ministerial

---

marine resources of the world's oceans and seas; and fisheries by-catch and discards and their impact on the sustainable use of the world's living marine resources, *A/RES/51/36,* 9 December 1996; UNGA, Fifty-second Session, Agenda item 39(c), Resolution adopted by the General Assembly [without reference to a Main Committee (A/52/L.30 and Add.1)], Large-scale pelagic drift-net fishing; unauthorised fishing in zones of national jurisdiction and on the high seas; fisheries by-catch and discards; and other developments, *A/RES/52/29,* 26 January 1998; UNGA, Fifty-third session, Agenda item 38(b), Large-scale pelagic drift-net fishing; unauthorised fishing in zones of national jurisdiction and on the high seas; fisheries by-catch and discards, and other developments, *A/RES/53/33,* 24 November 1998.

[34] UNGA, Fifty-fourth Session, Agenda Items 40(a) and (c), Oceans the Law of the Sea; Law of the Sea; Results of the Review by the Commission on Sustainable Development of the Sectoral Theme of "Oceans and Seas", Oceans and the Law of the Sea, *Report of the Secretary-General, A/54/429,* 30 September 1999, para. 249.

[35] *Ibid.,* paras. 255–257.

[36] FAO, Committee on Fisheries, Twenty-fourth Session, Rome, Italy, 26 February–2 March 2001, *Illegal, Unreported and Unregulated Fishing: A Proposal for a Draft International Plan of Action*, COFI 2001/7.

[37] FAO, Report of the Twenty-Third Session of the Committee on Fisheries, Rome, Italy, 15–19 February 1999, *FAO Fisheries Report No. 595* (Rome: FAO, 1999), at paras. 72 and 73. Although there

Meeting on Fisheries. The Declaration raised concern about the growing incidents of IUU fishing activities, including fishing vessels under "flags of convenience".[38] The meeting declared that the FAO Ministers and their representatives would develop a global plan of action to address such issues.[39]

In April 1999, regional fisheries bodies and States were informed of the FAO mandate to develop an International Plan of Action to combat IUU fishing. States and regional fisheries organisations were invited to provide information on the types and impacts of IUU fishing activities in their respective jurisdictions. The global review of IUU fishing identified the most common aspects of IUU fishing, including the lack of effective control of fishing vessels by some flag States; difficulty experienced by regional fisheries bodies in applying fisheries management and conservation measures to vessels of non-members, especially those flying "flags of convenience"; ineffectiveness of measures implemented by coastal States; and the lack of adequate human and financial resources.[40]

The Report of the Twenty-third Session of COFI was adopted at the 116th Session of the FAO Council in June 1999. The Council stressed the importance of implementing the FAO Code of Conduct for Responsible Fisheries in promoting the sustainability of global fisheries,[41] and noted that illegal, unauthorised and unreported fishing, including fishing by vessels flying "flags of convenience", undermine the conservation and management of fisheries resources.[42] It is worth noting that in the latter remark made by the Council, "unauthorised fishing" was used instead of "unregulated fishing." It was agreed that the FAO should develop an international plan of action to address the problem of IUU fishing within the framework of the FAO Code of Conduct.[43]

are no legally accepted definitions of both terms, "flag of convenience" may be distinguished from "open registry" in that flags of convenience are characterised as involving open registry States which are unable or unwilling to assume responsibility for the vessels it flags and thereby allow vessels to avoid compliance with conservation and management measures, whereas open register States is simply one that flies vessels owned by nationals of other countries. *See* Judith Swan, Fishing Vessels Operating Under Open Registers and the Exercise of Flag State Responsibilities: Information and Options, *FAO Fisheries Circular No. 980* (Rome: FAO, 2002).

[38] FAO, *The Rome Declaration on the Implementation of the Code of Conduct for Responsible Fisheries*, Adopted by the FAO Ministerial Meeting on Fisheries, Rome, Italy, 10–11 March 1999, para. 2.

[39] *Rome Declaration on the Implementation of the Code of Conduct for Responsible Fisheries*, para. 12(j).

[40] Kevin Bray, 'A Global Review of Illegal, Unreported, and Unregulated (IUU) Fishing', in FAO, Report of and Papers Presented at the Expert Consultation on Illegal, Unreported and Unregulated Fishing, Sydney, Australia, 15–19 May 2000, *FAO Fisheries Report No. 666* (Rome: FAO, 2001), at 91.

[41] FAO Council, Hundred and Sixteenth Session, Rome, Italy, 14–19 June 1999, Current World Food Situation, *CL 116/2* (Rome: FAO, 1999), para. 28.

[42] *Ibid.*, para. 30.

[43] *Ibid.*

2.1.5. *UN Commission on Sustainable Development*

The UN Commission on Sustainable Development, in its Seventh Session in 1999 declared that IUU fishing and unsustainable and uncontrolled distant water fishing constitute threats to the sustainable use of the oceans and seas.[44] The Commission supported the Rome Declaration endorsing the need to develop a global plan of action to deal with IUU fishing.[45] According to the Commission, the plan of action should include the following elements: effective flag State jurisdiction, control over vessels which operate in a manner that undermines international law or conservation and management measures, as well as coordinated efforts among States, FAO, RFMOs and other relevant international agencies such as the IMO.[46] The Commission specifically called on the IMO to ensure that all flag States comply with relevant international rules and standards and to develop effective port State control measures.[47]

Delegates to the Seventh Session of the Commission recognised the increasing problem of IUU fishing and fishing activities of fishing vessels under "flags of convenience" and identified the urgent need to eradicate IUU fishing.[48] Similar to the discussions in CCAMLR, UNGA and the FAO, the Seventh Session of the UN Commission on Sustainable Development highlighted the unregulated fishing activities of "flags of convenience" vessels.

2.1.6. *IMO Sessions*

The results of the Twenty-third session of COFI were discussed at the Seventy-first Session of the IMO's Maritime Safety Committee (MSC). In this session, the FAO invited the Committee to note concerns about the increasing reports of IUU fishing activities and their negative impact on the management and conservation of living resources on the high seas.[49]

IUU fishing was also discussed during the Eighth Session of the IMO's Sub-Committee on Flag State Implementation (FSI) in January 2000 where it was agreed that effective cooperation between the FAO and IMO to address the problem was needed. It was stressed that the impacts of IUU fishing transcend fisheries management.[50] There is a direct linkage between IUU fishing and issues such

---

[44] UN, Commission on Sustainable Development, *Report on the Seventh Session*, New York, 01 May and 27 July 1998, and 19–30 April 1999, Economic and Social Council Official Records, 1999 Supplement No. 9 (E/1999/29), United Nations, New York, 1999, Chapter I, Decision 7/1, Part II, para. 3.

[45] *Ibid.*, Chapter I, Decision 7/1, Part II, para. 18.

[46] *Ibid.*

[47] *Ibid.*, para. 35.

[48] *Ibid.*, Annex, Part III, para. 10.

[49] Doulman, *IUU Fishing: Mandate for an International Plan of Action, supra* note 16, at 51.

[50] International Maritime Organization (IMO), Sub-committee on Flag State Implementation (FSI), Eighth Session, Agenda Item 19, *Report to the Maritime Safety Committee and the Marine Environment Protection Committee*, FSI 8/19, 14 February 1999, para. 6.10.

as vessel safety, fishing vessel casualties and pollution of the marine environment, which are well within the mandate and jurisdiction of the IMO.[51] Most importantly, the Eighth Session of FSI emphasised the lack of implementation of flag State duties and vessel reflagging as the root causes of IUU fishing.[52] The formation of a joint IMO/FAO *Ad Hoc* Working Group to further examine the matter was proposed.[53]

The Forty-fourth Session of IMO's Marine Environment Protection Committee (MEPC) considered the recommendation of the Report of the FSI at its Eighth Session and agreed that the IMO should provide assistance to the FAO in dealing with IUU fishing, especially with respect to safety and pollution prevention from fishing vessels and other related issues.[54] The MSC also supported the recommendation of the FSI for the creation of a joint ad hoc working group and proposed that the group prepare a checklist of the necessary elements for effective flag State control over fishing vessels.[55]

2.1.7. *Sydney Experts Consultation Meeting*

Following consultations with RFMOs, the government of Australia, in cooperation with the FAO, convened an FAO Expert Consultation on IUU Fishing in May 2000. The report of the Expert Consultation appended the text of a draft plan entitled the "International Plan of Action to Prevent, Deter, and Eliminate Illegal, Unreported and Unregulated Fishing".[56]

The draft text of the International Plan of Action provided a vague scope of IUU fishing. Paragraph 3 of the draft stated that IUU fishing encompasses "fishing and related activities, including fishing in areas under national jurisdiction without the authorisation of the coastal State; fishing which contravenes or undermines conservation and management measures; failure to effectively exercise the required jurisdiction or control over vessels and nationals; and failure to fully and accurately meet fishery and fishing vessel data collection and reporting requirements."[57] Key measures for combating IUU fishing were also identified in the draft International Plan of Action. These include all State, flag State, and port

---

[51] *Ibid.*, para. 6.11.

[52] *Ibid.*, para. 6.10.

[53] *Ibid.*, para. 6.13.1.

[54] IMO, Marine Environment Protection Committee (MEPC), Forty-fourth Session, Agenda Item 20, *Report of the Marine Environment Protection Committee on Its Forty-fourth Session*, MEPC 44/20, 12 April 2000, para. 8.10.

[55] IMO, Maritime Safety Committee (MSC), 72nd Session, Agenda Item 23, *Report of the Maritime Safety Committee on Its Seventy-Second Session*, MSC 72/23, 31 May 2000, para. 7.25.

[56] *See* FAO, 'Report of the Expert Consultation,' in FAO, Report of and Papers Presented at the Expert Consultation on Illegal, Unreported and Unregulated Fishing, Sydney, Australia, 15–19 May 2000, *FAO Fisheries Report No. 666* (Rome: FAO, 2001), at 1–23.

[57] *Ibid.*, Appendix D, 'Preliminary Draft of the International Plan of Action to Prevent, Deter and Eliminate Illegal, Unreported and Unregulated Fishing,' para. 3.

State measures, internationally-agreed market measures, research, the role of RFMOs, special requirements of developing States, reporting, and the role of FAO.[58]

While discussion among States at the UNGA reflected the general view that illegal fishing primarily involves activities of foreign fishing vessels in the EEZs of coastal States,[59] the Expert Consultation Meeting alluded to IUU fishing in small scale fisheries which is normally conducted by fishing vessels in the internal, territorial, or archipelagic waters of their own States or neighbouring States.[60]

2.1.8. *FAO Technical Consultations*

The draft prepared at the Sydney Experts Consultation Meeting provided the basis for discussion and negotiation at the First FAO Technical Consultation on IUU fishing which was held in Rome in October 2000.[61] The Technical Consultation improved on the draft from the Sydney Experts Meeting and noted that further work was needed on two sections, namely, port State measures and the role of RFMOs, and proposed that a section on coastal State measures be added.[62] A second session was convened to refine the draft International Plan of Action prior to its submission to the Twenty-fourth Session of the COFI. The most notable addition to the draft text was the scope of IUU fishing. The nature and scope of each component of IUU fishing was refined and the phrase "related activities" which was included in the earlier draft was removed.

2.1.9. *Joint FAO/IMO Ad Hoc Working Group on IUU Fishing*

The Joint FAO/IMO Ad Hoc Working Group on IUU Fishing was established in October 2000.[63] At its first meeting, the Ad Hoc Working Group focused the scope of IUU fishing under two themes: the identification of the necessary

---

[58] *Ibid.*, para. 4.

[59] *See* UNGA, Reports of the Secretary-General from the Fifty-first to the Fifty-fifth Sessions, A/51/404, A/52/557, A/53/473, A/54/429, and A/55/386, 1996–2000. In these reports, the UNGA called upon States to adopt measures to ensure that no fishing vessels entitled to fly their flags fish in areas under the national jurisdiction of other States unless duly authorised by the competent authorities of the coastal States or States concerned. Such authorised fishing operations should also be carried out in accordance with the conditions set out in the authorisation.

[60] *See* Ousman K.L. Drammeh, 'Illegal, Unreported and Unregulated Fishing in Small-scale Marine and Inland Capture Fisheries,' in FAO, Report of and Papers Presented at the Expert Consultation on Illegal, Unreported and Unregulated Fishing, Sydney, Australia, 15–19 May 2000, *FAO Fisheries Report No. 666* (Rome: FAO, 2001), at 139.

[61] *See* FAO, Report of the Technical Consultation on Illegal, Unreported and Unregulated Fishing, Rome, Italy, 02–06 October 2000, *FAO Fisheries Report No. 634* (Rome: FAO, 2000).

[62] *Ibid.*, paras. 9–10.

[63] IMO, FSI, 9th Session, Agenda Item 15, 'Illegal, Unregulated, and Unreported Fishing and Related Matters,' *Outcome of the First Meeting of the Joint FAO/IMO Ad Hoc Working Group on IUU Fishing and Related Matters*, FSI 9/15, 08 November 2000.

elements for effective flag State control over fishing vessels and the review of measures that may be taken by port States for the inspection of foreign fishing vessels entering their ports.[64] A checklist for flag State control was also formulated which covered not only fishing vessels but also transport and support vessels.[65] The checklist included the registration of fishing vessels, authorisation to fish, records of fishing vessels, position reporting, and catch data reporting.

2.1.10. *FAO Twenty-Fourth Session*

The IPOA-IUU was adopted on 02 March 2001 by COFI and endorsed by the FAO Council at its 120th Session in June 2001. The IPOA-IUU provides the scope and various components of IUU fishing which are discussed in greater detail in the following section.

2.2. Scope of Illegal, Unreported, and Unregulated Fishing

2.2.1. *Definitions of "Fishing" and "Fishing Vessel"*

The concept of IUU fishing needs to be understood in the wider international legal context of the terms "fishing vessel" and "fishing". The definitions of these terms are not provided in the IPOA-IUU or in most international agreements. Various provisions of the LOSC refer to fishing,[66] fishing vessel,[67] and fisheries laws and regulations[68] without any definition of these terms. However some of the post-LOSC instruments and regional fisheries agreements have incorporated varying definitions of "fishing" and "fishing vessel".

2.2.1.1. *Fishing Vessel*

The FAO Compliance Agreement defines a fishing vessel as "any vessel used or intended for use for the purposes of the commercial exploitation of living marine resources, including mother ships and any other vessels directly engaged in such fishing operations."[69] This definition may be interpreted to cover vessels used for

---

[64] FAO, *Report of the Joint FAO/IMO Ad Hoc Working Group on Illegal, Unreported and Unregulated Fishing and Related Matters*, Rome, Italy, 09–11 October 2000, *FAO Fisheries Report No. 637* (Rome: FAO, 2001), at Appendix F.

[65] *Ibid.*

[66] See for example, United Nations Convention on the Law of the Sea, Montego Bay, Jamaica, concluded on 10 December 1982, in force 16 November 1994, 1833 UNTS 3; 21 ILM 1261 (1982). *Hereinafter referred to as LOSC*, Arts. 1(4), 19(2)(i), 42(1)(i), 51, 60(3), 61(3), 61(5), 62(4), 6392), 66(2), 66(3), 66(5), 67(2), 116, 147(2)(b), 221(1), and 297(3)(ii).

[67] See for example, *LOSC*, Arts. 42(1)(c) and 62(4)(a), (c) and (e).

[68] See for example, *LOSC*, Arts. 21(1)(e) and 73.

[69] FAO, *Agreement to Promote Compliance with International Conservation and Management Measures by Fishing Vessels on the High Seas*, Rome, Italy, concluded on 24 November 1993,

the harvesting of fish using different gears,[70] fish factory ships, fish carriers, and other vessels engaged in fishing. Because the definition focuses on the exploitation of fisheries resources, it may not necessarily include support vessels for the purpose of resupplying fishing vessels, as well as reefer vessels or other refrigerated cargo vessels carrying fish products that have not been previously landed in ports. Indeed, transshipment operations were implicitly excluded from the FAO Compliance Agreement. The most that could be achieved was a passing reference to vessels engaged in the transshipment of fish in the preamble. However, two years later when the UN Fish Stocks Agreement was finalised, no definition of "fishing vessel" was provided, while there were several references to "transshipment".[71]

The definition of "fishing vessel" provided in the FAO Compliance Agreement has been adopted in the FAO Technical Guidelines for Responsible Fisheries[72] and further expanded on in the FAO Model Scheme on Port State Measures to Combat IUU Fishing to specifically include support ships and carrier vessels.[73] This definition has also been incorporated into a number of regional fisheries management agreements, which made specific reference to vessels engaged in the transshipment of fish.[74] The European Union (EU) regulation on IUU fishing of 2008 (EC Regulation No 1005/2008) provides the most explicit description of the term "fishing vessel" consistent with the definition provided in international and regional fisheries agreements. This regulation defines a fishing vessel as "any vessel of any size used or intended for use for the purposes of commercial exploitation of fishery resources, including support ships, fish processing vessels, vessels engaged in transshipment and carrier vessels equipped for the transportation of fishery products, except container vessels."[75]

---

entered into force 24 April 2003, 33 ILM 968. *Hereinafter referred to as* FAO Compliance Agreement, Art. 1(a).

[70] Examples of gears that may be used for commercial exploitation of fish are purse seine, seine nets, trawls, hooks and lines, lift nets, gillnets, and entangling nets and traps.

[71] UN, *Agreement for the Implementation of the Provision of the United Nations Convention on the Law of the Sea of 10 December 1982 Relating to the Conservation and Management of Straddling Fish Stocks and Highly Migratory Fish Stocks*, New York, 08 September 1995. *Hereinafter referred to as UN Fish Stocks Agreement,* Arts. 18(3)(f), 18(3)(h), 23(3), and Annex I Art. 6(c).

[72] FAO Fishing Technology Service, Fishing Operations, *FAO Technical Guidelines for Responsible Fisheries No. 1* (Rome: FAO, 1996), at 3.

[73] FAO, *Model Scheme on Port State Measures to Combat Illegal, Unreported and Unregulated Fishing* (Rome: FAO, 2007), para. 1.2.

[74] See for example Convention on the Conservation of Highly Migratory Fish Stocks in the Western and Central Pacific Ocean, Honolulu, Hawaii, concluded 5 September 2000, entry into force 19 June 2004, [2004] ATS 15. *Hereinafter referred to as* WCPF Convention. Art. 1(e). Convention on the Conservation and Management of Fishery Resources in the Southeast Atlantic Ocean, Windhoek, Namibia, concluded on 20 April 2001, in force 13 April 2003; 41 ILM 257. *Hereinafter referred to as* SEAFO Convention, Art. 1(i); *Southern Indian Ocean Fisheries Agreement*, Rome, Italy, concluded on 29 December 2006, ATNIF 31, Art. 1(g).

[75] European Union (EU), Council Regulation (EC) No 1005/2008 of 29 September 2008 establishing a Community system to prevent, deter and eliminate illegal, unreported and unregulated

2.2.1.2. *Fishing*

One of the earliest international attempts to define "fishing" was in the Convention for the Prohibition of Fishing with Long Driftnets in the South Pacific in 1989 (Wellington Convention). Under this convention, "driftnet fishing" is defined to mean:

> (i) catching, taking or harvesting fish with the use of a driftnet;
>
> (ii) attempting to catch, take or harvest fish with the use of a driftnet;
>
> (iii) engaging in any other activity which can reasonably be expected to result in the catching, taking or harvesting of fish with the use of a driftnet, including searching for and locating fish to be taken by that method;
>
> (iv) any operations at sea in support of, or in preparation for any activity described in this paragraph, including operations of placing, searching for or recovering fish aggregating devices or associated electronic equipment such as radio beacons;
>
> (v) aircraft use, relating to the activities described in this paragraph, except for flights in emergencies involving the health or safety of crew members or the safety of a vessel; or
>
> (vi) transporting, transhipping and processing any driftnet catch, and cooperation in the provision of food, fuel and other supplies for vessels equipped for or engaged in driftnet fishing.[76]

Since then, other regional fisheries instruments have defined "fishing" in terms similar to that of the Wellington Convention.[77] Definitions in many domestic fisheries legislation, specifically in Pacific Islands region and Australia and New Zealand have adopted similar expanded definition of "fishing".[78]

The terms "fishing" and "fishing vessel" in international and regional fisheries conventions provide the scope for what activities may be considered IUU fishing under these instruments. Similarly in domestic fisheries legislation, the terms "fishing" and "fishing vessel", if at all defined, will provide the starting point for identifying what constitutes IUU fishing. One of the main issues however is the lack of uniform definition which may create conflicting interpretation and application across jurisdictions.[79]

---

fishing, amending Regulations (EEC) No 2847/93, (EC) No 1936/2001 and (EC) No 601/2004 and repealing Regulations (EC) No 1093/94 and (EC) No 1447/1999, *OJ L 286/1 29.10.2008*, Art. 2(5).

[76] *Convention for the Prohibition of Fishing with Long Driftnets in the South Pacific*, Wellington, New Zealand, concluded on 24 November 1989, *and its Protocols*, Noumea, New Caledonia, concluded on 20 October 1990, in force 17 May 1991, 1899 UNTS 3; 29 ILM 1454 (1990). *Hereinafter referred to as the* Wellington Convention, Art. 1(c).

[77] See for example *WCPF Convention*, Art. 1(d); *SEAFO Convention*, Art. 1(h); *Southern Indian Ocean Fisheries Agreement*, Art. 1(g).

[78] For example, Vanuatu, *Fisheries Act No. 55 of 2005*, Art. 1; Tonga, *Fisheries Management Act 2002*, Art. 1; New Zealand, *Fisheries Act 1996 No. 88*, Part I Art. 2(d); Australia, *Fisheries Management Act 1991* (Cth), Sec. 4.

[79] See for example International Tribunal for the Law of the Sea (ITLOS), *The M/V Saiga Case (Saint Vincent and the Grenadines v. Guinea) (Prompt Release)*, Case No. 1, Judgment of 04 December 1997, para. 64.

2.2.2. *IUU Fishing*

The term "IUU fishing" was not defined in the IPOA-IUU; however, the scope of each of the components of the term has been identified. Collectively, IUU fishing encompasses a wide range of fishing activities which can be considered in violation of or without regard to, applicable international, regional, or national fisheries regulations and standards. IUU fishing activities may be conducted by all fishing vessels, with or without nationality in internal waters, archipelagic waters, territorial seas, exclusive economic zones (referred to as "waters under the jurisdiction of a State" in paragraph 3.1 of the IPOA-IUU) and on the high seas, whether nor not under the competence of RFMOs.[80]

By referring to "Nature and Scope of IUU Fishing", the IPOA-IUU appears merely to offer a description or an explanation of what may constitute "illegal fishing", "unreported fishing", or "unregulated fishing", and not a strict definition of these terms.[81] Indeed, some of the terms are explained by using the very term that is being explained, such as in "unreported fishing", which is not normal drafting practice for a formal legal definition. As will be discussed in the sections that follow, the interpretation of these terms varies in the laws, regulations, and measures adopted by States and regional organisations. While the IPOA-IUU distinguished between the three components of IUU fishing, activities related to each element were not enumerated. A broad classification has the advantage of giving States the flexibility to determine which activities may fall within the scope of IUU fishing under the IPOA-IUU, rather than providing an exhaustive list of activities which may be considered as IUU fishing.

The components of IUU fishing are not explicitly identified nor defined in other fisheries related international instruments such as the LOSC, UN Fish Stocks Agreement, FAO Compliance Agreement, and FAO Code of Conduct for Responsible Fisheries.[82] However, some of the provisions of these international instruments, when considered together with the IPOA-IUU, assist in identifying which activities may be characterised as constituting illegal, unreported, or unregulated.

---

[80] The use of "waters under the jurisdiction of a State" in the IPOA-IUU with respect to addressing IUU fishing should, however, be distinguished from the specific and limited usage of the term "areas under national jurisdiction" under Article 3(1) of the UN Fish Stocks Agreement in terms of the obligations of States to cooperate and achieve compatibility of measures in respect of conserving and managing straddling fish stocks and highly migratory species. Given that the duty to cooperate under the LOSC applies only to the EEZ and the high seas, it can be argued that this narrows the scope of "areas under national jurisdiction" under the UN Fish Stocks Agreement to Part V of the LOSC, i.e the EEZ.

[81] *See* Part II heading of the IPOA-IUU. The paragraphs that followed which state that illegal fishing, unreported fishing, and unregulated fishing "refers to activities…" also support the argument that the IPOA-IUU does not provide a strict definition of IUU fishing.

[82] FAO, Code of Conduct for Responsible Fisheries, Adopted at the 28th Session of the FAO Conference, Rome, Italy, 31 October 1995. *Hereinafter referred to as FAO Code of Conduct.*

2.2.2.1. *Illegal Fishing*

"Illegal fishing" is the most common aspect of the IUU fishing terminology. Paragraph 3.1 of the IPOA-IUU provides:

> Illegal fishing refers to activities:
>
> 3.1.1 conducted by national or foreign vessels in waters under the jurisdiction of a State, without the permission of that State, or in contravention of its laws and regulations;
>
> 3.1.2 conducted by vessels flying the flag of States that are parties to a relevant regional fisheries management organisation but operate in contravention of the conservation and management measures adopted by that organisation or by which the States are bound, or relevant provisions of the applicable international law; or
>
> 3.1.3 in violation of national laws or international obligations, including those undertaken by cooperating States to a relevant regional fisheries management organisation.

"Illegal fishing" has three main components: the scope of illegal fishing activities, areas of jurisdiction, and application to different types of fishing vessels. Illegal fishing includes all fishing activities conducted in contravention of national and international laws as well as agreed regional fisheries management and conservation measures. In terms of areas of jurisdiction, paragraph 3.1.1 of the IPOA-IUU refers to illegal fishing activities being conducted in waters under the jurisdiction of a State while paragraph 3.1.2 refers to illegal fishing in areas within the competence of RFMOs. The vague reference to "international obligations" could also mean that any fishing activity on the high seas, as long as they are conducted contrary to internationally agreed fisheries conservation and management measures, may be considered illegal fishing.

The application of the concept of "illegal fishing" to fishing vessels varies depending on the maritime zone the fishing activity takes place. Paragraph 3.1.1 of the IPOA-IUU, which applies to waters under the jurisdiction of a State places no restriction on the types of vessels which may be considered to be conducting illegal fishing. Therefore, illegal fishing within national waters may be conducted by all types of vessels, both national and foreign. Within areas of competence of RFMOs, paragraphs 3.1.2 and 3.1.3 of the IPOA-IUU respectively only apply the concept to "vessels flying the flag of States that are Parties to a relevant regional fisheries management organisation" and "cooperating States to a relevant fisheries management organisation." This limits the scope and application of the term illegal fishing and suggests that the activities of fishing vessels flying the flags of States which are not members of relevant RFMOs would not be considered illegal. However, the fishing activities of third party vessels in the area of competence of an RFMO can be classified as "unregulated fishing", as discussed in Section 2.2.2.3.

Illegal fishing may take various forms. In waters under the jurisdiction of a State, the scope of illegal fishing will be determined by national legislation.

It may include fishing without licence or appropriate fishing vessel registration papers.[83] It may include fishing beyond allocated catch limits,[84] taking of juvenile fish,[85] taking of prohibited fish species,[86] fishing during closed seasons, or fishing in closed areas.[87] Illegal fishing may also include the use of unauthorised fishing gears or fishing contrary to the regulations on the types, sizes, and number of vessels allowed.[88] It may pertain to the use of destructive fishing methods such as explosives and poisons, use of small mesh fishing nets, highly destructive fishing gears, methods and techniques, use of unauthorised traps and weirs, wilful destruction of corals in reef fisheries, and catching juvenile fish. These activities are prohibited under the FAO Code of Conduct[89] and may be conducted by both national and foreign vessels.

The fisheries legislation of many States prohibit most of the activities identified above. For example, illegal fishing under the Australian *Fisheries Management Act 1991 (Cth)* include engaging in commercial fishing without a permit or concession, fishing contrary to the condition of the licence, using a foreign boat for recreational fishing in the Australian Fishing Zone, landing of fish by a foreign vessel without permit, and large-scale driftnet fishing.[90] Canada also identifies a number of prohibitions in its *Fisheries Act,* including the use of explosives, use of seines and nets that obstruct navigation or the passage of fish, unlawful sale or possession of fish, processing or exporting of fish without permit, and harmful alteration of fish habitat.[91] The *Fisheries Act 1996* of New Zealand prohibits similar fishing activities and further considers as an offence knowingly taking, possessing, receiving, procuring, processing, conveying, or selling any fish or aquatic life contrary to the Act to obtain benefit.[92] These fisheries offences and prohibitions are very similar to those identified under international and regional fisheries instruments.

Paragraph 3.1.1 of the IPOA-IUU includes the activities of fishing vessels conducted contrary to the provisions of bilateral fishing access agreements. Most fishing access agreements stipulate the terms and conditions within which vessels may conduct their operations. For example, fisheries partnership agreements entered into by the EU contain provisions relating to the total allowable catch, number of fishing vessels, declaration of catches, by-catch, mesh size, position

---

[83] *LOSC*, Art. 62(4)(a); *FAO Code of Conduct*, para. 8.2.2.
[84] *LOSC*, Art. 62(4)(c).
[85] *LOSC*, Art. 62(4)(d).
[86] *LOSC*, Art. 62(4)(b).
[87] *LOSC*, Art. 62(4)(c).
[88] *LOSC*, Art. 62(4)(c).
[89] *FAO Code of Conduct*, para. 8.4.2.
[90] Australia, *Fisheries Management Act 1991 (Cth)* as Amended, Part 2 Secs. 15 and 15A.
[91] Canada, *Fisheries Act*, R.S., c. F–14, s.1, Arts. 23–33 and 35.
[92] New Zealand, *Fisheries Act 1996*, Part 13, Art. 233.

reporting, and landing of catches.[93] In addition, some States make it an offence for any foreign licensed fishing vessels to fish in specified areas,[94] undertaking transshipment at sea,[95] fishing during closed seasons,[96] and non-compliance with vessel monitoring system requirements.[97] Failure to observe the conditions of these access agreements constitutes illegal fishing.

In the Pacific Islands region, fishing vessels of the United States (U.S.) which are permitted to engage in fishing under the Treaty on Fisheries Between the Governments of Certain Pacific Islands States and the Government of the United States of America (U.S. Treaty) are required to do so in accordance with the terms and conditions stipulated in the Treaty, as well as the applicable national laws of the Pacific Island parties involved.[98] These conditions apply with respect to licensing, fishing in closed areas, transshipment of fish, position and catch reporting, and placement of observers on board U.S. vessels. Any activity contrary to these agreed conditions may be considered illegal fishing under the U.S. Treaty.

On the basis of paragraph 3.1.2 of the IPOA-IUU, conservation and management measures adopted by RFMOs may also determine what activities constitute illegal fishing on the high seas. In general, RFMOs have adopted conservation

---

[93] *See* Annexes of bilateral protocols and administrative arrangements to fisheries partnership agreements entered into by the EU with third countries such as Cape Verde, Comoros, Côte d'Ivoire, Gabon, Gambia, Greenland, Guinea, Guinea Bissau, Equatorial Guinea, Kiribati, Madagascar, Mauritania, Mauritius, Federated States of Micronesia, Morocco, Mozambique, São Tomé and Príncipe, Senegal, Seychelles, and Solomon Islands. For example Protocol setting out the fishing opportunities and financial contribution provided for in the Fisheries Partnership Agreement between the European Community and the Republic of Cape Verde on fishing off the Coast of Cape Verde for the period from 1 September 2006 to 31 August 2011, 30.12.2002 *EN OJ L 414/8*; Protocol setting out the fishing opportunities and financial contribution provided for in the Agreement between the European Economic Community and the Islamic Federal Republic of the Comoros on fishing off the Comoros for the period from 1 January 2005 to 31 December 2010; Protocol setting out the fishing opportunities and financial contribution provided for in the Agreement between the European Community and the Republic of Guinea on fishing off the coast of Guinea for the period from 1 January 2009 to 31 December 2012, 19.6.2009 *EN L 156/40.*

[94] Protocol setting out the fishing opportunities and financial contribution provided for the in the Fisheries Partnership Agreement between the European Community and Republic of Kiribati for the period from 16 September 2006 to 15 September 2012, *7.8.2007 EN OJ L 205/8*, Annex Chapter II(1); Protocol setting out the fishing opportunities and financial contribution provided for the in the Partnership Agreement between the European Community and the Federated States of Micronesia on fishing in the Federated States of Micronesia, *6.6.2006 EN OJ L 151/8,* Annex Chapter II(1).

[95] Protocol setting out the fishing opportunities and payments provided for in the Partnership Agreement between the European Community and Solomon Islands on fishing off Solomon Islands, 13.4.2006 *EN OJ L 105/39*, Annex Chapter IX(9).

[96] Council Regulation (EC) No 41/2006 of 21 December 2006 fixing for 2007 the fishing opportunities and associated conditions for certain fish stocks and groups of fish stocks, applicable in Community waters and, for Community vessels, in waters where catch limitation are required, *20.1.2007 EN OJ L 15/1*, Art. 26.

[97] Fisheries Partnership Agreement between the Republic of Madagascar and the European Community, *17.12.2007 EN J OL 331/7*, Appendix IV.

[98] *See* Treaty on Fisheries Between the Governments of Certain Pacific Islands States and the Government of the United States of America, Port Moresby, concluded on 2 April 1987, in force 15 June 1988, ATS 1988 No 42, with subsidiary agreements in force 15 June 1993.

and management measures relating to the quantity of fish that may be caught, periods in which fishing may occur, size and sex of any species which may be taken, by-catch, fishing gear and technology which may be used, level of fishing effort, vessel types, and other measures protecting specific fisheries. Any activities contrary to these measures may be considered illegal. Some RFMOs have specifically defined what constitutes illegal fishing activity.

The International Commission for the Conservation of Atlantic Tunas (ICCAT) defined IUU fishing in general as "any fishing not in compliance with relevant ICCAT conservation and management measures, in the Convention Area or other areas."[99] The Northeast Atlantic Fisheries Commission (NEAFC) defines "serious infringements" to include fishing without a valid authorisation to fish issued by the flag State member, fishing without a quota or after exhausting quota limits, or use of prohibited gear.[100] Other types of serious infringements include directed fishing for a stock which is subject to a moratorium or for which fishing is prohibited; falsifying or concealing of the markings, identity or registration of a fishing vessel; preventing an inspector from carrying out his or her duties; and concealing, tampering with or disposing of evidence relating to an investigation.[101] The Northwest Atlantic Fisheries Organization (NAFO) defines fishing activities which are considered serious infringements[102] similar to those provided by the NEAFC. In addition, NAFO considers "fishing on an Other's quota"[103] without prior notification to the NAFO Executive Secretary, or more than seven days after fishing on such quota has been closed, as serious infringements,[104] and therefore illegal.

The Indian Ocean Tuna Commission (IOTC), ICCAT and CCAMLR also provide a list of fishing activities which may be considered illegal. These activities include harvesting of species without being registered or licensed, violation of the conditions of licence, fishing without quotas or effort allocation, fishing contrary to the laws and regulations of member States, or fishing with a vessel without nationality,[105] the taking or landing of undersized fish, fishing

---

99 International Commission for the Conservation of Atlantic Tunas (ICCAT), *Resolution by ICCAT Further Defining the Scope of IUU Fishing*, Resolution 01–18GEN, 22 March 2002.

100 Northeast Atlantic Fisheries Commission (NEAFC), *Scheme of Control and Enforcement*, 2009, Art. 29.

101 NEAFC, *Scheme of Control and Enforcement 2009*, Art. 29.

102 Northwest Atlantic Fisheries Organization (NAFO), *Conservation and Enforcement Measures*, NAFO/FC Doc. 09/1, Serial No. N5614, 2009, Art. 37.

103 Article 3(3) of the NAFO Conservation and Enforcement Measures provides that each Contracting Party which has not been allocated a quota of a particular stock listed may be allowed to fish on the quota allocated to "Others" with the permission of the organisation.

104 NAFO, *Conservation and Enforcement Measures 2009*, Art. 37(1).

105 Indian Ocean Tuna Commission (IOTC), *Resolution 09/03 on Establishing a List of Vessels Presumed to Have Carried Out Illegal, Unregulated and Unreported Fishing in the IOTC Area*, 2009, para. 1; ICCAT, *Recommendation by ICCAT to Establish a List of Vessels Presumed to Have*

during closed fishing periods or in closed areas, use of prohibited gear, transshipment with vessels included in the IUU list,[106] and engaging in at sea or port transshipment activities without authorisation.[107] Similar activities are considered IUU fishing under CCAMLR conservation measures.[108] In addition, CCAMLR has developed various regulations prohibiting directed fishing of certain types of fish in particular parts of its management area,[109] the violation of which constitutes an act undermining the Commission's conservation measures.

There are parallel measures against illegal fishing activities on the high seas, which reflect paragraph 3.1.3 of the IPOA-IUU. Under the LOSC, it can be argued that fishing which undermines high seas conservation and management measures with respect to the total allowable catch, fishing patterns, interdependence of stocks, and dependent species constitute illegal fishing.[110] Similarly, under the UN Fish Stocks Agreement, any activity of a fishing vessel undertaken contrary to established conservation and management measures would constitute a violation of the agreement.[111] The UN Fish Stocks Agreement defines a serious violation to include fishing without a valid licence; fishing in a closed area; fishing during a closed season or fishing without a quota established by relevant RFMOs; fishing for a stock which is subject to moratorium or for which fishing is prohibited; using prohibited gears; or other multiple violations which constitute a disregard of conservation and management measures.[112]

The FAO Compliance Agreement requires vessels fishing on the high seas not to engage in any activity that undermines the effectiveness of international conservation and management measures.[113] Although the FAO Compliance Agreement does not enumerate these activities, it provides that any act

---

Carried Out Illegal, Unreported and Unregulated Fishing Activities in the ICCAT Convention Area, 02–23 GEN, 04 June 2003, para. 1.

[106] Some RFMOs have created an IUU list or a record of vessels engaged in IUU fishing. *See* Chapter 9.

[107] IOTC, *Resolution 09/03 on IUU Vessel Listing,* para. 1; ICCAT, *Recommendation on IUU Vessel Listing,* para. 1;

[108] CCAMLR, *Conservation Measure 10–06 (2008)*, Scheme to Promote Compliance by Contracting Party Vessels with CCAMLR Conservation Measures, para. 5; CCAMLR, *Conservation Measure 10–07 (2009)*, Scheme to Promote Compliance by Non-Contracting Party Vessels with CCAMLR Conservation Measures, para. 9.

[109] For example, CCAMLR, Conservation Measure 32–02 (1998) on the Prohibition of Directed Fishing for Finfish in Statistical Subarea 48.1; Conservation Measure 32–09 (2007) on the Prohibition of Directed Fishing for *Dissotichus spp*. Except in Accordance with Specific Conservation Measures in the 2007/08 Season; Conservation Measure 32–17 (2003) on the Prohibition of Directed Fishing for *Electrona carlsbergi* in Statistical Subarea 48.3.

[110] *LOSC*, Art. 119.

[111] *UN Fish Stocks Agreement*, Art. 21(5).

[112] *UN Fish Stocks Agreement*, Arts. 18(3) and 21(11).

[113] *FAO Compliance Agreement*, Arts. III(1), V(1), and V(2).

undermining conservation and management measures is considered illegal.[114] Similarly, the FAO Code of Conduct refers to the contravention of applicable conservation and management measures as "illegal activities".[115] The FAO Compliance Agreement and the FAO Code of Conduct also emphasise that only fishing vessels with licences or authorisation to fish and carrying out their activities in accordance with such licences or authorisation may have the right to fish on the high seas.[116]

Under the UNGA resolutions and other international and regional instruments, the operation of large scale pelagic driftnet fishing is classified as illegal.[117] Chapter 17 of Agenda 21 and the FAO Code of Conduct call on States to prohibit dynamiting, poisoning, and other destructive fishing practices by vessels flying their flags and fishing on the high seas.[118] These activities may be classified as illegal fishing under paragraph 3.1.3 of the IPOA-IUU.

The Western and Central Pacific Fisheries Commission (WCPFC) provides a good illustration of RFMO implementation of the provisions of the UN Fish Stocks Agreement in respect of activities constituting serious violation, and which are therefore illegal. The WCPFC High Seas Boarding and Inspection Procedures identify a range of activities which constitute "serious violation on the high seas." These include: fishing without a licence, permit or authorisation issued by the flag State member; failure to maintain sufficient records of catch and catch related data in accordance with the Commission's reporting requirements or significant misreporting of such catch and/or catch related data; fishing in a closed area; fishing during a closed season; intentional taking or retention of species in contravention of any applicable conservation and management measure adopted by the Commission; significant violation of catch limits or quotas in force pursuant to the Convention; using prohibited fishing gear; falsifying or intentionally concealing the markings, identity or registration of a fishing vessel; concealing, tampering with or disposing of evidence relating to investigation of a violation; multiple violations which taken together constitute a serious disregard of measures in force pursuant to the Commission; refusal to accept a boarding and inspection, other than as provided in the High Seas Boarding and Inspection

---

114 *FAO Compliance Agreement*, Art. III(8).

115 *FAO Compliance Agreement*, Art. III(8); *FAO Code of Conduct*, Art. 8.2.7.

116 *UN Fish Stocks Agreement*, Arts. 18(3)(a) and (b)(ii); *FAO Compliance Agreement*, Art. III(2); *FAO Code of Conduct*, Art. 8.2.2.

117 UNGA, Large-scale pelagic drift-net fishing and its impact on the living marine resources of the world's oceans and seas, 79th Plenary Meeting, *A/RES/46/215*, 20 December 1991, para. 3(c). *See also* Agenda 21, para. 17.54; *Wellington Convention, supra* note 76; ICCAT, *Resolution by ICCAT on Large-scale Pelagic Driftnets*, 96–15 GEN, 03 February 1997.

118 *Agenda 21*, para. 17.53; *FAO Code of Conduct*, Art. 8.4.2.

Procedures; assaulting, resisting, intimidating, sexually harassing, interfering with, or unduly obstructing or delaying an authorised inspector; intentionally tampering with or disabling the vessel monitoring system; and such other violations as may be determined by the Commission.[119] This list of violations of WCPFC conservation and management measures is supported by IUU fishing activities listed under the WCPFC Conservation and Management Measure 2007–03 on IUU Vessel Listing.[120]

2.2.2.2. *Unreported Fishing*

According to paragraph 3.2 of the IPOA-IUU:

> Unreported fishing refers to fishing activities:
>
> 3.2.1 which have not been reported, or have been misreported, to the relevant national authority, in contravention of national laws and regulations; or
>
> 3.2.2 undertaken in the area of competence of a relevant regional fisheries management organisation which have not been reported or have been misreported, in contravention of the reporting procedures of that organisation.

Under the IPOA-IUU, therefore, unreported fishing can be categorised as non-reporting, misreporting, or under reporting of fisheries data that a State or an RFMO requires under its laws, regulations, or adopted conservation and management measures. Unreported fishing may also include reporting of data contrary to the reporting procedures of a State or an RFMO.

In practice, the distinction between "illegal fishing" and "unreported fishing" may not be easy to draw. The use of phrases "in contravention of national laws and regulations" and "in contravention of the reporting procedures of that organisation" in paragraph 3.2 of the IPOA-IUU can result in considerable overlap between "unreported fishing" and "illegal fishing". Since any fishing activity by national or foreign vessels in contravention of national laws and regulations may be considered illegal fishing under paragraph 3.1.1 of the IPOA-IUU, it follows that the reporting of catch contrary to national laws and regulations is also illegal. Similarly, under paragraph 3.2.2 of the IPOA-IUU, any fishing activity undertaken in areas of competence of a relevant RFMO which have not been reported or have been misreported, in contravention of the reporting procedures of that organisation may be illegal and at the same time unreported. In certain

[119] Western and Central Pacific Fisheries Commission (WCPFC), Boarding and Inspection Procedures, *Conservation and Management Measure 2006–08*, 15 December 2006, para. 37.

[120] Western and Central Pacific Fisheries Commission (WCPFC), Conservation and Management Measure to Establish a List of Vessels Presumed to Have Carried Out Illegal, Unreported and Unregulated Fishing Activities in the WCPO, *Conservation and Management Measure 2007–03,* 7 December 2007, para. 3.

circumstances, therefore, unreported fishing may be simply a subset of illegal fishing.

In areas under the competence of RFMOs, unreported fishing applies to fishing activities which are undertaken by vessels flying the flag of members, vessels without nationality and vessels flying the flag of non-members contrary to the reporting procedures of the relevant RFMO. In this case, unreported fishing under paragraph 3.2.2 of the IPOA-IUU may overlap with unregulated fishing under paragraph 3.3.1 of the IPOA-IUU. In practice, failure by vessels flagged to members of the RFMO and cooperating non-members to report, contrary to the measures instituted by the RFMO, will also constitute illegal fishing under paragraph 3.1.2 of the IPOA-IUU.

Similar to illegal fishing, prohibition on unreported fishing is well established in international law. In the EEZ, the LOSC permits coastal States to make laws and regulations "specifying information required of fishing vessels, including catch and effort statistics and vessel position reports."[121] The FAO Compliance Agreement has a broader provision on reporting of fisheries data and mandates a State to ensure that fishing vessels flying its flag provide all information on their operations as may be necessary to fulfil the State's obligations under the agreement.[122]

The UN Fish Stocks Agreement emphasises the importance of the timely collection and reporting of relevant fisheries data in accordance with subregional, regional, and global standards.[123] Relevant reporting requirements under the UN Fish Stocks Agreement include the total catch in number and/or weight (for both target and non-target species, by fishery, fleet, and fishing method); fishing location; date and time fished; composition of catch according to length, size and weight; and vessel related information (identification, vessel type, fishing gear description).[124] States are also required to ensure that vessels flying their flag submit logbook data on catch and effort, including data on fishing operations on the high seas at sufficiently frequent intervals to meet national requirements and regional and international obligations.[125] The UN Fish Stocks Agreement categorises some activities related to fisheries data reporting as serious violations. These include the failure to maintain accurate records of catch and catch related data required by subregional and regional fisheries management organisations and serious misreporting of catch contrary to the reporting requirements of such organisations.[126] Fishing vessels are not only required to report fisheries related

[121] *LOSC*, Art. 62(4)(e).
[122] *FAO Compliance Agreement*, Art. III(7).
[123] *UN Fish Stocks Agreement*, Art. 18(3)(e) and Annex I, Art. 1(1).
[124] *UN Fish Stocks Agreement*, Annex I, Arts. 3 and 4.
[125] *UN Fish Stocks Agreement*, Annex I, Art. 5.
[126] *UN Fish Stocks Agreement*, Art. 21(11)(b).

data and information, but also to make sure that such data is accurate and reported in a timely manner.

Chapter 17 of Agenda 21 requires "a full, detailed, accurate and timely reporting" of catches and effort.[127] The FAO Code of Conduct builds on the requirements under Chapter 17 of Agenda 21 and elaborates on the type of information needed to generate a sound fisheries statistical analysis.[128] Any reporting that does not fulfil this description may be considered unreported fishing.

On a regional level, catch reporting is based on various conservation measures relating to data reporting.[129] RFMOs have a number of catch reporting requirements for fishing vessels. The basic requirements include the maintenance of fishing logbooks containing detailed record of catches (by species, management area, type of gear) and other information such as the number of operations, entry into and exit from the regulatory area, transshipment and the daily or periodic reporting of such information to RFMOs.[130] Some RFMOs require fishing vessels to comply with catch documentation requirements. The RFMOs in this category include the CCAMLR *Dissostichus* catch document (DCD),[131] the Inter-American Tropical Tuna Commission (IATTC) Bigeye Tuna Statistical Document,[132] the Commission for the Conservation of Southern Bluefin Tuna (CCSBT) Catch Documentation Scheme,[133] and the IOTC Bigeye Tuna Statistical Document Programme.[134] Similar to international fisheries instruments, RFMOs emphasise

---

[127] *Agenda 21*, para. 17.51.

[128] *FAO Code of Conduct*, para. 7.4.4.

[129] *See* CCAMLR, Conservation Measure 23–01 (2005) on Five-day Catch and Effort Reporting System; Conservation Measure 23–02 (1993) on Ten-day Catch and Effort Reporting System; Conservation Measure 23–03 (1991) on Monthly Catch and Effort Reporting System; *The Convention for the Conservation of Salmon in the North Atlantic Ocean*, Reykjavik, Iceland, concluded on 01 October 1983, in force 2 March 1982, TIAS No 10789; 35 UST 2284; 1338 UNTS 33, Art. 15; IOTC, *Resolution 08/01*, Mandatory Statistical Requirements for IOTC Members and Cooperating Non-Contracting Parties (CPC's).

[130] NAFO, *Conservation and Enforcement Measures 2009*, Arts. 24 and 25; NEAFC, *Scheme of Control and Enforcement 2009*, Arts. 9 and 10; Inter-American Tropical Tuna Commission (IATTC), *Resolution on Data Provision*, Resolution C–03–05, 27 June 2003; International Pacific Halibut Commission (IPFHC), *Pacific Halibut Fishery Regulations* 2008, Sec. 16; ICCAT, *Resolution by ICCAT Concerning a Management Standard for the Large-scale Tuna Longline Fishery*, 01–20 GEN, 22 February 2002, Attachment 1; ICCAT, *General Outline of Integrated Monitoring Measures Adopted by ICCAT*, 02–31 GEN; ICCAT, *Recommendation by ICCAT Concerning Unreported Catches of Bluefin Tuna, Including Catches Classified as Not Elsewhere Included (NEI)*, 97–03 BFT, 13 June 1998, para. 1; IATTC, *Resolution on At-Sea Reporting*, Resolution C–03–04, 27 June 2003.

[131] CCAMLR, *Conservation Measures 10–05 (2009)*, Catch Documentation Scheme for *Dissostichus* spp.

[132] IATCC, *Resolution C–03–01*, Resolution on IATTC Bigeye Tuna Statistical Document Program, 24 June 2003.

[133] Commission for the Conservation of Southern Bluefin Tuna (CCSBT), *Resolution on the Implementation of a CCSBT Catch,* Adopted at the Fifteenth Annual Meeting, 14–17 October 2008.

[134] IOTC, *Resolution 01/06*, Recommendation by IOTC Concerning the IOTC Bigeye Tuna Statistical Document Programme, 2001.

the need for the recording and timely reporting of vessel position, catch of target and non-target species, fishing effort and other relevant fisheries data including an estimate of discards.[135]

Misrecording of catches and violations of catch communication requirements are listed as serious infringements under NAFO and NEAFC regulations.[136] Non-compliance with catch reporting requirements may also be a ground for a fishing vessel to be included on the IUU lists of various RFMOs. ICCAT, CCAMLR, IATTC and IOTC consider failure to record or report catches or the provision of false reports in their areas of competence as clear evidence that a vessel is engaged in IUU fishing.[137]

There are a few gaps in the scope of unreported fishing under the IPOA-IUU. As shown above, under the IPOA-IUU unreported fishing is confined to activities within waters under the jurisdiction of a State and RFMO areas. This would exclude activities on the high seas not regulated by any RFMO. Therefore, unless there is a legal requirement under domestic laws for vessels flying the flag of a State to report their high seas fishing activities, such activities would fall outside the scope of unreported fishing. Further, the IPOA-IUU limits unreported fishing to those activities conducted contrary to national and regional fisheries regulations and does not consider unreported fishing as an activity contrary to the reporting requirements of international fisheries instruments. The scope of unreported fishing under the IPOA-IUU does not seem to take into account the possibility that States and RFMOs may fail to promulgate appropriate reporting regulations, thus encouraging their fishing vessels not to report, misreport, or under-report their catches. The failure to establish adequate regulations on reporting may result in a State or RFMO sponsoring or condoning unreported fishing, which may also become unregulated fishing.

2.2.2.3. *Unregulated Fishing*

Under paragraph 3.3 of the IPOA-IUU unregulated fishing includes fishing activities:

> 3.3.1 in the area of application of a relevant regional fisheries management organisation that are conducted by vessels without nationality, or by those flying the flag of a State not

---

[135] ICCAT, *General Outline of Integrated Monitoring Measures Adopted by ICCAT*, 02–31 GEN; NEAFC, *Scheme of Control and Enforcement 2009*, Art. 10; *SEAFO Convention*, Art. 14(e).

[136] NAFO, *Conservation and Enforcement Measures 2009*, Art. 25; NEAFC, *Scheme of Control and Enforcement 2009*, Arts. 10, 12, and 13.

[137] ICCAT, *IUU Vessel Listing*, para. 1; CCAMLR, *Conservation Measure 10–06 (2008)*, Scheme to Promote Compliance by Contracting Party Vessels, para. 5(ii); IOTC, *Resolution 09/03 on IUU Vessel Listing*, para. 1(c); IATTC, *Resolution C-05-07*, Resolution to Establish a List of Vessels Presumed to Have Carried Out Illegal, Unreported and Unregulated Fishing in the Eastern Pacific Ocean, 24 June 2005, para. 1(b).

party to that organisation, or by a fishing entity, in a manner that is not consistent with or contravenes the conservation and management measures of that organisation; or

3.3.2 in areas or for fish stocks in relation to which there are no applicable conservation or management measures and where such fishing activities are conducted in a manner inconsistent with State responsibilities for the conservation of living marine resources under international law.

The main characteristic of unregulated fishing is the lack of regulations governing a particular area, fish stock, or type of vessel. Thus, the cause of unregulated fishing can either be the failure of governance by States and RFMOs or the failure by flag States to discharge their flag State duties to control the activities of their vessels on the high seas, such as through high seas fishing authorisation or permits.

Unlike illegal and unreported fishing, the term "unregulated fishing" is directly referred to in international instruments. The UN Fish Stocks Agreement, Agenda 21, and FAO Code of Conduct all identified unregulated fishing as one of the problems affecting the conservation and management of fisheries on the high seas.[138] Unregulated fishing is also referred to as "unauthorised fishing" by the UNGA and regional fisheries bodies.[139]

The analogous reference to "unregulated fishing" and "unauthorised fishing" in waters under the jurisdiction of a State and the high seas may lead to some confusion between the terms "unregulated fishing" and "illegal fishing." Paragraphs 3.1.2 and 3.3.1 of the IPOA-IUU provide a clear distinction between illegal and unregulated fishing. Illegal fishing outside areas under national jurisdiction takes place when vessels flying the flags of members of an RFMO fail to comply with conservation and management measures adopted by the RFMO; while unregulated fishing occurs when vessels flying the flags of non-members conduct fishing operations in the management areas of RFMOs. Unregulated fishing may also be undertaken by fishing vessels flying the flag of a member of an RFMO which then subsequently reflags to a non-member State to avoid compliance with RFMO regulations.[140]

The prevalence of unregulated fishing on the high seas may be partly attributable to the limited rules of international law governing the activities of non-members of RFMOs. Under international law, treaties do not create obligations or rights for third States without their consent.[141] Although the LOSC and the UN Fish Stock Agreement provide for the duty of all States to cooperate in the management

---

[138] *UN Fish Stocks Agreement*, Preamble; *Agenda 21*, para. 17.45; *FAO Code of Conduct*, Preface.

[139] *See* Section 2.1 of this chapter.

[140] *UNGA A/54/429*, *supra* note 34, para. 249.

[141] *Vienna Convention on the Law of Treaties*, Vienna, Austria, concluded 23 May 1969, in force 27 January 1980, UN Doc. A/Conf.39/27; 1155 UNTS 331; 8 ILM 679 (1969); 63 AJIL 875 (1969), Art. 36.

of fisheries resources on the high seas,[142] the extent to which vessels of non-members of RFMOs are bound by conservation and management measures is not clear. According to Article VIII of the FAO Compliance Agreement, Parties must "cooperate" and "exchange information" with respect to vessels of non-Parties, or "encourage" any non-Party to accept or adopt laws and regulations consistent with the Agreement. These provisions are ambiguous[143] and provide very little clarification of the legal measures that may be undertaken to regulate the fishing activities by vessels flying the flags of non-Parties to the FAO Compliance Agreement.

The UN Fish Stocks Agreement contains few provisions with respect to non-members of RFMOs or non-participants in subregional or regional fisheries management arrangements. A State which is not a member of an RFMO is not discharged from the obligation to cooperate in the conservation and management of straddling and highly migratory fish stocks on the basis of the LOSC and the UN Fish Stocks Agreement.[144] Such a State is required not to authorise vessels flying its flag to engage in fishing activities in the areas of competence of RFMOs.[145] However, since this provision is not binding on non-parties to the UN Fish Stocks Agreement, it is not clear how the provision will be implemented in practice.

In order to address issues related to the law of treaties, some regional fisheries management agreements have incorporated provisions on unregulated fishing by non-Contracting Party vessels. For example, the Convention on Future Multilateral Cooperation in the Northwest Atlantic Fisheries "invites the attention of any Non-Contracting Party to the Convention relating to fisheries activities in the Convention area which appear to affect adversely the attainment of the objectives of the Convention."[146] Similarly, the Convention for the Conservation of Anadromous Stocks in the North Pacific Ocean "encourages any State or entity not Party to the Convention to adopt laws and regulations consistent with the provisions of the Convention and invite the attention of such States and entities which could adversely affect the conservation of anadromous stocks within the Convention area."[147] The Southeast Atlantic Fisheries Organisation (SEAFO) agreement contains a provision requesting non-Parties to cooperate fully with the

---

[142] *LOSC*, Arts. 63 and 64; *UN Fish Stocks Agreement*, Art. 17(1).

[143] Patricia Birnie, 'New Approaches to Ensuring Compliance At Sea: The FAO Agreement to Promote Compliance with International Conservation and Management Measures by Fishing Vessels on the High Seas', *RECIEL* 8:1(1999), at 54.

[144] *UN Fish Stocks Agreement*, Art. 17(1).

[145] *UN Fish Stocks Agreement*, Art. 17(2).

[146] *Convention on Future Multilateral Cooperation in the Northwest Atlantic Fisheries*, Ottawa, Canada, concluded 24 October 1978, in force 01 January 1979, 1135 UNTS 269; 34 ILM 1452 (1995), Art. XIX.

[147] *Convention for the Conservation of Anadromous Stocks in the North Pacific Ocean*, Moscow, concluded 11 February 1992, in force 16 February 1993, TIAS No 11465, Art. IV.

organisation either by becoming party to the Convention or by agreeing to apply the conservation and management measures adopted by the Commission.[148]

The weak requirements relating to unregulated fishing activities in regional fisheries agreements are remedied by specific enforcement measures adopted by RFMOs. For example, ICCAT, IATTC, and NEAFO have procedures to identify vessels flying the flag of non-members that diminish the effectiveness of their conservation measures and actions that may be taken against these vessels.[149] Measures adopted by NEAFC, CCAMLR, IOTC, IATTC, and NAFO provide that non-member vessels sighted fishing in their regulatory areas would be presumed to be undermining the effectiveness of conservation and enforcement measures adopted by the respective organisations.[150] Although the activities of non-member vessels in RFMO areas of competence may not be illegal under the IPOA-IUU, based on paragraph 3.3.1 of the IPOA-IUU, such activities may be classified as unregulated fishing. In this case, it is the failure of flag States to regulate their vessels which turns the activities of their vessels into unregulated fishing.

According to paragraph 3.3.1 of the IPOA-IUU, unregulated fishing includes activities of vessels without nationality or stateless vessels in RFMO areas. Fishing vessels may become stateless when it sails under the flag of two or more States, using them according to convenience,[151] or where a flag State revokes the registration of the vessel because of the vessel's continued disregard of the laws of the flag State.[152] Since vessels without nationality are outside the management framework of RFMOs, their activities are considered unregulated.

Another gap in the current international regulatory framework is the non-representation of all actors in RFMOs. A particular case in point is the difficulty that some RFMOs face in including Taiwan (or Chinese Taipei) as a fishing entity in their respective organisations, largely because of international relations

---

[148] *SEAFO Convention*, Art. 22.

[149] ICCAT, *Resolution by ICCAT on Compliance with the ICCAT Conservation and Management Measures (Including Addendum)*, 23 January 1995, 94–09 GEN; ICCAT, *Recommendation by ICCAT on Transshipments and Vessel Sightings*, 13 June 1998, 97–11 GEN; IATTC, Resolution C–04–03, *Resolution on a System of Notification of Sightings and Identification of Vessels Operating in the Convention Area*; NEAFC, *Scheme of Control and Enforcement 2009*, Art. 37.

[150] NEAFC, *Scheme of Control and Enforcement 2009*, Art. 37(2); CCAMLR, *Conservation Measure 10–07 (2009)*, para. 9; IATTC, *IUU Vessel Listing*, para. 1; IOTC, Resolution 01/03, *Establishing a Scheme to Promote Compliance by Non-Contracting Party Vessels with Resolutions Established by IOTC*, para. 2; IOTC, *Resolution 09/03 on IUU Vessel Listing*, para 2; ICCAT, *Recommendation by ICCAT Concerning the Ban on Landing and Transshipments of Vessels from Non-Contracting Parties Identified as Having Committed a Serious Infringement*, 21 June 1999, 98–11 GEN, para. 1; NAFO, *Conservation and Enforcement Measures 2009*, Art. 52(1); and IATTC, *Resolution C–05–07 on IUU Vessel Listing*, para. 1.

[151] *LOSC*, Art. 92(1).

[152] Ted L. McDorman, 'Stateless Fishing Vessels, International Law, and the UN High Seas Fisheries Conference', 25 *Journal of Maritime Law and Commerce* (1994), at 531; *See also* Deirdre M. Warner-Kramer and Krista Canty, 'Stateless Fishing Vessels: The Current International Regime and a New Approach', 5 *Ocean and Coastal Law Journal* (2000), at 230.

concerns and politics within the UN concerning the status of Taiwan.[153] RFMOs which operate under the framework of FAO such as the IOTC are only open to Members and Associate Members of the FAO and those which are members of the United Nations.[154] Such membership conditions limit the participation of Taiwan in the IOTC to "Invited Experts", and not as members, despite its significant fishing activities in the management area, its expression of interest to become a member of the Commission, and participation in the Commission meetings.[155]

Recent developments in international fisheries law have provided the legal basis for, and the rights and responsibilities of fishing entities.[156] The UN Fish Stocks Agreement provides that the agreement applies *mutatis mutandis* to fishing entities whose vessels fish on the high seas.[157] Implementation of this provision has resulted in Taiwan enjoying certain rights and assuming obligations almost similar to those of State Parties of the Agreement.[158] Some RFMOs which operate outside the FAO framework have developed schemes to provide for the participation of Taiwan. Resolutions have been adopted by ICCAT,[159] CCSBT,[160] and IATTC[161] to allow Taiwan to participate as a fishing entity in their respective commissions. Taiwan is also a member of the WCPFC as a fishing entity, having expressed its legal commitment to be bound by the regime established by the Convention for the Conservation and Management of Highly Migratory Fish Stocks in the Western and Central Pacific (WCPF Convention).[162] Unless the participation of Taiwan in all RFMOs is addressed as a global fisheries management issue, the international political system itself will be creating conditions for unregulated fishing.

---

[153] UN, *Resolution 2758 on Restoring the Rights of the PRC*, 25 October 1971.

[154] *Agreement of the Establishment of the Indian Ocean Tuna Commission*, Rome, Italy, concluded on 25 November 1993, in force 27 March 1996, ATS 1996 No 20, Art. IV and XVII.

[155] *See* IOTC, *Report of the Seventh Session of the Scientific Committee*, Victoria, Seychelles, 08–12 November 2004, para. 6.34.

[156] *See* Warwick Gullett, 'Taiwan's Engagement in the Developing International Legal Regime for Fishing: Implications for Taiwanese Officials and Fishers,' 3 *Taiwan Ocean Law Review* (2004) 163–216.

[157] *UN Fish Stocks Agreement*, Art. 1(3).

[158] Martin Tsamenyi, 'The Legal Substance and Status of Fishing Entities in International Law: A Note', 37 *Ocean Development and International Law* (2006), at 126.

[159] ICCAT, *Criteria for Attaining the Status of Cooperating Non-Contracting Party, Entity, or Fishing Entity in ICCAT*, 03–20 GEN, 2003.

[160] CCSBT, *Resolution to Establish the Status of Co-operating Non-member of the Extended Commission and the Extended Scientific Committee*, 2003, para. 3; *See also* Convention for the Conservation of Southern Bluefin Tuna, Canberra, Australia, concluded on 10 May 1993, in force 20 May 1994, *ATS 1996 No 16*, Art. 14.

[161] IATTC, *Resolution on the Participation of Taiwan in the Work of the Inter-American Tropical Tuna Commission*, Adopted During the 62nd Meeting of the IATTC, La Jolla, California, October 1998; IATTC, Resolution C–07–02, *Resolution on Criteria for Attaining the Status of Cooperating Non-Party or Fishing Entity in IATTC*.

[162] WCPFC, *Arrangement for the Participation of Fishing Entities*, Honolulu, USA, 05 September 2000.

Paragraph 3.3.2 of the IPOA-IUU may be interpreted to apply to high seas areas where there exists no conservation and management measure. Any fish stocks in RFMO areas falling outside the scope of species managed by that RFMO or on the high seas where no regulations have been established may be considered unregulated.

The so-called discrete fish stocks, which are not covered under the UN Fish Stocks Agreement, may also be unregulated. Examples of such discrete stocks include orange roughy and alfonsino in the high seas of the Indian Ocean.[163] This gap in the management of fisheries resources on the high seas is partially being addressed through the adoption of the Southern Indian Ocean Fisheries Agreement (SIOFA).[164] Similarly, the management of other marine living resources in the Indian Ocean which are not covered under existing regional fisheries agreements is being dealt with through the Southwest Indian Ocean Fisheries Commission (SWIOFC).[165] In addition to the deep sea stocks of the Southern Indian Ocean, the UNGA has identified the Southeast Pacific Ocean beyond the limits of the IATTC area as a possible management gap,[166] and consequently unregulated. In 2006, Australia, Chile, and New Zealand initiated a process to establish the South Pacific Regional Fisheries Management Organisation (SPRFMO) which will focus on the discrete high seas stocks and some pelagic and demersal stocks which straddle the high seas and the EEZs of coastal States.[167] Over 20 States are

---

[163] FAO, Report of the Ad hoc Meeting on Management of Deepwater Fisheries Resources of the Southern Indian Ocean, Swakopmund, Namibia, 30 May–01 June 2001, *FAO Fisheries Report No. 652* (Rome: FAO, 2001), at para. 13.

[164] Southern Indian Ocean Fisheries Agreement, *supra* note 74.

[165] Statutes of the South West Indian Ocean Fisheries Commission, *Resolution 1/127*, Adopted at the Hundred and Twenty-seventh Session of the FAO Council, Rome, 22–27 November 2004. *See also* reports of the First Intergovernmental Consultation on the Establishment of a Southwest Indian Ocean Fisheries Commission, Saint-Denis, Reunion, 06–09 February 2001; First Ad Hoc Meeting on Management of Deepwater Fisheries Resources of the Southern Indian Ocean, Swakopmund, Namibia, 30 May–01 June 2001; Second Intergovernmental Consultation on the Establishment of a Southwest Indian Ocean Fisheries Commission, Antananarivo, Madagascar, 25–28 September 2001; Second Ad Hoc Meeting on Management of Deepwater Fisheries Resources of the Southern Indian Ocean, Fremantle, Western Australia, 20–22 May 2002; Third Intergovernmental Consultation on the Establishment of a Southwest Indian Ocean Fisheries Commission, Nairobi, Kenya, 27–30 January 2004; and Fourth Intergovernmental Consultation on the Establishment of a Southwest Indian Ocean Fisheries Commission, Mahe, Seychelles, 13–16 July 2004; Fifth Intergovernmental Consultation on the Southern Indian Ocean Fisheries Agreement, Mombassa, Kenya, 20–23 April 2005.

[166] UNGA Fifty-eighth Session, Item 53(b) of the Provisional Agenda, Oceans and the Law of the Sea: Sustainable Fisheries, Including Through the 1995 Agreement for the Implementation of the Provisions of the United Nations Convention on the Law of the Sea of 10 December 1982 Relating to the Conservation and Management of Straddling Fish Stocks and Highly Migratory Fish Stocks, and Related Instruments, *Report of the Secretary-General, A/58/215*, 05 August 2003, para. 38(a).

[167] First International Meeting on the Establishment of the South Pacific Regional Fisheries Management Organisation, Wellington, New Zealand, 14–17 February 2006, Information Paper Submitted by the Co-sponsors Australia, Chile, New Zealand, *SP/01/Inf3_rev1*, para. 3. *See also*

involved in the process of establishing the new RFMO which has adopted interim measures to be applied to vessels flying the flag of negotiating members fishing for non-highly migratory species on the high seas of the South Pacific ocean.[168]

The continued existence of unregulated fishing on the high seas illustrates a gap in the overall global governance of fisheries.[169] This gap is recognised in paragraph 3.4 of the IPOA-IUU, which provides that "certain unregulated fishing may take place in a manner which is not in violation of applicable international law, and may not require the application of measures envisaged under the International Plan of Action." However, until specific measures are adopted by RFMOs and implemented by States, unregulated fishing will continue to exist.

The IUU fishing terminology is now firmly established in international fisheries law and practice. Since the adoption of the IPOA-IUU, a number of States, regional organisations and RFMOs have elaborated on the concept and have applied it in different ways. Other terminologies to describe IUU fishing, coined largely by environmental non-governmental organisations, have emerged since the adoption of the IPOA-IUU. These terms include: "pirate fishers" or "pirate fishing-flags of convenience", "rogue fishing", "predatory fishing," "plundering of ocean stocks", and "bad actors of ocean fisheries".[170]

Although as illustrated in this Chapter, there are three distinct and often interrelated components of the term "IUU fishing," what has emerged since the adoption of the IPOA-IUU is a tendency to equate IUU fishing with illegal fishing. This tendency may result in the failure to develop comprehensive measures and processes to address all components of the IUU fishing problem. Another danger in equating IUU fishing with illegal fishing is that States and RFMOs will conveniently fail to recognise their significant roles and responsibilities in developing governance measures to address unregulated fishing. Without a comprehensive and holistic approach to IUU fishing, global, regional and national measures to address IUU fishing will remain half-baked.

International Consultations on the Establishment of the South Pacific Regional Fisheries Management Organisation, www.southpacificrfmo.org. Accessed on 10 September 2009.

[168] See Interim Measures Adopted by Participants in Negotiations to Establish the South Pacific Regional Fisheries Management Organisation, *Third International Meeting on the Establishment of the South Pacific Regional Fisheries Management Organisation*, Reñaca, Chile, 30 April– 04 May 2007.

[169] *See* High Seas Task Force, 'High Seas Governance,' *Presented at the Meeting of the High Seas Task Force*, Paris, France, 09 March 2005, HSTF/09, February 2005. www.high-seas.org. Accessed on 12 January 2006.

[170] *See* Hélène Bours, Matthew Gianni, and Desley Mather, 'Pirate Fishing Plundering the Oceans', *Greenpeace International Campaign Against Pirate Fishing*, February 2001, www.greenpeace.org. Accessed on 31 March 2005; Greenpeace International, *Pirate Fishing-Stolen Fish, Stolen Futures*, www.oceans.greenpeace.org. Accessed on 14 May 2006; and *See* David Balton, 'Dealing with the "Bad Actors" of Ocean Fisheries', in OECD, *Fish Piracy: Combating Illegal, Unreported and Unregulated Fishing* (Paris: OECD, 2004), at 57–64.

Chapter Three

The International Legal and Policy Framework for Sustainable Fisheries

3.1. Introduction

The prevention, deterrence, and elimination of IUU fishing is a key component of promoting the sustainable management of fisheries resources. Because of this, the international legal framework to address IUU fishing justifiably lies within the scope of the international framework for sustainable fisheries as a whole. This framework comprises a very diverse range of instruments embracing legally binding treaties and non-binding, voluntary instruments adopted both within the fisheries sector and in related sectors such as shipping, trade, environment, and labour relations.

The complex nature of fisheries entails the adoption of management measures that may vary from one type of fishery to another. Sustainable fisheries management is also influenced by the social and economic conditions of a State, available scientific data, and administrative structures in place. Given this complexity, fisheries management is conditioned by a number of factors including resource considerations, environmental constraints, biodiversity and ecological factors, technological development and socio-economic considerations.[1] Accordingly, the legal framework for achieving sustainable fisheries management not only includes fisheries specific instruments but also non-fisheries specific agreements. These non-fisheries specific instruments provide additional tools which address a broad range of issues and challenges relating to the conservation and management of fisheries resources and are environment, trade, and maritime safety related. Together, these instruments provide the wider legal and policy context for the implementation of all State, flag State, coastal State, port State, internationally agreed market State and RFMO measures to prevent, deter and eliminate IUU fishing as discussed in Chapters 4 to 9. Some of the emerging instruments relevant for addressing IUU fishing are also examined in this chapter.

---

[1] Food and Agriculture Organization (FAO), Fisheries Management, *FAO Technical Guidelines for Responsible Fisheries No. 4* (Rome: FAO, 1997), at 14.

3.2. International Fisheries Specific Instruments Supporting the Implementation of the IPOA-IUU

Developments in the international law regulating fisheries may be characterised by a continuous pursuit of measures that were intended to effectively conserve and manage global fisheries resources. The adoption of measures to control fishing access started when the classical doctrine of the freedom of the high seas was slowly eroded and replaced by the principles of sustainable and shared use under the LOSC.[2] Restrictions on access to fisheries, particularly on the high seas, were deemed to be necessary to prevent the occurrence of a 'tragedy of the commons.'[3]

The modern international legal and policy framework regulating fisheries can be divided into two categories: legally binding multilateral agreements and non-binding instruments, or so-called "soft law" instruments. Major fisheries related legal instruments adopted prior to the IPOA-IUU such as the FAO Compliance Agreement[4] and the UN Fish Stocks Agreement,[5] as well as non-binding instruments such as the FAO Code of Conduct for Responsible Fisheries[6] and UN resolutions on driftnet fishing[7] do not deal directly with IUU fishing. However, these

---

[2] United Nations Convention on the Law of the Sea, Montego Bay, Jamaica, concluded on 10 December 1982, in force 16 November 1994, 1833 UNTS 3; 21 ILM 1261 (1982). *Hereinafter referred to as* LOSC; *See* Gary Knight and Hungdah Chiu, *The International Law of the Sea: Cases, Documents and Readings* (London: Elsevier Applied Science and UNIFO Publishers, Inc., 1991), at 319. *See also* Dolliver Nelson, 'The Development of the Legal Regime of High Seas Fisheries', in Alan Boyle and David Freestone, (eds.), *International Law and Sustainable Development* (Oxford: Oxford University Press, 1999), at 113–134; Francisco Orrego Vicuña, *The Changing International Law on High Seas Fisheries* (Cambridge: University Press, 1999), at 8–13; Shigeru Oda, *International Control of Sea Resources* (Dordrecht: Martinus Nijhoff Publishers, 1989).

[3] Douglas M. Johnston, *The International Law of Fisheries: A Policy-Oriented Inquiry* (New Haven: New Haven Press, 1987), at xxxvi.

[4] FAO, Agreement to Promote Compliance with International Conservation and Management Measures by Fishing Vessels on the High Seas, Rome, Italy, concluded on 24 November 1993, entered into force 24 April 2003, 33 ILM 968. *Hereinafter referred to as* the FAO Compliance Agreement.

[5] Agreement for the Implementation of the Provisions of the United Nations Convention on the Law of the Sea of 10 December 1982 relating to the Conservation and Management of Straddling Fish Stocks and Highly Migratory Fish Stocks, concluded on 4 September 1995, in force 11 December 2001, 34 ILM 1542 (1995); 2167 UNTS 88. *Hereinafter referred to as* UN Fish Stocks Agreement.

[6] FAO, *Code of Conduct for Responsible Fisheries,* Adopted at the 28th Session of the FAO Conference, Rome, Italy, 31 October 1995. *Hereinafter referred to as* FAO Code of Conduct.

[7] UN General Assembly, 79th Plenary meeting, *A/RES/46/215* Large-scale pelagic drift-net fishing and its impact on the living marine resources of the world's oceans and seas, 20 December 1991; UNGA, Forty-seventh Session, *Decision 47/443* Large-scale pelagic drift-net fishing and its impact on the living marine resources of the world's oceans and seas, adopted on 22 December 1992; UNGA, Forty-fourth Session, *Decision 44/445* Large-scale pelagic drift-net fishing and its impact on the living marine resources of the world's oceans and seas, adopted on 21 December 1993; UNGA, Forty-ninth Session, *Decision 49/436* Large-scale pelagic drift-net fishing and its impact on the living marine resources of the world's oceans and seas, adopted on 19 December 1994; UNGA,

instruments attempt to address numerous fisheries management concerns, some of which are direct forms of or issues related to IUU fishing[8] and have provided the catalyst for the development of international regulations to address IUU fishing. International instruments, including declarations and resolutions adopted after the IPOA-IUU, such as various UN resolutions on sustainable fisheries, the Rome Declaration on IUU Fishing,[9] the FAO Model Scheme on Port State Measures,[10] and the Agreement on Port State Measures to Combat IUU Fishing[11] provide direct reference to the need to address IUU fishing and also form part of the broader framework for the implementation of the IPOA-IUU.

3.2.1. *Legally Binding Instruments*

There are four major global agreements directly related to fisheries: the LOSC, FAO Compliance Agreement, UN Fish Stocks Agreement, and the Port State Measures Agreement.

Fiftieth Session, Agenda item 96(c), *A/RES/50/25* Large-scale pelagic drift-net fishing and its impact on the living marine resources of the world's oceans and seas; unauthorised fishing in zones of national jurisdiction and its impact on the living marine resources of the world's oceans and seas; and fisheries by-catch and discards and their impact on the sustainable use of the world's living marine resources, adopted on 5 December 1995; UNGA, Fifty-first session, Agenda item 24(c), *A/RES/51/36* Large-scale pelagic drift-net fishing and its impact on the living marine resources of the world's oceans and seas; unauthorised fishing in zones of national jurisdiction and its impact on the living marine resources of the world's oceans and seas; and fisheries by-catch and discards and their impact on the sustainable use of the world's living marine resources, adopted on 9 December 1996; UNGA, Fifty-second session, Agenda item 39(c), *A/RES/52/59* Large-scale pelagic drift-net fishing and its impact on the living marine resources of the world's oceans and seas; unauthorised fishing in zones of national jurisdiction and its impact on the living marine resources of the world's oceans and seas; and fisheries by-catch and discards and their impact on the sustainable use of the world's living marine resources, adopted on 26 November 1997; UNGA, Fifty-third session, Agenda item 38(b), *A/RES/53/33* Large-scale pelagic drift-net fishing; unauthorised fishing in zones of national jurisdiction and on the high seas; fisheries by-catch and discards, and other developments, adopted on 24 November 1998; UNGA, Fifty-fifth session, Agenda item 34(b), *A/RES/55/8* Large-scale pelagic drift-net fishing and its impact on the living marine resources of the world's oceans and seas; unauthorised fishing in zones of national jurisdiction and its impact on the living marine resources of the world's oceans and seas; and fisheries by-catch and discards and their impact on the sustainable use of the world's living marine resources, adopted on 30 October 2000; UNGA, Fifty-seventh session, Agenda item 25(b), *A/RES/57/412* Large-scale pelagic drift-net fishing; unauthorised fishing in zones of national jurisdiction and on the high seas; fisheries by-catch and discards, and other developments, adopted on 12 December 2002.

[8] These IUU fishing issues were discussed in Chapter 2.

[9] *The 2005 Rome Declaration on Illegal, Unreported and Unregulated Fishing*, Adopted by the FAO Ministerial Meeting on Fisheries, Rome, 12 March 2005,

[10] FAO, *Model Scheme on Port State Measures to Combat Illegal, Unreported and Unregulated Fishing* (Rome: FAO, 2007).

[11] Agreement on Port State Measures to Prevent, Deter and Eliminate Illegal, Unreported and Unregulated Fishing, Adopted in November 2009, Appendix V of the FAO Council, Hundred and Thirty-seventh Session, Rome, 28 September-02 October 2009, Report of the 88th Session of the Committee on Constitutional and Legal Matters (CCLM), 23–25 September 2009, *CL 137/5*, September 2009.

3.2.1.1. *United Nations Convention on the Law of the Sea*

The United Nations Convention on the Law of the Sea which came into force on 16 November 1994 provides a general framework for the management of ocean spaces and its resources. One of the principles adopted under the LOSC which provides the basis for all fisheries provisions of the Convention is the equitable, efficient utilisation, and conservation of living resources.[12] These principles are most applicable to the management of fisheries resources in the EEZ and on the high seas. In the case of the internal waters, archipelagic waters, and the territorial sea,[13] the LOSC does not contain specific provisions on the management of fisheries resources primarily because these zones fall exclusively under coastal State sovereignty. On this point, Hey (1999) correctly observed:

> The LOS Convention does not contain provisions that require the coastal state or archipelagic state to conserve or optimally utilise the fisheries resources that are subject to its sovereignty. On the basis of the LOS Convention, coastal states and archipelagic states thus have a wide margin of discretion in regulating the use of the resources in maritime internal waters, archipelagic waters and the territorial sea. They also are entitled to the benefits to be obtained from the fisheries resources in these zones, subject to the explicit condition that in archipelagic waters existing treaties and traditional fishing rights are to be respected.[14]

The development of the legal regime of the EEZ under the LOSC marked a significant change in the management of fisheries resources as it placed about 90 per cent of marine living resources under the jurisdiction of coastal States.[15] The regime of the EEZ recognises the sovereign rights of coastal States in conserving and managing the living resources in the zone.[16] Coastal States are given sovereign rights in respect of the living resources in their EEZs and are subject to the basic duties of ensuring that the fisheries resources therein are not endangered by over-exploitation[17] and promoting the objective of optimum utilisation of such resources.[18] The LOSC confers on coastal States the discretion to formulate conservation measures to implement those obligations, such as adopting laws and regulations with regard to access by foreign fishing vessels in the EEZ. These laws and regulations may relate to the licensing of fishers, fishing vessels and equipment; determining the type and amount of species to be caught; seasons and areas of fishing; the types, sizes, and amount of gear, and the types, sizes, and number of fishing vessels that may be used; fixing the age and size of fish and other

---

[12] *LOSC*, Preamble.

[13] *LOSC*, Arts. 2 and 49(1).

[14] Ellen Hey, 'The Fisheries Provisions of the LOS Convention' in Ellen Hey, (ed.), *Developments in International Fisheries Law* (Hague: Kluwer Law International, 1999), at 20.

[15] *Ibid.*, at 27.

[16] *LOSC*, Part V.

[17] *LOSC*, Art. 61(2).

[18] *LOSC*, Art. 62(1).

species that may be caught; specifying information required of fishing vessels; placing of observers on board vessels; landing of catch; terms and conditions on joint ventures and other cooperative arrangements; and enforcement procedures.[19] The determination of the effectiveness of such regulations is left solely to coastal States. While the regime of the EEZ benefited a significant number of coastal States, it also created some negative impacts on other States. The EEZ resulted in fleet dislocation in distant water fishing States which aggravated the problem of overcapacity,[20] a major issue associated with IUU fishing.

The LOSC also contains provisions on fishing on the high seas.[21] When the freedom of the high seas was codified under the 1958 Convention on the High Seas[22] and eventually superseded by the LOSC, the right of States for their nationals to engage in fishing on the high seas was already subject to some form of regulation. Limitations on the right to fish on the high seas were primarily based on treaty obligations, interests and rights of other coastal States, and conservation and management measures embodied in these international conventions.[23] Under the LOSC, States are required to fulfil three general obligations. The first obligation is to adopt measures for the conservation of the living resources of the high seas.[24] The second obligation is to ensure that the nationals of States adhere to those measures and the third is to cooperate with other States to achieve the same objective.[25]

Another significant provision in the LOSC which links fisheries conservation and management provisions in the EEZ and the high seas relates to cooperation to manage transboundary stocks.[26] The LOSC not only provides for the duty to cooperate among States on the conservation and management of living resources in areas of the high seas,[27] but also stipulates the obligation of States to cooperate more specifically on the management of straddling stocks,[28] highly migratory species,[29] anadromous stocks,[30] and catadromous species.[31] Such duty to cooperate underpins the establishment of regional fisheries management organisations (RFMOs), as well as the development of the UN Fish Stocks Agreement.

---

[19] *LOSC*, Art. 62(4).

[20] Douglas M. Johnston, *The International Law of Fisheries*, *supra* note 3, at LXXVII.

[21] *LOSC*, Part VII.

[22] *Convention on the High Seas,* Geneva, Switzerland, concluded on 29 April 1958, in force 30 September 1962, 13 UST 2312; 450 UNTS 11.

[23] *LOSC*, Art. 116; *See also* Convention on Fishing and Conservation of Living Resources of the High Seas, Geneva, Switzerland, concluded on 29 April 1958, in force 20 March 1966, 17 UST 138; 559 UNTS 285. *Hereinafter referred to as High Seas Fishing Convention*. Art. 1(1).

[24] *LOSC*, Arts. 117 and 118; *See also High Seas Fishing Convention*, Arts. 1(2), 3, and 4.

[25] *LOSC*, Arts. 117 and 118; *See also High Seas Fishing Convention*, Arts. 1(2), 3, and 4.

[26] *LOSC*, Arts. 63 and 64; *High Seas Fishing Convention*, Arts. 4 and 5.

[27] *LOSC*, Art. 118.

[28] *LOSC*, Art. 63.

[29] *LOSC*, Art. 64.

[30] *LOSC*, Art. 66.

[31] *LOSC*, Art. 66.

3.2.1.2. *FAO Compliance Agreement*

The Agreement to Promote Compliance with International Conservation and Management Measures by Fishing Vessels on the High Seas was adopted on 24 November 1993 within the framework of the FAO, and it entered into force on 24 April 2003. The FAO Compliance Agreement began as an effort to solve the problem of "reflagging".[32] This issue was taken up by the FAO Technical Consultation Council on High Seas Fishing in September 1992 which requested the FAO Secretariat to bring the matter to the attention of the FAO Committee on Fisheries.[33] At the 102nd Session of the FAO Council, it was "agreed that the issue of reflagging of fishing vessels into flags of convenience to avoid compliance with agreed conservation and management measures…should be addressed immediately by FAO, with a view to finding a solution which could be implemented in the near future."[34]

The FAO Compliance Agreement applies to all fishing vessels that are used or intended for fishing on the high seas,[35] except for vessels of 24 metres or less in length[36] subject to certain conditions.[37] The agreement emphasises the obligation of States to take effective compliance measures with respect to exempt vessels.[38]

The FAO Compliance Agreement reiterates the provisions of the LOSC with respect to the effective control of fishing vessels on the high seas; and specifies measures that flag States are required to implement to ensure the compliance of their vessels with international conservation and management measures on the high seas. These international conservation and management measures not only pertain to fisheries, but also all measures needed to conserve and manage one or more species of living marine resources in accordance with the LOSC and other relevant rules of international law.[39] These measures include the issuance of

---

[32] David Balton, 'The Compliance Agreement', in Ellen Hey, (ed.), *Developments in International Fisheries Law* (Hague: Kluwer Law International, 1999), at 34.

[33] FAO, Report of the Technical Consultation on High Seas Fishing, *FAO Fisheries Report No. 484* (Rome: FAO, 1992), at para. 45.

[34] FAO, *Report of the FAO Council*, 102nd Session, Rome, Italy 9–20 November 1992 (Rome: FAO, 1992), at para 58. For a discussion of the provisions of the FAO Compliance Agreement and the UN Fish Stocks Agreement, see William Edeson, David Freestone and Elly Gudmundsdottir, *Legislating for Sustainable Fisheries: A Guide to Implementing the 1993 FAO Compliance Agreement and 1995 UN Fish Stocks Agreement* (Washington, D.C.: The World Bank, 2001).

[35] *FAO Compliance Agreement*, Art. II(1).

[36] *FAO Compliance Agreement*, Art. II(2).

[37] *FAO Compliance Agreement*, Art. II(2)(a) and (c) and III. Article III provides that in a fishing region where bordering coastal States have not yet declared EEZs, or equivalent zones of national jurisdiction over fisheries, such coastal States as Parties to the Agreement may agree to establish a minimum length of fishing vessels below which the Agreement shall not apply.

[38] *FAO Compliance Agreement*, Art. III(1)(b). The definition of fishing vessels only include those engaged in fishing operations, and does not include support vessels Art. I(a). See Chapter 2 Section 2.2.1.

[39] *FAO Compliance Agreement*, Art. I(b).

authorisations to fish,[40] maintenance of records of fishing vessels[41] and cooperation among States for the exchange of information,[42] which are critical elements for addressing IUU fishing.

3.2.1.3. *UN Fish Stocks Agreement*

The Agreement for the Implementation of the Provisions of the United Nations Convention on the Law of the Sea of 10 December 1982 relating to the Conservation and Management of Straddling Fish Stocks and Highly Migratory Fish Stocks outlines international standards for the conservation and management of straddling and highly migratory fish stocks to ensure that there are effective mechanisms for compliance and enforcement of those measures on the high seas and compatibility of measures between the EEZ and the high seas.[43] Unlike the FAO Compliance Agreement, the UN Fish Stocks Agreement aims to facilitate the implementation of the provisions of the LOSC only with regard to straddling fish stocks and highly migratory species. The UN Fish Stocks Agreement generally applies to the high seas, although some of its provisions are also applicable to the EEZ, based on the principle of compatibility of conservation and management measures.[44] In addition to the flag State duties stipulated in the FAO Compliance Agreement, the UN Fish Stocks Agreement enumerates additional flag State duties such as the marking of fishing vessels and gear, vessel monitoring systems, observer programmes, boarding and inspection and port State measures.[45] Part III of the UN Fish Stocks Agreement provides mechanisms for international cooperation, particularly through the creation of subregional and regional fisheries management organisations and arrangements. Provisions in the UN Fish Stocks Agreement on high seas boarding and inspection[46] are major developments in international fisheries law that allow States other than the flag State to take limited action against fishing violations on the high seas.[47] The UN Fish Stocks Agreement also recognises the special requirements of developing States in implementing their obligations under the agreement.[48]

---

[40] *FAO Compliance Agreement*, Art. III(2).

[41] *FAO Compliance Agreement*, Art. IV.

[42] *FAO Compliance Agreement*, Art. V.

[43] UN Conference on Straddling and Highly Migratory Fish Stocks, Fourth Session, *Statement Made by the Chairman of the Conference at the Closing of the Fourth Session, on 26 August 1994*, A/CONF.164/24, para 5, in Jean-Pierre Lévy and Gunnar G Schram, *United Nations Conference on Straddling Fish Stocks and Highly Migratory Fish Stocks: Selected Documents* (Hague: Martinus Nijhoff Publishers, 1996).

[44] *UN Fish Stocks Agreement*, Art. 7.

[45] *UN Fish Stocks Agreement*, Art. 18.

[46] *UN Fish Stocks Agreement*, Arts. 21 and 22

[47] David Balton, 'The Compliance Agreement,' *supra* note 32, at 33.

[48] *UN Fish Stocks Agreement*, Part VII.

Another significant aspect of the UN Fish Stocks Agreement is the emphasis on cooperation among States to conserve and manage straddling and highly migratory fish stocks through appropriate regional or sub-regional fisheries management organisations and arrangements.[49] Cooperation in the conservation and management of resources on the high seas is not only an obligation of RFMO members, but is also extended to non-members of or non-participants in such organisations. In particular, Article 8(3) of the UN Fish Stocks Agreement stipulates that States fishing for the stock on the high seas may choose not to join or participate in relevant RFMOs but are still obligated not to undermine the conservation and management measures adopted by RFMOs. Under Article 17(1) of the UN Fish Stocks Agreement, States which are not members of RFMOs or which do not agree to apply the conservation and management measures established by such RFMOs, are not discharged from the obligation to cooperate in the conservation and management of relevant straddling fish stocks and highly migratory fish stocks.[50] The UN Fish Stocks Agreement further provides that only States which are members of relevant RFMOs or agree to apply the conservation and management measures of such RFMOs are entitled to have access to the fisheries resources to which the measures apply.[51]

The provisions of the UN Fish Stocks Agreement noted above qualify the traditional freedom of fishing on the high seas. The linkage between access to fisheries and membership in RFMOs or acceptance to apply regional conservation and management measures adopted through RFMOs suggests that RFMOs have the exclusive competence to regulate straddling and highly migratory fish stocks on the high seas.[52] Thus, States cannot simply escape their obligation to cooperate by not participating in RFMOs while their fishing vessels continue to fish on the high seas.[53]

The provisions of the UN Fish Stocks Agreement on the duty to cooperate also seek to address, to a certain extent, the legal problems associated with the application of treaties to non-parties under Article 34 of the Vienna Convention on the Law of Treaties. As explained in Section 2.2.2.3, under international law, a non-party to a regional fisheries management agreement cannot be bound by such agreement without its express consent. However, if a non-member of an RFMO is a party to the UN Fish Stocks Agreement, it is bound to cooperate under the agreement; hence, it would need to become a member of or participate in the

---

[49] *UN Fish Stocks Agreement*, Art. 8.

[50] Similar provisions are also found in FAO Compliance Agreement, Preamble; FAO Code of Conduct, Article 7.1.5; Agenda 21, para. 17.60.

[51] *UN Fish Stocks Agreement*, Art. 8(4).

[52] Tore Henriksen, Geir Hønneland, and Are Sydnes, *Law and Politics in Ocean Governance: The UN Fish Stocks Agreement and Regional Fisheries Management Regimes* (Leiden: Martinus Nijhoff Publishers, 2006), at 16.

[53] *Ibid.*

organisation or arrangement (for example, as a cooperating non member), or apply the conservation and management measures adopted by RFMOs in their areas of competence.

3.2.1.4. *FAO Port State Measures Agreement*

Under international law, States have full sovereignty over their ports, with minor exceptions such as non-discriminatory treatment of vessels of different nationalities and admission of vessels in distress.[54] Because the development and implementation of port State measures is essentially within the sovereign discretion of each State, international conventions refrain from prescribing any generally applicable standards for port States. This is confirmed by Article 23(4) of the UN Fish Stocks Agreement which provides that "(n)othing in this article affects the exercise by States of their sovereignty over ports in their territory in accordance with international law."

In the context of global efforts to combat IUU fishing, the discretion States have to impose conditions of entry into their ports give rise to differential standards among members of RFMOs in terms of inspection procedures, information to be provided by vessels intending to enter into port and penalties imposed. This is generally perceived to weaken the effectiveness of international conservation and management measures to combat IUU fishing.

To address this gap in the regulatory framework, the Twenty-seventh Session of the FAO Committee on Fisheries agreed to the development of a legally binding agreement on port State measures to combat IUU fishing.[55] Following an Expert Consultation and several Technical Consultations,[56] a final text of the Agreement on Port State Measures to Prevent, Deter and Eliminate Illegal,

---

[54] *LOSC*, Arts. 25 and 218.

[55] FAO, Committee on Fisheries, Report of the Twenty-seventh Session of the FAO Committee on Fisheries, Rome, Italy, 5–9 March 2007, *FAO Fisheries Report No. 830* (Rome: FAO, 2007), at para. 68.

[56] FAO, Report of the Expert Consultation to Draft a Legally-binding Instrument on Port State Measures, Washington, D.C., USA, 4–8 September 2007, *FAO Fisheries Report No 846* (Rome, FAO, 2007). See also Technical Consultation to Draft a Legally-binding Instrument on Port State Measures to Prevent, Deter, and Eliminate Illegal, Unreported and Unregulated Fishing, Rome, Italy, 23–27 June 2008; Informal Open-ended Technical Meeting to Review the Annexes of the Draft Legally-binding Instrument on Port State Measures to Prevent, Deter and Eliminate IUU Fishing, Rome, Italy, 25–27 November 2008; and a resumed session on Technical Consultation to Draft a Legally-binding Instrument on Port State Measures to Prevent, Deter, and Eliminate Illegal, Unreported and Unregulated Fishing from 26–29 January 2009; Technical Consultation to Draft a Legally-binding Instrument on Port State Measures to Prevent, Deter and Eliminate Illegal, Unreported and Unregulated Fishing (Second Resumed Session), Rome, Italy, 4–8 May 2009; Technical Consultation to Draft a Legally-binding Instrument on Port State Measures to Prevent, Deter and Eliminate Illegal, Unreported and Unregulated Fishing (Third Resumed Session), Rome, Italy, 24–28 August 2009.

Unreported and Unregulated Fishing (FAO Port State Measures Agreement)[57] was agreed to in August 2009, under Article XIV of the FAO Constitution. The preamble to the FAO Port State Measures Agreement recognises "that port State measures provide a powerful and cost-effective means of preventing, deterring and eliminating illegal, unreported and unregulated fishing." The FAO Port State Measures Agreement reiterates the sovereignty of port States in adopting measures to combat IUU fishing. In particular, Article 4(1) of the FAO Port State Measures Agreement provides that

> 1. Nothing in this Agreement shall prejudice the rights, jurisdiction and duties of Parties under international law. In particular, nothing in this Agreement shall be construed to affect:
>
> (a) the sovereignty of Parties over their internal waters, archipelagic and territorial waters or their sovereign rights over their continental shelf and in their exclusive economic zones;
>
> (b) the exercise by Parties of their sovereignty over ports in their territory in accordance with international law, including their right to deny entry thereto as well as to adopt more stringent port State measures than those provided for in this Agreement, including such measures adopted pursuant to a decision of a regional fisheries management organization."

The FAO Port State Measures Agreement is based largely on the FAO Model Scheme on Port State Measures and contains provisions on the integration of port State measures with other relevant measures to address IUU fishing, cooperation and exchange of information among national authorities and States, requirements for prior entry into port, use of ports, port inspection procedures, training of inspectors, the role of flag States, requirements of developing States, dispute settlement, dealing with non-parties, and monitoring and review of the implementation of the agreement.

The FAO Port State Measures Agreement will enter into force thirty days after the date of deposit with the Depositary of the twenty-fifth instrument of ratification, acceptance, approval or accession.[58] If it gains wide ratification, the FAO Port State Measures Agreement will help to establish common procedures for inspection and agreed measures against IUU fishing vessels. The implementation of the measures provided under the FAO Port State Measures Agreement will also promote harmonisation of port State measures across RFMOs and strengthen global measures to address IUU fishing.

3.2.2. *Non-binding Instruments*

A range of non-binding fisheries instruments also provide the policy context for modern international fisheries law and general framework to combat IUU fishing.

---

[57] *FAO Port State Measures Agreement*, *supra* note 11.

[58] *FAO Port State Measures Agreement*, Art. 29.

3.2.2.1. *UN Resolutions on Driftnet Fishing*

International efforts to regulate large scale pelagic driftnet fishing started with concerns about the negative impacts of driftnet fishing on targeted and non-targeted species of fish, marine mammals, seabirds, and other living marine resources which become entangled in such driftnets.[59] Large scale pelagic driftnet fishing proliferated in the North Pacific, Central and South Pacific, Indo-Pacific and Indian Ocean, Atlantic and the Mediterranean in the late 1980s and early 1990s.[60] Regional initiatives developed in the North Pacific Ocean,[61] Caribbean,[62] and South Pacific[63] in response to the problem all emphasised the need for strong and concerted action by the international community to address the problem.

As a result, in 1991, the United Nations General Assembly (UNGA) adopted Resolution 46/215 which imposed a global moratorium on large scale pelagic driftnet fishing.[64] The UNGA Resolution specifically called on States to reduce large scale pelagic driftnet fishing effort by 50 per cent and to ensure that the areas of operation for driftnet fishing vessels did not expand.[65] UNGA Resolution 46/215 further encouraged States to take measures, individually or collectively, to prevent large scale pelagic driftnet fisheries on the high seas and enclosed and semi-enclosed seas.[66] The need for the full implementation of the global moratorium on large scale pelagic high seas driftnet fishing, together with associated issues such as unauthorised fishing in zones of national jurisdiction and by-catch and discards, was reiterated in succeeding UN General Assembly resolutions.[67] The UNGA resolutions urged all States to take greater enforcement responsibility to ensure compliance with UNGA Resolution 46/215 and to impose appropriate sanctions against acts contrary to the terms of the resolution. The issue of large scale pelagic driftnet fisheries has been included in further resolutions of the UN General Assembly on sustainable fisheries.[68]

---

[59] UNGA, 85th Plenary Meeting, Large-scale pelagic driftnet fishing and its impact on the living marine resources of the world's oceans and seas, *A/RES/44/225*, adopted on 22 December 1989; UNGA, 71st Plenary meeting, Large-scale pelagic driftnet fishing and its impact on the living marine resources of the world's oceans and seas, *A/RES/45/197*, adopted on 21 December 1990.

[60] *See* S.P. Northridge, 'Driftnet Fisheries and their Impacts on Non-target Species: A Worldwide Review', *FAO Fisheries Technical Paper No. 320* (Rome: FAO, 1991).

[61] Earthtrust.org Website *North Pacific Joint Observer Program 1989–1991* http://earthtrust.org/dnpaper/npac.html. Accessed on 22 April 2008.

[62] *Castries Declaration*, St Lucia, 20–24 November 1989.

[63] *Convention for the Prohibition of Fishing with Long Driftnets in the South Pacific,* Wellington, New Zealand, concluded on 24 November 1989, *and its Protocols*, Noumea, New Caledonia, concluded on 20 October 1990, in force 17 May 1991, 1899 UNTS 3; 29 ILM 1454 (1990).

[64] UNGA, 79th Plenary Meeting, Large-scale pelagic drift-net fishing and its impact on the living marine resources of the world's oceans and seas, *UNGA 46/215*, 20 December 1991.

[65] *Ibid.*, para 3.

[66] *Ibid.*, para 4.

[67] *UNGA A/RES/50/25*, para 1; *UNGA A/RES/51/36*, para 1; *UNGA A/RES/52/29*, para 1; *UNGA A/RES/53/33*, para 2; *UNGA A/RES/55/8*, para 2; *UNGA A/RES/57/142*, para 5. See *supra* note 7.

[68] UNGA, Fifty-eighth session, Agenda Item 52(b), Resolution adopted by the General Assembly, Sustainable fisheries, including through the 1995 Agreement for the Implementation of the

3.2.2.2. *The Declaration of Cancun*

The Declaration of Cancun was adopted on 8 May 1992 and called upon States to draft an International Code of Conduct for Responsible Fishing, as well as convene an intergovernmental conference on high seas fisheries.[69] This declaration also highlighted the need to reach agreement on measures to promote international trade in fish products consistent with the WTO General Agreement on Tariffs and Trade and principles of responsible fishing.[70]

3.2.2.3. *The Rome Consensus on World Fisheries*

In March 1995, the Rome Consensus on World Fisheries reiterated the importance of bringing to conclusion the FAO Code of Conduct and UN Fish Stocks Agreement, as well as the acceptance of the FAO Compliance Agreement.[71] In addition, the Rome Consensus on World Fisheries also urged governments and international organisations to take prompt action to eliminate overfishing, minimise wasteful fisheries practices, rehabilitate fish habitats, review fishing fleet capacity, and ensure the effectiveness of fisheries conservation and management measures.[72]

Provisions of the United Nations Convention on the Law of the Sea of 10 December 1982 relating to the Conservation and Management of Straddling Fish Stocks and Highly Migratory Fish Stocks, and related instruments, *A/RES/58/14*, 21 January 2004, para. 33; UNGA, Fifty-ninth session, Agenda Item 49(b), Resolution adopted by the General Assembly, Sustainable fisheries, including through the 1995 Agreement for the Implementation of the Provisions of the United Nations Convention on the Law of the Sea of 10 December 1982 relating to the Conservation and Management of Straddling Fish Stocks and Highly Migratory Fish Stocks, and related instruments, *A/RES/59/25*, 17 January 2005, para. 44; UNGA, Sixtieth session, Agenda Item 75(b), Resolution adopted by the General Assembly, Sustainable fisheries, including through the 1995 Agreement for the Implementation of the Provisions of the United Nations Convention on the Law of the Sea of 10 December 1982 relating to the Conservation and Management of Straddling Fish Stocks and Highly Migratory Fish Stocks, and related instruments, 10 March 2006, *A/RES/60/31*, para. 50; UNGA, Sixty-first session, Agenda Item 71(b), Resolution adopted by the General Assembly, Sustainable fisheries, including through the 1995 Agreement for the Implementation of the Provisions of the United Nations Convention on the Law of the Sea of 10 December 1982 relating to the Conservation and Management of Straddling Fish Stocks and Highly Migratory Fish Stocks, and related instruments, 6 March 2007, *A/RES/61/105*, para. 59; UNGA, Sixty-second session, Agenda Item 77(b), Resolution adopted by the General Assembly, Sustainable fisheries, including through the 1995 Agreement for the Implementation of the Provisions of the United Nations Convention on the Law of the Sea of 10 December 1982 relating to the Conservation and Management of Straddling Fish Stocks and Highly Migratory Fish Stocks, and related instruments, 28 February 2008, *A/RES/62/177*, para. 72.

[69] *Declaration of the International Conference on Responsible Fishing*, Cancun, Mexico, 8 May 1992, paras. I and IV.

[70] *Declaration of the International Conference on Responsible Fishing*, para. V.

[71] FAO Ministerial Conference on Fisheries, *The Rome Consensus on World Fisheries*, Rome, 14–15 March 1995, para. 10.

[72] *Ibid.*

3.2.2.4. *FAO Code of Conduct*

A significant development in the field of the international management of fisheries is the FAO Code of Conduct for Responsible Fisheries, which is a voluntary instrument adopted on 31 October 1995 to provide internationally agreed principles and standards applicable to the conservation, management and development of all fisheries.[73] The FAO Code of Conduct not only deals with the management of capture fisheries but also covers other aspects of fisheries development such as aquaculture, processing, and trade of fish. Because of its comprehensive nature, the FAO Code of Conduct contains applicable principles and measures addressing a wide range of IUU fishing related activities, particularly illegal fishing such as destructive fishing practices[74] and fishing without authorisation.[75] The FAO Code of Conduct is global in scope and is directed at members and non-members of the FAO, fishing entities, subregional, regional, and global governmental or nongovernmental organisations, and all persons concerned with the management and development of fisheries.[76]

Apart from its all encompassing nature, there are two other reasons for the international significance of the FAO Code of Conduct. First, it highlights the precautionary approach to fisheries as well as emphasises the need for the introduction of new understanding and methods of dealing with the uncertainties inherent to fisheries management.[77] The other significance of the FAO Code of Conduct relates to its legal effect. Voluntary instruments such as the FAO Code of Conduct are considered to be an important supplement to treaties and customary international law. They provide useful and provisional norms which, by the consent of States, can later be adopted as binding measures in treaties.[78] The widespread acceptance and implementation of the FAO Code of Conduct may provide evidence of the development of new norms or principles of international law applicable to sustainable fisheries management.[79]

3.2.2.5. *IPOA-Capacity*

The International Plan of Action for the Management of Fishing Capacity (IPOA-Capacity) is a voluntary instrument developed to implement the FAO Code of

[73] *FAO Code of Conduct, supra* note 6.

[74] *See* FAO Code of Conduct, paras. 8.4.2 and 8.4.5.

[75] *See* FAO Code of Conduct, para. 8.2.2.

[76] *FAO Code of Conduct*, Art. 1.4.

[77] Gerald Moore, 'The Code of Conduct for Responsible Fisheries', in Ellen Hey (ed.), *Developments in International Fisheries Law* (Hague: Kluwer Law International, 1999), at 95.

[78] Ingrid Detter, *The International Legal Order* (Hants: Dartmouth Publishing Company, 1994), at 251.

[79] William Edeson, 'Closing the Gap: The Role of "Soft" International Instruments to Control Fishing', 20 *Australian Yearbook of International Law* (1999), at 21. See also discussion in Chapter 1 Section 1.2.1.

Conduct in order to address the problem of excess fishing capacity.[80] The IPOA-Capacity recognises that excess fishing capacity threatens the long term sustainability of fisheries by contributing substantially to overfishing, degradation of marine resources, and decline of food production potential.[81] The objective of the IPOA-Capacity is for States and regional fisheries organisations to achieve an efficient, equitable and transparent management of fishing capacity at the global level.[82] The IPOA-Capacity contains "urgent actions" and identifies mechanisms to promote its implementation. Some of these "urgent actions" include the assessment and monitoring of fishing capacity and preparation and implementation of national plans of action.[83] The specific actions stated in the IPOA-Capacity are similar to those adopted in the IPOA-IUU, such as the development of national plans, establishment of records of fishing vessels, and elimination of subsidies and economic incentives that build up excessive fishing capacity.

3.2.2.6. *IPOA-Seabirds*

The International Plan of Action for Reducing Incidental Catch of Seabirds in Longline Fisheries (IPOA-Seabirds) is another instrument elaborated within the framework of the FAO Code of Conduct.[84] It applies to States in the waters of which longline fishing is being conducted by their own or foreign vessels and to States whose vessels conduct longline fishing on the high seas and the EEZ of other States.[85] Key longline fisheries in which incidental catch of seabirds are known to exist are tuna, swordfish and billfish, Patagonian toothfish in the Southern Ocean, and halibut and cod in the Pacific and Atlantic Oceans,[86] most of which fall under the competence of RFMOs.

The IPOA-Seabirds provides a guide in developing a national plan of action and adopts technical and operational measures for reducing the incidental catch of seabirds in longline fisheries. States may prescribe appropriate mitigation measures provided in the IPOA-Seabirds and include them in their national plans of action based on their assessment of the extent and nature of the problem. Other

---

[80] FAO, *International Plan of Action for the Management of Fishing Capacity*, Adopted at the Twenty-third Session of the FAO Committee on Fisheries, Rome, Italy, 1999. *Hereinafter referred to as* IPOA-Capacity. Article 6(3) of the FAO Code of Conduct provides that "States should prevent overfishing and excess fishing capacity and should implement management measures to ensure that fishing effort is commensurate with the productive capacity of the fishery resources and their sustainable utilisation.

[81] *IPOA-Capacity*, para. 1.

[82] *IPOA-Capacity*, para. 7.

[83] *IPOA-Capacity*, Part III.

[84] FAO, *International Plan of Action for Reducing Incidental Catch of Seabirds in Longline Fisheries*, Adopted at the Twenty-third Session of the FAO Committee on Fisheries, Rome, Italy, 1999. *Hereinafter referred to as* IPOA-Seabirds.

[85] *IPOA-Seabirds*, para. 9.

[86] *IPOA-Seabirds*, para. 2.

elements of the national plan of action may include plans for research and development to improve technologies and practices in reducing the incidental capture of seabirds and data collection.

3.2.2.7. *IPOA-Sharks*

Similar to the IPOA-Capacity and IPOA-Seabirds, the International Plan of Action for the Conservation and Management of Sharks (IPOA-Sharks) is a non-binding instrument that implements the FAO Code of Conduct.[87] It applies to all States with shark fisheries, including those whose vessels catch sharks on the high seas.[88] For the purpose of the IPOA-Sharks, shark catch is taken to include directed, bycatch, commercial, recreational, and other forms of taking sharks.[89]

The IPOA-Sharks provides for the development of a national plan of action for the conservation and management of sharks (Sharks-plan). The Sharks-plan should aim to ensure that shark catches, whether directed or non-directed fisheries, are sustainable. It would also need to assess threats to sharks population and determine critical habitats, and facilitate the identification, monitoring, and reporting of shark catches and species specific biological and trade data.[90]

3.2.2.8. *UN Resolutions on Sustainable Fisheries*

Since 2004, the UNGA has been adopting resolutions on sustainable fisheries which contain specific provisions on IUU fishing.[91] These resolutions emphasise that the FAO Code of Conduct, together with its associated international plans of action, set out principles and global standards for responsible fishing practices.[92] The resolutions also express concern that IUU fishing threatens to seriously deplete populations of certain fish species and significantly damage marine ecosystems to the detriment of sustainable fisheries and food security.[93] The UNGA resolutions support some of the measures adopted under the FAO Code of

---

[87] FAO, *International Plan of Action for the Conservation and Management of Sharks*, Adopted at the Twenty-third Session of the FAO Committee on Fisheries, Rome, Italy, 1999. *Hereinafter referred to as* IPOA-Sharks.

[88] *IPOA-Sharks*, para. 17.

[89] *IPOA-Sharks*, para. 10. The term "shark" includes all species of sharks, skates, rays and chimaeras (Class *Chondrichthyes*).

[90] *IPOA-Sharks*, para. 10.

[91] UN resolutions adopted earlier than 2004 referred to "unauthorised fishing in zones of national jurisdiction and its impact on the living marine resources of the world's oceans and seas" which has been subsequently subsumed under the heading of IUU fishing. *See* A/RES/49/116, A/RES/50/25, A/RES/51/36, A/RES/52/29, A/RES/53/33, A/RES/55/8, and A/RES/57/142, *supra* note 7.

[92] *UNGA A/RES/58/14*, Preamble; *UNGA A/RES/59/25*, Preamble and para. 26; *UNGA A/RES/60/31*, Preamble and para. 33; *UNGA A/RES/61/105*, Preamble and para. 33; and *UNGA A/RES/62/177,* Preamble and para. 33. *See supra* note 68.

[93] *Ibid*. *See* Preamble of UNGA *A/RES/58/14*, *A/RES/59/25*, *A/RES/60/31*, *A/RES/61/105*, and *A/RES/62/177*.

Conduct, the FAO Compliance Agreement, and the UN Fish Stocks Agreement, such as effective control over nationals, monitoring and control of transshipments on the high seas, vessel monitoring systems, record of fishing vessels, port State measures, intergovernmental cooperation, and development and implementation of national and regional plans of action.[94] In these resolutions, the IMO has been called upon to examine and clarify the concept of genuine link in relation to flag State duties, prevent the operation of sub-standard vessels, and eliminate subsidies that contribute to IUU fishing.

The UNGA Resolution 60/31 adopted in 2006 encourages States and other international organisations to develop more effective measures to trace fish and fishery products to enable importing States to identify fish caught in a manner that undermines international conservation and management measures.[95] It also urges States, individually and through RFMOs to adopt and implement internationally agreed market related measures to address IUU fishing.[96] This resolution further supports the development of a comprehensive global record of fishing vessels within the FAO.[97] UNGA resolutions adopted in 2007 and 2008 promote the development of a legally binding instrument on minimum standards for port State measures.[98] These resolutions also encourage flag and port States to share data on landing and catch quotas.[99] Furthermore, UNGA Resolution 62/177 supports actions of RFMOs against IUU fishing such as the development of a common list of vessels identified as having conducted IUU fishing, as well as the denial of port access to vessels appearing on such list.[100]

In addition to these measures, the UNGA Resolution 61/105 adopted in 2006 contains some ambitious goals. It includes an important new section on responsible fisheries in the marine ecosystem. The UNGA Resolution 61/105 sought to build upon the achievements of the Reykjavik Declaration on Responsible Fisheries in the marine ecosystem. However, its most significant component is the specific call for action to be taken with respect to bottom fisheries. In particular, paragraph 83 of the resolution called upon RFMOs, with competence to regulate bottom fisheries "to adopt and implement measures in accordance with precautionary approach, ecosystem approaches and international law" by 31 December 2008.[101] UNGA Resolution 61/105 also calls for the identification of

[94] *Ibid. See UNGA A/RES/58/14,* paras. 19–29; *UNGA A/RES/59/25,* para 27–38; *UNGA A/RES/60/31*, paras. 34–46; *UNGA A/RES/61/105*, paras. 34–46; *UNGA A/RES/62/177,* para. 38–55.

[95] *Ibid., UNGA A/RES/60/31*, para. 40.

[96] *Ibid., UNGA A/RES/60/31*, para. 46; *UNGA A/RES/61/105*, para. 46; *UNGA A/RES/62/177*, para. 55.

[97] *Ibid., UNGA A/RES/60/31*, para. 45.

[98] *Ibid., UNGA A/RES/61/105*, para. 43; *UNGA A/RES/62/177*, para. 51.

[99] *Ibid., UNGA A/RES/61/105*, para. 44; *UNGA A/RES/62/177*, para. 53.

[100] *Ibid., UNGA A/RES/62/177*, paras. 44 and 50.

[101] *See also Ibid.*, UNGA A/RES/62/177, paras. 96–97.

vulnerable marine ecosystems and an assessment of the impacts of bottom fishing activities on such ecosystems and to the long term sustainability of deep sea fish stocks. Paragraph 83(c) stated:

> (c) In respect of areas where vulnerable marine ecosystems, including seamounts, hydrothermal vents and cold water corals, are known to occur or are likely to occur based on the best available scientific information, to close such areas to bottom fishing and ensure that such activities do not proceed unless conservation and management measures have been established to prevent significant adverse impacts on vulnerable marine ecosystems.

In addition to calling on RFMOs, the UNGA resolution in paragraph 83(d) calls on members of RFMOs to require vessels flying their flag to cease bottom fishing activities in areas where vulnerable marine ecosystems are encountered in the course of fishing operations, and to report the encounter so that appropriate measures can be adopted in respect of the relevant site. Similar requirements are also imposed on flag States whose vessels are engaged in bottom fisheries on the high seas where there are no RFMOs or regional fisheries management arrangement with the competence to regulate such fisheries. In this case flag States are called upon either to adopt and implement interim measures or cease to authorise bottom fishing activities in such areas.[102]

3.2.2.9. *FAO Model Scheme on Port State Measures to Combat IUU Fishing*

A Model Scheme on Port State Measures to Combat IUU Fishing was adopted by the FAO in 2004[103] which outlines principles and guidelines to be used by States as a reference for the negotiation and adoption of regional memoranda of understanding (MOUs), adoption of resolutions within RFMOs, or measures to be adopted at the national level.[104] The FAO Model Scheme is the first international policy document to outline a list of activities considered as IUU fishing based on the IPOA-IUU. Paragraph 4 of the FAO Model Scheme provides that a foreign fishing vessel is believed to have engaged in, or supported IUU fishing activities, if there is reasonable evidence that the vessel has, among other things, fished without a valid authorisation issued by the flag State or the relevant coastal State, fished in a closed area, fished during a closed season, or fished without or after

---

[102] *Ibid., UNGA A/RES/61/105*, para. 86.

[103] This Model Scheme is similar to the Memorandum of Understanding on Port State Measures to Combat IUU Fishing which was proposed in 2002 by Mr. Terje Lobach. At that time, it was considered impractical to adopt an MOU on Port State Measures at the global level. The substantive aspects of both documents are very similar. *See* FAO, Report of the Expert Consultation to Review Port State Measures to Combat Illegal, Unreported, and Unregulated Fishing, Rome, Italy, 04–06 November 2002, *FAO Fisheries Report No. 692* (Rome: FAO, 2002). *See also* Terje Lobach, 'Port State Control of Foreign Fishing Vessels', *FAO Legal Papers Online No. 29* (Rome: FAO, 2002).

[104] FAO, Report of the Technical Consultation to Review Port State Measures to Combat Illegal, Unreported and Unregulated Fishing, Rome, Italy, 31 August–02 September 2004, *FAO Fisheries Report No. 759* (Rome: FAO, 2004), at para. 16.

attainment of a quota, engaged in directed fishing for a stock which is subject to a moratorium or for which fishing is prohibited, used prohibited fishing gear, and falsified or concealed the markings, identity or registration of the vessel. The FAO Model Scheme also provides that IUU fishing includes multiple violations which together constitute a serious disregard of relevant conservation and management measures.[105]

In addition, the FAO Model Scheme provides guidelines for carrying out inspections of foreign vessels in ports, a list of information that should be provided by vessels in advance to port States before entry, expected results from port inspections, training of port inspectors, and information sharing system among port States.[106] The Model Scheme conforms to the measures adopted under the IPOA-IUU and all relevant rules of international law. The universal adoption of the Model Scheme would help establish common procedures for port inspections of fishing vessels globally and strengthen measures against IUU fishing vessels.

3.2.2.10. *Rome Declaration on IUU Fishing*

The Rome Declaration on IUU Fishing was adopted in 2005 to enlist the commitment of States to fully implement international fisheries instruments for the sustainable use of marine living resources. This declaration reiterated the need to implement measures adopted in the IPOA-IUU, such as the review of national legislation to ensure compliance with fisheries management measures, implementation of internationally agreed boarding and inspection procedures, and implementation of vessel marking requirements.[107] The Rome Declaration on IUU Fishing called for the adoption of new actions, such as measures to address fleet overcapacity, and exchange of VMS and observer data.[108]

The Rome Declaration further called on States "to develop a comprehensive record of fishing vessels within FAO, including refrigerated transport vessels and supply vessels, that incorporates available information on beneficial ownership, subject to confidentiality requirements in accordance with national law."[109] This call for action led to the study by FAO to determine the feasibility and viability of developing a comprehensive record of fishing vessels. The study confirmed that there is neither a complete record of the number of fishing vessels in the world,

---

[105] *FAO Model Scheme*, para. 4(h). A similar list of activities which may be considered IUU fishing has been deliberated on in the drafting of the legally binding agreement on port State measures to combat IUU fishing. However, such list was not adopted and only the general scope and nature of IUU fishing under paragraph 3 of the IPOA-IUU has been included in the final text of the FAO Port State Measures Agreement.

[106] *FAO Fisheries Report No. 759*, Appendix E, *supra* note 104.

[107] *The 2005 Rome Declaration on Illegal, Unreported and Unregulated Fishing*, Adopted by the FAO Ministerial Meeting on Fisheries, Rome, 12 March 2005, paras. 3, 5, and 6.

[108] *Rome Declaration on IUU Fishing,* para. 4

[109] *Ibid.*

nor a complete, single source of information from which it would be possible to trace individual vessels and ownership.[110]

3.2.2.11. *FAO Guidelines*

International guidelines relevant to sustainable fisheries that also address IUU fishing have been formulated under the FAO after the adoption of the IPOA-IUU.[111] An example is the FAO Guidelines for the Eco-labelling of Fish and Fishery Products from Marine Capture Fisheries[112] which provides the principles, minimum substantive requirements and criteria, and procedural and institutional aspects of eco-labelling of fish and fishery products. The FAO Guidelines on Eco-labelling of Fish and Fish Products also provide procedural guidelines for setting standards for sustainable fisheries, accreditation of independent certifying bodies, and certification that a fishery and the product chain of custody are in conformity with required standards and procedures. These guidelines provide a framework for ensuring that fish and fishery products which are awarded eco-labels are obtained from well managed marine capture fisheries.

The International Guidelines for the Management of Deep-sea Fisheries in the High Seas[113] were adopted in September 2008 and are directly related to UNGA resolutions on large scale driftnet fishing and sustainable fisheries and aim to implement UNGA Resolution A/RES/61/105. In general, the FAO Guidelines on Deep-sea Fisheries call for the establishment and implementation of national legal and institutional frameworks for the effective management of deep sea fisheries and their impacts on vulnerable marine ecosystems, particularly through flag, port, and market State measures and by exercising jurisdiction over nationals.[114] The guidelines also provide for the need to strengthen the capacity of existing RFMOs with the competence to manage deep sea fisheries, as well as the need to cooperate in the establishment of new RFMOs or arrangements to regulate bottom fisheries.[115]

[110] FAO, Committee on Fisheries, Twenty-seventh Session, Rome, Italy, 5–9 March 2007, Report on the Development of a Comprehensive Record of Fishing Vessels, *COFI/2007/Inf.12* (Rome: Italy, 2007), at para. 6.

[111] There are currently 12 technical guidelines supporting responsible fisheries. These technical guidelines focus on ecosystem approach to fisheries, precautionary approach to capture fisheries and species introductions, managing fishing capacity, inland fisheries, small scale fisheries, fishing operations, implementation of the IPOA-IUU, responsible fish trade, responsible fish utilisation, conservation and management of sharks, aquaculture development, indicators for sustainable development of marine capture fisheries, integration of fisheries into coastal area management, vessel monitoring systems, and information knowledge and sharing.

[112] FAO, *Guidelines for the Ecolabelling of Fish and Fishery Products from Marine Capture Fisheries* (Rome: FAO, 2005).

[113] FAO, *International Guidelines for the Management of Deep-sea Fisheries in the High Seas* (Rome: FAO, 2008).

[114] *Ibid.*, para. 5.

[115] *Ibid.*, para. 35.

3.3. Environment Related Instruments

Traditionally, global international environmental instruments dealing with renewable natural resources were designed to operate as checks against the over-exploitation of species or the destruction of habitats of high conservation value. This is achieved through, for example, the regulation and prohibition of the taking of designated species, the protection of habitat by creating protected areas and the regulation of international trade in endangered species. During the 1980s a change in the approach to conservation began to occur as it became apparent that effective conservation depends on the adoption of strategies which take into account the interrelationship between individual species, their gene pools and the ecosystems on which they depend. There was a corresponding international acceptance and promotion of the value of components of biological diversity, not only as resources to be exploited, but as part of the network that is necessary to sustain the quality of life on earth for current and future generations.

This sub-section describes global environmental instruments most relevant to sustainable fisheries management in general. The relevant instruments include the Convention on Wetlands of International Importance Especially as Waterfowl Habitat (Ramsar Convention),[116] the Convention on International Trade in Endangered Species of Wild Flora and Fauna (CITES),[117] and the Convention on the Conservation of Migratory Species of Wild Animals (Bonn Convention)[118] and the Convention on the Conservation of Biological Diversity (CBD).[119] Although these Conventions are not specifically considered in the context of global efforts to combat IUU fishing, their effective implementation will promote a sustainable fishing culture and indirectly contribute to the eradication of IUU fishing. Some of these instruments, especially CITES, can also play a positive role in supporting the development and application of internationally agreed trade measures to combat IUU fishing.

---

[116] *Convention on Wetlands of International Importance especially as Waterfowl Habitat*, Ramsar, Iran, concluded on 2 February 1971, in force 21 December 1975, 996 UNTS 245; TIAS 11084; 11 ILM 963 (1972). *Hereinafter referred to as* Ramsar Convention. As amended by the Paris Protocol, 3 December 1982 and Regina Amendments, 28 May 1987.

[117] *Convention on International Trade in Endangered Species of Wild Fauna and Flora*, Washington D.C., USA, concluded on 3 March 1973, in force 1 July 1975, 27 UST 1087; TIAS 8249; 993 UNTS 243. *Hereinafter referred to as* CITES. Amended at Bonn, on 22 June 1979 and Gaborone on 30 April 1983.

[118] *Convention on the Conservation of Migratory Species of Wild Animals,* Bonn, concluded on 23 June 1979, in force 01 November 1983, 19 ILM 15 (1980); ATS 1991/32; BTS 87 (1990), Cm. 1332. *Hereinafter referred to as* Bonn Convention.

[119] *Convention on Biological Diversity*, Rio de Janeiro, Brazil, concluded on 5 June 1992, in force 29 December 1993, 1760 UNTS 79; 31 ILM 818 (1992).

3.3.1. *Ramsar Convention*

The Convention on Wetlands of International Importance Especially as Waterfowl Habitat provides the legal framework for the conservation and sustainable use of wetlands and their resources through local, regional and national actions and international cooperation. The Convention provides for the designation of wetlands within national territory for inclusion on the List of Wetlands of International Importance,[120] as well as the formulation and implementation of plans for the conservation of such wetlands.[121] Part of the responsibilities of the Ramsar Convention is based on the expectation of common and equitably shared transparent accountability.[122] Some States have included Ramsar obligations in national law and policy.[123]

The Ramsar Convention has adopted a Classification of Wetland Type which includes 42 types, grouped into three categories, namely marine and coastal wetlands, inland wetlands, and human-made wetlands. The Conference of Contracting Parties has developed criteria for identifying wetlands of international importance.[124] Group A of the Criteria includes sites containing representative, rare or unique wetland types. Group B of the Criteria includes sites of international importance for conserving biological diversity. There are several criteria for Group B wetland types, including criteria based on species and ecological communities and specific criteria based on waterbirds, fish, and other taxa. Because of the clear relationship between wetlands and biodiversity concerns, the Conference of the Contracting Parties of the Ramsar Convention agreed to develop ways and means at the national level for improving liaison between the authorities implementing the Ramsar Convention and focal points for related conventions and agreements such as biodiversity related conventions.[125]

Resolutions have been adopted by the Conference of the Parties of the Ramsar Convention recognising the importance that marine wetlands play in supporting aquatic species populations and resources. Resolution IX.4 adopted in 2005 on fisheries resources urges Ramsar Convention Contracting Parties and invites

---

[120] *Ramsar Convention*, as amended, Art 2(1).

[121] *Ramsar Convention*, as amended, Art 3(1).

[122] *Ibid.*

[123] *Ibid.*

[124] Ramsar Convention, *Criteria for Identifying Wetlands of International Importance*, Adopted by the 7th (1999) and 9th (2005) Meetings of the Conference of the Contracting Parties, superseding earlier Criteria adopted by the 4th and 6th Meetings of the COP (1990 and 1996), to guide implementation of Article 2.1 on designation of Ramsar sites. www.ramsar.org. Accessed on 28 April 2008.

[125] Ramsar Convention, Resolution IX.5, *Synergies with other international organisations dealing with biological diversity; including collaboration on, and harmonisation of, national reporting among biodiversity-related conventions and agreements*, Adopted at the Ninth Conference of the Contracting Parties, Kampala, Uganda, 8–15 November 2005.

relevant organisations to use the habitat and species conservation provisions of the Ramsar Convention to support the introduction and implementation of management measures that mitigate the environmental impacts of fishing.[126] The resolution is particularly concerned with ecologically damaging fishing gears used in many fisheries and significantly alter habitat structure and prevent the movement of species.[127]

3.3.2. *CITES*

The Convention on International Trade in Endangered Species of Wild Fauna and Flora aims to ensure that the international trade in specimens of wild animals and plants does not threaten their survival. CITES lists species according to the degree of the protection they need[128] and provides for the control of the international trade in these endangered species through a system of permits and certificates.[129] Appendix I of CITES includes species threatened with extinction, in which trade is permitted only in exceptional circumstances.[130] Appendix II of CITES includes species not necessarily threatened with extinction, but in which trade must be controlled in order to avoid utilisation incompatible with their survival.[131] Appendix III contains species that are protected in at least one country, which has requested other CITES Parties for cooperation in controlling the trade.[132] CITES regulates the trade of species listed under these appendices in terms of export, re-export, import, and introduction from the sea.[133] The Conference of Parties of CITES, which is the decision-making body of the Convention and comprises all its member States,[134] has agreed to a set of biological and trade related criteria to help determine whether a species should be included in Appendices

---

[126] Ramsar Convention, Resolution IX.4, *The Ramsar Convention and conservation, production and sustainable use of fisheries resources*, Kampala, Uganda, 8–15 November 2005, para. 26.

[127] Ramsar Convention, *Resolution IX.4*, Annex, Issue 3.

[128] *CITES*, Art II and Appendices I, II, and III.

[129] *CITES*, Art III-V.

[130] *CITES*, Art II(1).

[131] *CITES*, Art II(2).

[132] *CITES*, Art II(3).

[133] Article I(e) of CITES defines "introduction from the sea" as "transportation into a State of specimens of any species which were taken in the marine environment not under the jurisdiction of any State." CITES Conference of Parties agreed that the 'the marine environment not under the jurisdiction of any State' means those marine areas beyond the areas subject to the sovereignty or sovereign rights of a State consistent with international law, as reflected in the United Nations Convention on the Law of the Sea. *See* CITES Conf.14.6, Introduction from the Sea, Adopted at the Fourteenth meeting of the Conference of the Parties, Hague, Netherlands, 3–15 June 2007. For further discussions on the matter, *see also* Erik Franckx Applications of the Term "Introduction from the Sea", Background Paper for the FAO, Report of the Expert Consultation on Legal Issues Related to CITES and Commercially-exploited Aquatic Species, Rome, 22–25 June 2004, *FAO Fisheries Report No. 746* (Rome: FAO, 2004).

[134] *CITES*, Art XI.

I and II.[135] There are about 5,000 species of animals and 28,000 species of plants currently protected by CITES, including 15 species of fish listed in Appendix I and 71 species in Appendix II, as well as whales, turtles, and corals.[136]

One of the basic issues about the implementation of CITES with respect to fisheries is its application to commercially exploited species.[137] CITES primarily focused on the listing of species other than exploited aquatic fishes or invertebrates until the early 1990s.[138] Since then, concerns have been raised about the appropriateness of the CITES listing criteria for exploited and managed fishery resources.[139] Some of the criteria for listing species under Appendices I and II are deemed to be only applicable to a few exploited marine species.[140] There are also cases where little is known about the sustainable catch of a particular species,[141] and it may be difficult to ascertain if trade in such species would be detrimental to their survival.[142] Concerns have also been expressed about possible "false alarms" (*i.e.* classification of species which are not at risk of extinction to a category requiring trade restrictions) and "misses" (*i.e.* failure to classify species at risk under categories offering them necessary protection from trade) in applying the present CITES criteria for the listing of commercially exploited aquatic species.[143] This may not be perceived as a critical error for purposes of conservation; however, the unnecessary prevention of trade may have severe economic consequences, particularly for harvesting developing States.[144] This may, in turn, inadvertently promote or encourage IUU fishing.

A number of legal issues are associated with the application of CITES to commercially exploited aquatic species. One issue is whether a CITES listing, particularly for Appendix I or Appendix II, would contravene the LOSC and related

---

[135] CITES, Conference of Parties, Conf. 9.24 (Rev. CoP14), Criteria for amendment of Appendices I and II.

[136] CITES Website, The CITES Species. www.cites.org. Accessed on 22 April 2008.

[137] "Commercially-exploited aquatic species" in relation to CITES has been agreed within FAO to encompass resources exploited by fisheries in marine and large freshwater bodies. *See* FAO, Report of the Technical Consultation on the Sustainability of the CITES Criteria for Listing Commercially-Exploited Aquatic Species, Rome, Italy, 28–30 June 2000, *FAO Fisheries Report No. 629* (Rome: FAO, 2000), at para. 8; FAO, Report of the Technical Consultation on Implementation Issues Associated with Listing Commercially-Exploited Aquatic Species on CITES Appendices, Rome, 25–28 May 2004, *FAO Fisheries Report No. 741* (Rome: FAO, 2004), at para. 11.

[138] FAO, An appraisal of the suitability of the CITES criteria for listing commercially-exploited aquatic species, *FAO Fisheries Circular No. 954* (Rome: FAO, 2000), at 1.

[139] *Ibid.*

[140] FAO, Report of the Second Technical Consultation on the Sustainability of the CITES Criteria for Listing Commercially-exploited Aquatic Species, Windhoek, Namibia, 22–25 October 2001, *FAO Fisheries Report No. 667* (Rome: FAO, 2001), at Appendix F.

[141] *FAO Fisheries Report No. 741*, *supra* note 137, at para 24.

[142] Article IV(2)(a) of CITES provides the requirement for a scientific authority of the State of export to advise that such export will not be detrimental to the survival of that species as one of the conditions for the granting of an export permit for a species listed in Appendix II.

[143] *FAO Fisheries Circular No. 954*, *supra* note 138, at 32.

[144] *Ibid.*, at 33.

instruments by diminishing the rights and obligations to fish, including the freedom to fish on the high seas.[145] More particularly, the concern relates to the harvesting of species in the EEZ or on the high seas which could not be subsequently introduced into the ports of a CITES Party because of the listing of such species in Appendix I.[146] Other issues include the tedious process and high administrative cost of reversing decisions on listing of species under Appendices I and II.[147] These issues are continuously being addressed in expert consultations conducted within the FAO.[148]

3.3.3. *Bonn Convention*

The Convention on the Conservation of Migratory Species requires its Parties to conserve migratory species and their habitat by taking action, either individually or in cooperation with other States.[149] The Bonn Convention provides for measures that may be applied by range States[150] in conserving migratory species listed under Appendices I and II of the Convention. Appendix I of the Convention lists migratory species which are endangered throughout all or a significant proportion of their range, and for which States are required to strive for strict protection of such species.[151] Appendix II lists migratory species with an unfavourable conservation status or which would benefit significantly from international cooperation, such as global and regional agreements that would conserve and manage individual or a group of species.[152]

The Conference of Parties of the Bonn Convention has recognised by-catch as a major cause of mortality of migratory species listed on Appendices I and II of the Convention, including seabirds, sharks, turtles, marine mammals and sturgeons.[153]

---

[145] FAO, Report of the Expert Consultation on Legal Issues Related to CITES and Commercially-exploited Aquatic Species, Rome, 22–25 June 2004, *FAO Fisheries Report No. 746* (Rome: FAO, 2004), at para 36.

[146] *Ibid.*

[147] *See* Robert W G Jenkins, 'The Significant Trade Process: Making Appendix II Work', in Jon Hutton and Barnabas Dickson (eds.), *Endangered Species Threatened Convention: The Past, Present and Future of CITES* (London: Earthscan Publications Ltd, 2000), at 49.

[148] Technical and experts consultation on the implementation of CITES criteria to commercially-exploited aquatic species can be traced back to the call of FAO to establish an ad hoc group to discuss the matter in 1998. This idea was further endorsed at the Twenty-third Session of COFI in 1999 and the consultation meetings began in 2000. In 2003, an Ad Hoc Expert Advisory Panel for Assessment of Proposals to CITES was established, to provide information and advice on proposals to amend CITES Appendices I and II. *See* Appendix E, Report of the Twenty-fifth Session of COFI, FAO, Rome, 24–28 February 2003.

[149] *Bonn Convention,* Art. II.

[150] A Range State means any State that exercises jurisdiction over any part of the range of the migratory species, or a State, the flag vessels of which are engaged outside national jurisdiction limits in taking that migratory species. *See* Art. I.1(h) of the Bonn Convention.

[151] *Bonn Convention*, Art. III.

[152] *Bonn Convention*, Art. IV.

[153] *Bonn Convention*, By-catch, *UNEP/CMS/Resolution 8.14*, Adopted at the Conference of the Parties at its Eighth Meeting, Nairobi, 20–25 November 2005.

Consequently, the Conference of Parties to the Bonn Convention has adopted a resolution calling on its Parties to implement the IPOA-Seabirds and the IPOA-Sharks and develop and implement national plans required by these international plans of action.[154] The resolution urges States to implement measures adopted by regional environmental and fisheries organisations to address by-catch issues.[155] Because of their similar concerns on endangered species, the Secretariats of the Bonn Convention and CITES have entered into a memorandum of understanding to ensure compatibility of measures, in respect of certain species, such as marine turtles, whale sharks, and sturgeons.[156]

The Bonn Convention has broad applicability to fisheries issues through its framework, approach, and requirements for Parties and participating States to undertake conservation measures aimed not only at migratory species but also the habitats and species upon which they may be dependent. A number of commercially fished species could potentially be suitable for listing under the Bonn Convention, although efforts to have such species listed have so far failed. The flexibility of the Bonn Convention framework has however enabled the development of and adoption of a range of issue specific memoranda of understanding and subsidiary agreements. Of the thirteen memoranda of understanding and subsidiary agreements currently in place under the Bonn Convention umbrella, six apply to marine species including marine turtles, pinnipeds, cetaceans and seabirds.[157] There is potential for commercially fished species to also be covered by such agreements.

3.3.4. *Convention on the Conservation of Biological Diversity*

Biological diversity is defined by the Convention on Biological Diversity as "the variability among living organisms from all sources including, inter alia, terrestrial, marine and other aquatic ecosystems and the ecological complexes of which they are part: this includes diversity within species, between species

---

[154] *Ibid.*, para. 2(a).

[155] *Ibid.*, paras. 2(c) and 4.

[156] Thirteenth Meeting of the Conference of the Parties, Cooperation and Synergy with the Convention on the Conservation of Migratory Species of Wild Animals, Bangkok, Conf.13.3, Thailand, 2–14 October 2004.

[157] Subsidiary agreements include the *Agreement on the Conservation of Albatrosses and Petrels (ACAP)*, Cape Town, South Africa, concluded on 1 February 2001, in force 1 February 2004, 2258 UNTS 257; *African-Eurasian Waterbird Agreement (AEWA)*, Hague, the Netherlands, concluded 16 June 1995, in force 1 November 1999, 2365 UNTS I-42632; *Agreement for the Conservation of Cetaceans of the Black Sea, Mediterranean Sea and Contiguous Atlantic Area (ACCOBAMS)*, Monaco, concluded on 24 November 1996, in force 1 June 2001, 2183 UNTS 303; *Agreement on the Conservation of Small Cetaceans of the Baltic and North Seas (ASCOBANS)*, New York, United States, concluded on 13 September 1991, in force 29 March 1994, 1772 UNTS 217; *Agreement for the Conservation of Seals in the Wadden Sea*, Bonn, Germany, concluded on 16 October 1990, in force 1 October 1991, www.cms.int.

and of ecosystems."[158] The primary objective of the CBD is to conserve biological diversity, including the sustainable use of its components and the fair and equitable sharing of benefits arising out of the utilisation of genetic resources.[159] A basic principle under the CBD is that States have sovereign rights to exploit their own resources pursuant to their own environmental policies, and the responsibility to ensure that activities within their jurisdiction or control do not cause damage to the environment of other States or of areas beyond the limits of national jurisdiction.[160]

CBD provides for the application of in-situ and ex-situ conservation measures such as the establishment of a system of protected areas, protection of ecosystems, rehabilitation and restoration of degraded ecosystems, and recovery of threatened species.[161] The Convention also provides for the introduction of appropriate procedures requiring impact assessment of projects that are likely to have significant adverse effects on biological diversity.[162] Furthermore, it outlines the obligation of Contracting Parties to provide incentives that would promote the conservation and sustainable use of components of biological diversity.[163] However, none of the conservation obligations set out in the CBD are absolute in nature.[164] The obligations are qualified by the phrase "as far as possible and as appropriate" which weaken the obligations to conserve biological diversity to a large extent.[165]

As part of achieving its objective, the Conference of Parties has agreed to set targets and goals, some of which are fisheries related. The goals set out under Decision VIII/15 of the CBD Conference of Parties include the conservation of genetic diversity of harvested species of trees, fish and wildlife and other valuable species conserved and the sustainable use and consumption of all exploited fisheries products.[166] Some fisheries concerns are more specifically covered in the decisions adopted by the CBD Conference of Parties on marine and coastal biological diversity, particularly with respect to the elimination of destructive fishing practices and use of selective fishing gear in order to avoid

---

[158] *Convention on Biodiversity*, Art. 2.

[159] *Convention on Biodiversity*, Art. 1.

[160] *Convention on Biodiversity*, Art. 3.

[161] *Convention on Biodiversity*, Arts. 8 and 9

[162] *Convention on Biodiversity*, Art. 14.

[163] *Convention on Biodiversity*, Art. 11.

[164] Rüdiger Wolfrum, 'The Protection and Management of Biological Diversity', in Fred L Morrison and Rüdiger Wolfrum (eds.), *International, Regional and National Environmental Law* (Hague: Kluwer Law International, 2000), at 359.

[165] *Ibid. See Convention on Biodiversity*, Arts. 5, 6(b), 7, 8, 9, 10, 11, and 14.

[166] Convention on Biological Diversity, *Decision VIII/15*, Framework for monitoring implementation of the achievement of the 2010 target and integration of targets into the thematic programmes of work, Adopted at the Conference of Parties at its Eighth Meeting, Curitiba, 20–31 March 2006.

or minimise by-catch.[167] Similar to other relevant international conventions, the Secretariats of the CBD and CITES have entered into a memorandum of understanding to coordinate their programme activities and investigate opportunities whereby CITES can become a partner in the implementation of appropriate provisions of the CBD.[168]

3.3.5. *Johannesburg Plan of Implementation*

At the World Summit on Sustainable Development (WSSD) in 2002, the Johanne sburg Declaration on Sustainable Development and Plan of Implementation were adopted. The Johannesburg Declaration on Sustainable Development recognised the continuous depletion of fish stocks and loss of biodiversity as part of environmental deterioration.[169] This political declaration also affirmed the commitments of States under the Rio Declaration on Environment and Development and Agenda 21.[170]

The Johannesburg Plan of Implementation, which was adopted to build on the achievements made since the United Nations Conference on Environment and Development (UNCED), provides that sustainable fisheries and ensuring the sustainable development of the oceans require effective coordination and cooperation, including at the global and regional levels, between relevant bodies, and actions at all levels.[171] To achieve sustainable fisheries, the Johannesburg Plan of Implementation encourages States to ratify or accede to and effectively implement the relevant United Nations and, where appropriate, associated regional fisheries agreements or arrangements, the FAO Code of Conduct, as well as relevant IPOAs and technical guidelines of the FAO.[172] The Johannesburg Plan of Implementation further urges States to establish effective monitoring, reporting, enforcement and control of fishing vessels to further the IPOA-IUU and eliminate subsidies that contribute to IUU fishing and overcapacity.[173] This plan of implementation also promotes the conservation and management of the oceans with due regard to other relevant international instruments such as the CBD and the Ramsar Convention.[174]

---

[167] Convention on Biological Diversity, *Decision VII/5*, Marine and coastal biological diversity, Adopted at the Seventh Meeting of the Conference of Parties, Kuala Lumpur, 9–20 February 2004.

[168] Convention on Biological Diversity, Cooperation and synergy with the Convention on Biological Diversity, *Conf. 10.4 (Rev. 14)*, Adopted at the Tenth meeting of the Conference of the Parties, Harare, Zimbabwe, 9–20 June 1997.

[169] World Summit on Sustainable Development (WSSD), *Johannesburg Declaration on Sustainable Development*, Adopted at the 17th plenary meeting on 4 September 2002, para. 13.

[170] The relevance of UNCED and Agenda 21 in combating IUU fishing is discussed in Chapter 2.

[171] WSSD, *Johannesburg Plan of Implementation*, Adopted at the 17th plenary meeting on 4 September 2002, para. 30

[172] *Johannesburg Plan of Implementation*, para. 31(b), (c) and (d).

[173] *Johannesburg Plan of Implementation*, para. 31(d) and (f).

[174] *Johannesburg Plan of Implementation*, para. 32(b) and (e).

3.4. Trade Related Agreements

The WTO has not adopted a separate agreement with specific rules on fisheries matters. Fisheries remains covered by General Agreement on Tariffs and Trade (GATT) 1994, as well as specific agreements such as the Agreements on Technical Barriers to Trade, Sanitary and Phytosanitary Measures, Pre-shipment Inspection, Rules of Origin, Import Licensing Procedures, and Subsidies and Countervailing Measures.[175] These agreements elaborate on the use of trade related measures for conservation purposes subject to certain safeguards against abuse.[176] These agreements are most relevant in addressing IUU fishing.

3.4.1. *GATT 1994*

The relevant provisions of GATT 1994 in addressing IUU fishing are Articles III, VIII, X, XI, XIII, and XX. These provisions have a direct bearing on the adoption of measures by States covering fish and fishery products that may enter into international trade, including those which may have been caught through IUU means. Article III on National Treatment on Internal Taxation and Regulation provides that there should be no differentiation between domestic and imported products.[177] Article VIII on Fees and Formalities requires minimisation of complex import and export formalities and simplification of import and export documentation requirements.[178] Article X of GATT 1994 states that no measures imposing restrictions on imports may be enforced before such measures have been officially published. Furthermore, each State is required to administer in a uniform, impartial and reasonable manner its trade laws and regulations. Article XI of GATT 1994 prohibits the application of any restriction on the importation of any product from a Contracting Party of the agreement other than duties, taxes or other charges.[179] An exception to the general prohibition on quotas is with respect to import restrictions on any agricultural or fisheries product necessary for the enforcement of governmental measures that restrict the

---

[175] *See* Ruangrai Tokrisna, 'WTO-Consistent Trade-Related Measures to Address IUU Fishing-Developing Countries Issues', in FAO, Report of and Papers Presented at the Expert Consultation on Illegal, Unreported and Unregulated Fishing, Sydney, Australia, 15–19 May 2000, *FAO Fisheries Report No. 666* (Rome: FAO, 2001), at 282–293.

[176] Linda A. Chaves, 'Illegal, Unreported and Unregulated Fishing: WTO-Consistent Trade Related Measures to Address IUU Fishing', in FAO, Report of and Papers Presented at the Expert Consultation on Illegal, Unreported and Unregulated Fishing, Sydney, Australia, 15–19 May 2000, *FAO Fisheries Report No. 666* (Rome: FAO, 2001), at 269.

[177] General Agreement on Tariffs and Trade 1994, concluded on 15 April 1994, in force 1 January 1995, 1867 UNTS 187; 33 ILM 1153 (1994), in the *Final Act Embodying the Results of the Uruguay Round of Multilateral Trade Negotiations*, Annex 1A, 15 April 1994, Art. III(4). *Hereinafter referred to as* GATT 1994.

[178] *GATT 1994*, Art. VIII(1)(c).

[179] *GATT 1994*, Art. XI.

marketing or production of like domestic products, removal of a temporary surplus of a like domestic product, or restriction of the production of any animal product that is directly dependent on the imported commodity.[180] This exception allows WTO Members to ensure that imports of agricultural and fisheries products do not interfere with domestic programmes designed to support or allocate agriculture and fisheries production.[181] Article XIII also prohibits non-discriminatory quantitative restrictions.

Lastly, Article XX of GATT 1994 provides exceptions to GATT obligations. It says that nothing in GATT 1994 prevents the adoption or enforcement of measures "relating to the conservation of exhaustible natural resources if such measures are made effective in conjunction with restrictions on domestic production or consumption."[182] However, based on the chapeau of Article XX, even if a measure falls within the exception of the article, such measures are not to be applied in a manner that would constitute a means of arbitrary or unjustifiable discrimination, or be disguised as a restriction on international trade, between countries where the same conditions prevail. These provisions strengthen the relationship between trade and conservation and management of fisheries resources.

3.4.2. *Agreement on Technical Barriers to Trade*

The Agreement on Technical Barriers to Trade recognises that the problem of non-tariff barriers to trade is mainly due to the disparities between national standards, technical regulations and procedures for approval and certification, creating the need for the development of international standards.[183] Consistent with the provisions of the GATT 1994, the Agreement on Technical Barriers to Trade requires WTO Members to provide equal treatment to national and imported products[184] and that technical regulations adopted do not create unnecessary obstacles to international trade.[185] Technical regulations cannot be made trade restrictive except when required to fulfil legitimate objectives, such as national security requirements, the prevention of deceptive practices, protection of human health or safety, animal or plant life or health, or the environment.[186] Actions to address IUU fishing may be construed as a legitimate objective that would address deceptive practices and assist in protecting the environment. However, in order to

---

[180] *GATT 1994*, Art. XI(c).

[181] Mitsuo Matsushita, Thomas J. Schoenbaum and Petros C. Mavroidis, *The World Trade Organization: Law, Practice and Policy* (Oxford: Oxford University Press, 2006), at 273.

[182] *GATT 1994*, Art. XX.

[183] M Rafiqul Islam, *International Trade Law of the WTO* (Oxford: Oxford University Press, 2006), at 155.

[184] WTO, *Agreement on Technical Barriers to Trade*, Geneva, concluded on 12 April 1979, in force 1 January 1980, 1186 UNTS 276 Art. 2.1.

[185] *Agreement on Technical Barriers to Trade*, Art. 2.2.

[186] *Agreement on Technical Barriers to Trade*, Art. 2.2.

ensure that no unnecessary trade restrictions are applied, the Agreement on Technical Barriers to Trade provides that if a WTO Member requires, another Member would need to justify a technical regulation which may have an effect on trade.[187] Technical regulations shall not be maintained if the circumstances or objectives giving rise to their adoption no longer exist or if the changed circumstances or objectives can be addressed in a less trade restrictive manner.[188]

3.4.3. *Agreement on Pre-shipment Inspection*

The Agreement on Pre-shipment Inspection allows WTO Members to conduct pre-shipment inspections across all exporters affected by such programmes so long as they are carried out in a non-discriminatory manner.[189] Pre-shipment inspection activities include all activities relating to the verification of the quality, quantity, price, including currency exchange rate and financial terms, and/or the customs classification of goods to be exported to the territory of the user Member.[190] The Agreement requires pre-shipment inspections to be conducted in a transparent manner,[191] ensure confidentiality of business information,[192] and not cause unnecessary delays.[193] Relevant authorities can either issue a Clean Report of Findings or provide a detailed written explanation specifying the reasons for non-issuance following a pre-shipment inspection.[194] Such pre-shipment inspection may be used to certify that an export fish or fish product has not been caught through IUU fishing.

3.4.4. *Agreement on Rules of Origin*

Rules of origin are defined under the WTO Agreement on Rules of Origin as "those laws, regulations and administrative determinations of general application applied by any Member to determine the country of origin of goods provided such rules of origin are not related to contractual or autonomous trade regimes leading to the granting of tariff preferences going beyond the application of paragraph 1 of Article I of GATT 1994."[195] Rules of origin also apply within the

---

187 *Agreement on Technical Barriers to Trade*, Art. 2.5.

188 *Agreement on Technical Barriers to Trade*, Art. 2.3.

189 WTO, *Agreement on Pre-shipment Inspection*, http://www.wto.org/english/docs_e/legal_e/21-psi_e.htm, Art. 2.1.

190 *Agreement on Pre-shipment Inspection*, Art. 1.3.

191 *Agreement on Pre-shipment Inspection*, Art. 2.5.

192 *Agreement on Pre-shipment Inspection*, Art. 2.9–2.13.

193 *Agreement on Pre-shipment Inspection*, Art. 2.15.

194 *Agreement on Pre-shipment Inspection*, Art. 2.16.

195 WTO, *Agreement on Rules of Origin*, http://www.wto.org/english/docs_e/legal_e/22-roo.pdf, Art. 1.1. Paragraph 1 of Article I of GATT 1994 provides that "(W)ith respect to customs duties and charges of any kind imposed on or in connection with importation or exportation or imposed on the international transfer of payments for imports or exports, and with respect to the method of levying such duties and charges, and with respect to all rules and formalities in connection with importation

context of the most favoured nation treatment,[196] anti-dumping and countervailing duties,[197] safeguard measures,[198] origin marking requirements,[199] any discriminatory quantitative restrictions or tariff quotas, and rules of origin used for government procurement and trade statistics.[200] Similar to Clean Report of Findings issued under the Agreement on Pre-shipment Inspections, certificates of origin may also be used for preventing trade in IUU-caught fish.

Article 2 of the Agreement on Rules of Origin provides that rules of origin shall not themselves create restrictive, distorting or disruptive effects on international trade[201] and must be administered in a consistent, uniform, impartial and reasonable manner.[202] For the purpose of establishing harmonised rules of origin, the country determined to be the origin of a particular good is either the country where the good has been wholly obtained or, when more than one country is concerned in the production of the good, the country where the last substantial transformation has been carried out.[203] A Committee on Rules of Origin and a Technical Committee on Rules of Origin under the Customs Cooperation Council have been established to consult on matters relating to the harmonisation and application of rules of origin.[204]

3.4.5. *Agreement on Import Licensing Procedures*

Import licensing under the Agreement on Import Licensing Procedures involves the administrative procedures requiring the submission of an application or other documentation, other than that required for customs purposes, to the relevant administrative body as a prior condition for importation into the customs territory of the importing Member.[205] The Agreement requires Members to ensure that the administrative procedures used to implement import licensing rules are in conformity with the relevant provisions of GATT 1994, including its annexes and protocols. The objective is to prevent trade distortions that may arise from an inappropriate operation of those procedures, taking into account the economic development purposes and financial and trade needs of developing country Members.[206]

---

and exportation, and with respect to all matters referred to in paragraphs 2 and 4 of Article III, any advantage, favour, privilege or immunity granted by any contracting party to any product originating in or destined for any other country shall be accorded immediately and unconditionally to the like product originating in or destined for the territories of all other contracting parties."

[196] *GATT 1994*, Arts. I, II, III, XI and XIII.

[197] *GATT 1994*, Art. VI.

[198] *GATT 1994*, Art. XIX.

[199] *GATT 1994*, Art. IX.

[200] *Agreement on Rules of Origin,* Art. 1.2.

[201] *Agreement on Rules of Origin,* Art. 2(c).

[202] *Agreement on Rules of Origin,* Art. 2(d) and (e).

[203] *Agreement on Rules of Origin,* Art. 3(b).

[204] *Agreement on Rules of Origin,* Art. 4 and Annex I.

[205] WTO, *Agreement on Import Licensing Procedures*, Geneva, concluded on 12 April 1979, in force 1 January 1980, 1186 UNTS 372, Art. 1.1.

[206] *Agreement on Import Licensing Procedures*, Art. 1.2

The Agreement on Import Licensing Procedures requires WTO Members to publish rules and all information concerning procedures for the submission of applications, including the eligibility of persons, firms and institutions to make such applications, the administrative body or bodies to be approached, and the lists of products subject to the licensing requirement, at least 21 days prior to their effective date, in order to enable governments to be familiar with them.[207] Any exception, derogations or changes in or from the rules concerning licensing procedures or the list of products subject to import licensing are also required to be published in the same manner and within the same time period.[208]

The Agreement on Import Licensing Procedures contains provisions on automatic import licensing where approval of the application is granted in all cases.[209] It also provides for non-automatic import licensing. A WTO Member who has an interest in the trade of the product included in the non-automatic import licensing may require the information. Such information may include the administration of the restrictions, the import licences granted over a recent period, the distribution of such licences among supplying countries, and where practicable, import statistics with respect to the products subject to import licensing.[210] Import licensing procedures may include requirements to ascertain whether a fishery product being imported has not been obtained from IUU-related activities.

3.4.6. *Agreement on the Application of Sanitary and Phytosanitary Measures*

The Agreement on the Application of Sanitary and Phytosanitary Measures (SPS Agreement) aims to establish a multilateral framework for the development, adoption and enforcement of sanitary and phytosanitary measures in order to minimise their negative effects on trade and ensure that such measures are not applied in a manner which constitutes a disguised restriction on international trade. The SPS Agreement requires Members to ensure that any sanitary or phytosanitary measure is applied only to the extent necessary to protect human, animal or plant life or health, and should be based on scientific principles and is not maintained without sufficient scientific evidence.[211] In cases where relevant scientific evidence is insufficient, a WTO Member may provisionally adopt sanitary or phytosanitary measures on the basis of available pertinent information, including that from the relevant international organisation as well as from measures applied by other Members.[212] The application of such measures should take

---

[207] *Agreement on Import Licensing Procedures*, Art. 1.4(a).

[208] *Agreement on Import Licensing Procedures*, Art. 1.4(a).

[209] *Agreement on Import Licensing Procedures*, Art. 2.1.

[210] *Agreement on Import Licensing Procedures*, Art. 3.5.

[211] WTO, *Agreement on the Application of Sanitary and Phytosanitary Measures*, concluded on 15 December 1993, in force 01 January 1995, GATT Doc. MTN/FA Ii-A1A-4, 33 ILM XXX, Art. 2(2). *Hereinafter referred to as SPS Agreement.*

[212] *SPS Agreement*, Art. 7(5).

into account the special needs of developing WTO Members, particularly least developed Members.[213] As will be discussed in Chapter 8, although the SPS Agreement is currently not directly relevant to addressing IUU fishing, it is a critical component of the general framework for international trade in fish and fish products.

3.4.7. *Agreement on Subsidies and Countervailing Measures*

The WTO Agreement on Subsidies and Countervailing Measures (SCM Agreement) provides a framework for addressing subsidies, although it is not specific to fisheries. Part I of the SCM Agreement applies to subsidies that are specifically provided to an enterprise or industry group while Parts II and III contain rules and procedures on prohibited subsidies and actionable subsidies. Part V establishes the substantive and procedural requirements that must be fulfilled before a Member may apply a countervailing measure against subsidised imports. Parts VI and VII establish the institutional structure and notification/surveillance modalities for the implementation of the SCM Agreement. Part VIII contains special and differential treatment rules for various categories of developing country Members. Part IX contains transition rules for developed country and former centrally planned economy Members. Parts X and XI contain dispute settlement and final provisions.

Article 1 of the SCM Agreement states that a subsidy is deemed to exist if a financial contribution or income or price support is provided by a government, and a benefit is conferred, and that such subsidy is specific. Article 2 defines the concept of specificity, which is deemed to exist when accesses to the subsidy is limited, explicitly or in fact, to certain enterprises. The SCM Agreement classifies specific subsidies into three: prohibited, actionable (*i.e.,* permitted, but potentially subject to action), and non-actionable (*i.e.* permitted, and shielded from action). Prohibited subsidies are irrefutably presumed to distort trade. Certain kinds of subsidies within the actionable category were deemed, via rebuttable presumption, to cause serious prejudice to other members of the WTO.

The focus of the SCM Agreement is to ensure that the subsidies applied by States do not cause an injury to the domestic industry of another WTO Member and that such subsidy does not nullify nor impair any benefit accruing directly or indirectly to other WTO Members.[214] The SCM Agreement further prohibits subsidies which create serious prejudice to the interests of other Members.[215] Serious prejudice in this context includes effects of subsidies such as the displacement or

[213] *SPS Agreement,* Art. 10(1).

[214] WTO Agreement on Subsidies and Countervailing Measures, concluded on 15 April 1994, 1867 UNTS 14, GATT Doc. MTN/FA Ii-A1A-4, 33 ILM XXX. *Hereinafter referred to as* SCM Agreement, Art. 5.

[215] *SCM Agreement,* Art. 5.

impediment of imports or exports of a like product between WTO Members and a third country, and a significant price undercutting by the subsidised product in the same market.[216]

The WTO has acknowledged that fisheries subsidies are trade distorting and undermine the sustainable use of fisheries resources.[217] Within the WTO, there have been arguments that positive subsidies or subsidies that reduce excessive fishing capacity should be allowed.[218] However, there is still debate on how certain fisheries subsidies contribute to overcapacity in the fishing fleet and affect the sustainability of fisheries resources.[219] It is also difficult to find a WTO Member that could not be perceived as applying one or more types of fisheries subsidies that may fall within the ambit of the SCM Agreement.[220] While the SCM Agreement provides a framework to prevent industries from gaining unfair advantage over others through government subsidies, it is not yet clear how such disciplines on subsidies may be applied consistently to fisheries matters across WTO Members.

3.4.8. *Draft WTO Rules on Fisheries Subsidies*

The WTO conducts continuous discussions on subsidy reduction and elimination to identify measures that would achieve benefits for both trade and the environment.[221] Discussions on WTO disciplines on fisheries subsidies started in 2001 when WTO members agreed "…to clarify and improve WTO disciplines on fisheries subsidies, taking into account the importance of this sector to developing countries."[222] However, it was only in November 2007 that a draft text was circulated by the Chair of the Doha Round negotiations group on rules on fisheries subsidies. The draft text is proposed to be inserted as Annex VIII to the SCM Agreement. The proposed Annex VIII contains provisions on prohibited subsidies[223] and non-prohibited[224] subsidies. Examples of prohibited subsidies are those

---

[216] *SCM Agreement*, Art. 6.3.

[217] Beatrice Chaytor, 'An Examination of Trade and Environmental Policy Responses to the Challenge of Fisheries Subsidies', 8 *Review of European Community & International Environmental Law* (1999), at 280.

[218] Marc Benitah, 'Ongoing WTO Negotiations on Fisheries Subsidies,' *ASIL Insights*, The American Society of International Law Website, June 2004, www.asil.org. Accessed on 24 April 2008.

[219] *Ibid.*

[220] Marc Benitah, *The Law of Subsidies under the GATT/WTO System* (Hague: Kluwer Law International, 2001), at 84.

[221] Chaytor, *supra* note 217, at 280.

[222] WTO, Ministerial Conference, Fourth Session, Doha, 9–14 December 2001, Ministerial Declaration, adopted on 14 November 2001, *WT/MIN(01)/DEC/1*, 20 November 2001, para. 28.

[223] WTO, Negotiating Group on Rules, *Draft Consolidated Texts of the AD and SCM Agreements*, TN/RL/W/213, 30 November 2007, Annex VIII, Fisheries Subsidies. *Hereinafter referred to as* Draft Annex VIII to the SCM Agreement, Art. I.

[224] *Draft Annex VIII to the SCM Agreement*, Art. II(b), (c), and (d).

the benefits of which are conferred on any vessel engaged in IUU fishing.[225] It also provides certain exemptions from prohibited subsidies, such as least developed country members.[226] The proposed amendment to the SCM Agreement made reference to the FAO Code of Conduct, FAO Compliance Agreement, and UN Fish Stocks Agreement as basis for the provisions of Annex VIII which are aimed at preventing overfishing.[227] Furthermore, Article IV of the draft Annex VIII provides that no WTO Member shall cause, through the use of any subsidy, depletion of or harm to, or creation of overcapacity in respect of straddling or highly migratory fish stocks whose range extends into the EEZ of another member, or stocks in which another member has identifiable fishing interests such as quota allocations.

The draft Annex VIII on fisheries subsidies may provide a sound basis for the facilitation of discussions on the matter in WTO; however, the application of fisheries specific instruments and international regulations on subsidies to address fisheries concerns, including IUU fishing, is a complex issue that would require intensive negotiation by WTO members. There have been different positions on how to improve the application of the WTO discipline on fisheries subsidies.[228] These varying positions of States, together with different national fisheries management frameworks in place, are major impediments to concluding a trade related agreement on fisheries subsidies.

3.5. Maritime Safety and Labour Related Agreements

Fishing vessel safety has been a matter of concern for the International Maritime Organization (IMO). However, the differences in design and operation of fishing vessels prevent them from being regulated by the International Convention for the Safety of Life at Sea 1974, as amended (SOLAS Convention) and the International Convention on Load Lines 1966, as amended. The IMO and FAO are currently promoting the adoption of an international mandatory safety regime for fishing vessels, in the form of the 1993 Torremolinos Protocol for the Safety of Fishing Vessels,[229] the 1995 International Convention on Training, Certification

---

225 *Draft Annex VIII to the SCM Agreement*, Art. I.1(h).

226 *Draft Annex VIII to the SCM Agreement*, Art. III.1.

227 *See* Draft Annex VIII to the SCM Agreement, Art. III(3), IV, and V.

228 *See* for example, WTO, Negotiating Group on Rules, Submission of the European Communities to the Negotiating Group on Rules-Fisheries Subsidies, *TN/RL/W/82*, 23 April 2003; WTO, Negotiating Group on Rules, Communication from the Delegations of Antigua and Barbuda, Belize, Fiji Islands, Guyana, the Maldives, Papua New Guinea, Solomon Islands, St Kitts and Nevis, *TN/RL/W/136*, 14 July 2003; WTO, Negotiating Group on Rules, Fisheries Subsidies: Overcapacity and Overexploitation, Communication from New Zealand, *TN/RL/W/154*, 26 April 2004.

229 International Maritime Organization (IMO), *Torremolinos Protocol of 1993 relating to the Torremolinos International Convention for the Safety of Fishing Vessels 1977,* Torremolinos, Spain, concluded on 2 April 1993, www.imo.org. *Hereinafter referred to as* 1993 Torremolinos Protocol.

and Watchkeeping for Fishing Vessels,[230] as well as a number of fishing vessel safety codes and voluntary guidelines. Both of the international conventions and protocol are yet to enter into force.

In the case of protecting and promoting the rights of fishers onboard fishing vessels, the International Labour Organization (ILO), the global body responsible for formulating and overseeing international labour standards, has adopted The Work in Fishing Convention in 2007 (C188),[231] which is supplemented by the Work in Fishing Recommendation 2007 (R199).[232] The new standards revised existing ILO instruments addressing conditions of workers in the fishing sector.[233] Similar to the IMO Conventions directly affecting fishing vessels, the Work in Fishing Convention is yet to enter into force. It will enter into force after ratification by ten of ILO's 180 Member States, which should include eight coastal States.[234]

Other ILO Conventions that have provisions which apply to commercial maritime fishing include the Shipowners' Liability (Sick and Injured Seamen) Convention; 1936 (C55), Seafarers' Welfare Convention, 1987 (C163); Health Protection and Medical Care (Seafarers) Convention, 1987 (C164); Social Security (Seafarers) Convention (Revised), 1987 (C165); Repatriation of Seafarers Convention, (Revised), 1987 (No. 166); Labour Inspection (Seafarers) Convention, 1996 (No. 178); Recruitment and Placement of Seafarers Convention, 1996 (C179); and the Seafarers' Hours of Work and the Manning of Ships Convention, 1996 (No. 180).

3.5.1. *Torremolinos Convention and its Protocol*

The 1977 Torremolinos Convention for the Safety of Fishing Vessels was the first international convention on the safety of fishing vessels. The 1977 Torremolinos Convention applied to fishing vessels of 24 metres in length and over, including those vessels processing their catch.[235] It contained safety and stability

[230] IMO, *International Convention on Standards of Training, Certification and Watchkeeping for Fishing Vessel Personnel,* adopted on 7 July 1995, in force 1 February 1997, Select Documents on International Affairs No. 43 (1995) 12, p. 148. www.austlii.org. *Hereinafter referred to as* STCW-F.

[231] International Labour Organization (ILO), Convention Concerning Work in the Fishing Sector (C188), Geneva, Switzerland, adopted on 14 June 2007, www.ilo.org. *Hereinafter referred to as* Work in Fishing Convention.

[232] ILO, Recommendation Concerning the Work in the Fishing Sector (R199), Geneva, Switzerland, adopted on 14 June 2007, www.ilo.org. *Hereinafter referred to as* Work in Fishing Recommendation.

[233] The major ILO Conventions revised by the Work in Fishing Convention 2007 are the Minimum Age (Fishermen) Convention, 1959 (C112), the Medical Examination (Fishermen) Convention, 1959 (C113), Fishermen's Articles of Agreement Convention, 1959 (C114), Accommodation of Crews (Fishermen) Convention, 1966 (C126). The Work in Fishing Convention 2007 did not revise the Fishermen's Competence Certificates Convention, 1966 (C125).

[234] *The Work in Fishing Convention*, Art. 48.

[235] IMO, *Torremolinos International Convention for the Safety of Fishing Vessels,* Torremolinos, Spain, 2 April 1977. *Hereinafter referred to as* Torremolinos Convention.

requirements for the construction and equipment of new, decked, and seagoing fishing vessels.

The 1977 Torremolinos Convention was superseded by the 1993 Torremolinos Protocol which took into account the technological developments in the intervening years, particularly the large scale distant and deep water fishing operations. The 1993 Torremolinos Protocol applies to the same types of fishing vessels and contains provisions on machinery installations and spaces, lifesaving appliances and arrangements, vessel requirements, protection of the crew, and shipborne navigational equipment and arrangements.

3.5.2. *STCW-F*

The International Convention on Standards of Training, Certification and Watchkeeping for Fishing Vessel Personnel or STCW-F was adopted in 1995 as the first attempt to make standards of safety for crews of fishing vessels mandatory. Similar to the 1993 Torremolinos Protocol, the STCW-F is short and contains technical guidance on the certification of skippers, officers, engineer officers, and radio operators.

3.5.3. *The 2007 Work in Fishing Convention*

The 2007 Work in Fishing Convention applies to all fishers and all fishing vessels engaged in commercial fishing operations.[236] This Convention, together with the Work in Fishing Recommendation 2007 provide the minimum requirements for work onboard fishing vessels and conditions of service, particularly in respect of minimum age for work, requirement for medical examination, manning and hours of rest, maintenance of a crew list, repatriation of fishers, recruitment and placement of fishers, payment, and accommodation and food. Provisions on accommodation are adopted to ensure that fishing vessels are constructed in a manner that would ensure suitable living conditions for workers on board vessels. New standards have also been adopted under the Work in Fishing Convention on improved occupational health and safety, medical care at sea and ashore, protection of a work agreement, and social security protection. In addition to technical standards, the Work in Fishing Convention provides for the right of ILO

[236] *The Work in Fishing Convention*, Art. 2(1). The Convention defines commercial fishing as all fishing operations, including fishing operations on rivers, lakes or canals, with the exception of subsistence fishing and recreational fishing. Article 3 of the Convention, however, also provides that if there are special problems of a substantial nature in the light of particular conditions of fisheries or of fishing vessel operations, an ILO Member may exclude (a) fishing vessels engaged in fishing operations in rivers, lakes or canals and (b) limited categories of fishers and fishing vessels from the requirements of the Convention. Article 4 also provides for a progressive implementation of all or some of the provisions which affect some types of fishing vessels or fishing activities.

Members to establish a system of inspections, monitoring, complaint procedure, penalty and corrective measures to ensure compliance with the provisions of the Convention.[237]

3.5.4. *Voluntary Codes and Guidelines*

Apart from these international agreements, the IMO has also developed a number of non-mandatory instruments, in collaboration with FAO and the ILO. These include the FAO/ILO/IMO Document for Guidance on Fishermen's Training and Certification,[238] revised Code of Safety for Fishermen and Fishing Vessels,[239] and revised FAO/ILO/IMCO Voluntary Guidelines for the Design, Construction and Equipment of Small Fishing Vessels.[240]

The Document for Guidance on Fishermen's Training and Certification is a response to the call by the IMO Maritime Safety Committee to prepare guidelines and recommendation for the training and certification of personnel on board fishing vessels of 12 metres in length and over and for the prevention of fatigue of fishing vessel personnel.[241] This Guidance also respects the FAO Code of Conduct, particularly Article 8 on Fishing Operations which provides for the need for States to enhance the education, skills, and professional qualification of fishers through education and training.[242] The Code of Safety was formulated with the desire to coordinate facilities provided by various fishing nations to improve safety and health of fishermen and fishing vessels at sea.[243] Lastly, the Voluntary Guidelines for Small Fishing Vessels was developed to provide a generally applicable code of safe practice for the design, construction, and equipment of new decked fishing vessels of 12 metres in length and over but less than 24 metres in length.[244]

---

237 *The Work in Fishing Convention*, Part VII.

238 *See* IMO, *FAO/ILO/IMO Document for Guidance on Training and Certification of Fishing Vessel Personnel* (London: IMO, Second Edition, 2001).

239 *See* IMO, *Code of Safety for Fishermen and Fishing Vessels, Part A Safety and Health Practice for Skippers and Crews* (London: IMO, 2005); IMO, *Code of Safety for Fishermen and Fishing Vessels, Part B Safety and Health Requirements for the Construction and Equipment of Fishing Vessels* (London: IMO, 2005).

240 IMO, *FAO/ILO/IMCO Voluntary Guidelines for the Design, Construction and Equipment of Small Fishing Vessels* (London: IMO, 2005).

241 Reference was made to Resolution 3 of the 1995 STCW-F Convention in *FAO/ILO/IMO Document for Guidance on Training and Certification of Fishing Vessel Personnel*, *supra* note 238, at 1.

242 *FAO Code of Conduct*, Art. 8.1.7.

243 IMO, *Code of Safety for Fishermen and Fishing Vessels,* at 3.

244 *FAO/ILO/IMCO Voluntary Guidelines for the Design, Construction and Equipment of Small Fishing Vessels*, para. 1.1.

Chapter Four

All State Measures

4.1 General Framework for All State Responsibilities

IUU fishing is a global problem and therefore requires action by the international community, including coastal, port, flag and market States. In recognition of the global character and impact of IUU fishing, the IPOA-IUU identifies a range of measures that need to be implemented by "All States" to combat IUU fishing. These measures include the implementation of international instruments; enactment of national legislation; development of national plan of action; control of nationals; jurisdiction over vessels without nationality; imposition of sanctions of sufficient severity; actions against non cooperating States, avoidance of economic incentives to IUU fishing activities; monitoring, control and surveillance (MCS); cooperation between States; publicity; and provision of technical capacity and resources. In the implementation of the IPOA-IUU, it is important to bear in mind that all the actions listed under "All State Responsibilities" cut across and overlap with coastal State, port State, flag State, market State, and RFMO measures to combat IUU fishing. This Chapter focuses on aspects of "All State Measures" that are not discussed in other Chapters which are critical to combating IUU fishing. The measures analysed by way of examples include the ratification and implementation of international instruments, adoption of national plans of actions and legislation to combat IUU fishing, control of nationals, and elimination of economic incentives.

4.2 Ratification and Implementation of International Instruments

The IPOA-IUU provides for the responsibility of States to give effect to relevant norms of international law, as reflected in the LOSC, to combat IUU fishing.[1] It also encourages all States to ratify, accede to, accept, and implement international fisheries instruments such as the FAO Compliance Agreement, UN Fish Stocks Agreement, and the FAO Code of Conduct and its associated International Plans

---

[1] FAO, *International Plan of Action to Prevent, Deter, and Eliminate Illegal, Unreported, and Unregulated Fishing*, adopted on 23 June 2001 at the 120th Session of the FAO Council. *Hereinafter referred to as* IPOA-IUU.

of Action.[2] However, as seen from the discussion in Chapter 3, the international instruments relevant to combating IUU fishing are wide ranging and include fisheries specific instruments and non-fisheries specific instruments. Therefore in order to adequately address IUU fishing, States would need to take into account all of the fisheries, trade environment, and maritime safety related instruments. The important role that international instruments play in the global fight against IUU fishing has been emphasised by the FAO by stating that "IUU fishing contravenes existing international standards for fishery conservation and management. In other words, if all States were fully implementing those international standards, there would be much less IUU fishing."[3]

Some of the relevant international fisheries instruments fall squarely within the domain and competence of fisheries agencies, while others fall outside the area of responsibility of national fisheries authorities, thus requiring a wider inter-agency cooperation at the domestic level. At the end of the day, the effectiveness of international instruments to combat IUU fishing depends largely on the extent of their implementation at the national level. This requires the development of appropriate policies and national legislation to give effect to obligations under relevant international instruments.

Although many States are quick to ratify international instruments, ratification is only the start of addressing IUU fishing. The real challenge facing many States is not so much the ratification of relevant instruments, but the domestic implementation of the provisions of these instruments through legislation and management measures. Many States, particularly developing States, are handicapped in terms of capacity to translate international obligations into national implementation measures such as legislation. This challenge has been recognised by many of the international instruments which have made provisions for technical and financial assistance to developing States in implementing their obligations. A number of international institutions and developed States have programmes to provide such assistance to developing States to implement sustainable fisheries management practices.[4]

---

[2] *IPOA-IUU*, paras. 11–14.

[3] Food and Agriculture Organization (FAO), Fisheries Department, Implementation of the International Plan of Action to Prevent, Deter and Eliminate Illegal, Unreported and Unregulated Fishing, *FAO Technical Guidelines for Responsible Fisheries No. 9* (Rome: FAO, 2002), at 8.

[4] See for example, United Nations General Assembly (UNGA), Sixty-third session, Item 73(a) of the provisional agenda, Oceans and the Law of the Sea, Available assistance to and measures that may be taken by developing States, in particular the least developed States and Small Island Developing States (SIDS), as well as coastal African States, to realise the benefits of sustainable and effective development of marine resources and uses of the oceans within the limits of national jurisdiction, Study Prepared by the Secretariat, *A/63/342*, 3 September 2008.

4.3 Adoption of National Legislation and National Plan of Action

A fundamental obligation of States is to bring domestic law into conformity with obligations under international law.[5] This is reiterated in the IPOA-IUU which provides the need for States to give full effect to relevant norms of international law in order to prevent, deter, and eliminate IUU fishing.[6] There are two principal ways by which States can implement the IPOA-IUU: first is the adoption of measures in national legislation;[7] and second is the development and implementation of national plans of action (NPOA-IUU).[8] The IPOA-IUU provides for the development of NPOA-IUU by 2004, or three years after the adoption of the IPOA-IUU.[9] The IPOA-IUU also provides for the review of the implementation of the NPOA-IUU at least every four years after its adoption.[10]

In a review conducted in 2003 on the progress of implementing the FAO Code of Conduct for Responsible Fisheries and the international plans of action, 70 FAO Members indicated that they have taken steps in implementing the IPOA-IUU.[11] However, to date only a small number of States have finalised their national plans of action.[12] Some States are still in the process of drafting and finalising their NPOA-IUU.[13] While the limited number of NPOA-IUU may suggest a lack of compliance or inadequate support from States to implement the IPOA-IUU, the current trend in domestic legislation indicates the adoption of measures which are more stringent than those provided under the IPOA-IUU.

---

[5] Ian Brownlie, *Principles of Public International Law* (Oxford: Clarendon Press, 1998), at 35; *See also* William Edeson, 'Towards Long-term Sustainable Use: Some Recent Developments in the Legal Regime of Fisheries', in Alan Boyle and David Freestone, (eds.), *International Law and Sustainable Development: Past Achievements and Future Challenges* (Oxford: Oxford University Press, 1999), at 165.

[6] *IPOA-IUU*, para. 3.

[7] *IPOA-IUU*, para. 16.

[8] *IPOA-IUU*, para. 25.

[9] *IPOA-IUU*, para. 25.

[10] *IPOA-IUU*, para. 26.

[11] FAO, Committee on Fisheries, Twenty-fifth Session, Progress in the Implementation of the Code of Conduct for Responsible Fisheries and Related International Plans of Action, Rome, Italy, 24–28 February 2003, *COFI/2003/Rev.1* (Rome: FAO, 2003), at 9.

[12] Examples of States that have developed their NPOA-IUU published in the FAO website are Argentina, Australia, Canada, Chile, Japan, New Zealand, U.S., and South Korea. *See* FAO Website, *IPOA-IUU National Plans of Action*. www.fao.org. Accessed on 04 June 2009.

[13] Examples of States which reported to have draft NPOA-IUU are Cook Islands, Federated States of Micronesia, Fiji, Marshall Islands, Nauru, Niue, Palau, Samoa, Tonga, Tuvalu, the Philippines, Malaysia, Angola, Mozambique, and Tanzania. *See* David Lymer, Simon Funge-Smith, Jesper Clausen and Weimin Miao, Status and Potential of Fisheries and Aquaculture in Asia and the Pacific 2008, *RAP Publication 2008/15* (Bangkok: FAO-RAP, 2008), at 20; *See also* Stop Illegal Fishing Programme, IPOA-IUU Implementation Status: Report for the SADC Region, Members and RFBs, Volume 3, Annexes NPOA-IUU, December 2007, www.illegal-fishing.info, Accessed on 10 February 2009.

The implementation of the IPOA-IUU through NPOA-IUU and legislative measures entails a better understanding of the different forms and impacts of IUU fishing, a thorough review of relevant national, bilateral, and regional regulations and practices, and formulation of applicable flag, coastal, port, market, and all State measures. Such understanding will help strengthen the legal and policy framework to effectively combat IUU fishing. Ideally, NPOAs would need to be supported by national policy and adequate legislation. This will ensure that such measures are translated into enforceable measures for relevant government agencies, non-government entities, and most importantly the fishing sector.

4.3.1 *Legislative Measures*

A number of legislative measures have been adopted by States to address IUU fishing. In addition to flag, coastal, port, and market measures, some States have incorporated provisions in national legislation to strengthen measures related to addressing IUU fishing, including effective control over nationals. Some States have made it a violation in their laws for their nationals to engage in fishing activities that violate fisheries conservation and management laws of any other State or that undermine the effectiveness of conservation and management measures adopted by a relevant RFMO.[14] More recently, extensive legislative measures directly addressing IUU fishing have been adopted by the United States (U.S.) and the European Union (EU), which are discussed in Chapter 10.

4.3.2 *National Plans of Action*

Most NPOA-IUU submitted to FAO by States follow the structure of the IPOA-IUU and contain a number of components, namely the importance of fisheries to the economy of the State, scope and nature of IUU fishing, applicable national legislation and relevant institutions, ratification and adoption of relevant international instruments, specific measures being undertaken as a flag, coastal, port, and market State in combating IUU fishing, and participation in regional organisations.[15] The two areas of significant importance to States in the adoption of

---

[14] *FAO Technical Guidelines for Responsible Fisheries No. 9*, *supra* note 3, at 12–14. See for example, New Zealand, *Fisheries Act 1996 Amendment Act (No. 2) 1999*, Art. 113A; Australia, *Fisheries Management Act 1991* (Cth), Part 6, Division 5A.

[15] See for example, Republic of Chile, Ministry of Economy and Energy, *National Plan of Action to Prevent, Deter, and Eliminate Illegal, Unreported and Unregulated Fishing*, 2004; Ghana, *National Plan of Action to Prevent, Deter, and Eliminate Illegal, Unreported and Unregulated Fishing*, June 2004; Canada, *National Plan of Action to Prevent, Deter, and Eliminate Illegal, Unreported and Unregulated Fishing*, March 2005; Republic of Seychelles, *Fisheries Profile and National Plan of Action to Prevent, Deter, and Eliminate Illegal, Unreported and Unregulated Fishing*, 2004; Spain, *National Plan of Action to Prevent, Deter, and Eliminate Unreported and Unregulated Fishing*, November 2002; New Zealand, Ministry of Fisheries, *Plan of Action to Prevent, Deter, and Eliminate Illegal, Unreported and Unregulated Fishing*, May 2004; Republic of

national plans to combat IUU fishing are the scope and nature of IUU fishing and measures undertaken to address the problem.

Two trends can be observed in existing NPOA-IUU with respect to the use of the term IUU fishing. The first trend is the general reference to the scope and nature of IUU fishing provided under paragraph 3 of the IPOA-IUU.[16] The second trend is the adoption of a much wider definition of each of the components of the terminology based on the provisions of the IPOA-IUU. As an example, Chile has developed a definition of IUU fishing in its NPOA-IUU which is much broader than what is provided for under the IPOA-IUU. According to paragraph 3 of the Chilean NPOA-IUU:

> **Illegal fishing** means extractive fishing activities carried out by industrial or artisanal ship owners with or without a vessel or by domestic or foreign vessels, in waters under national jurisdiction or on the high seas, in violation of standing domestic regulations or those established by Regional or International Agencies, regardless of Chile's condition of party or non-party thereof.
>
> Also considered **illegal fishing** is the use of resources in violation of standing international and national regulations, in activities such as processing, trade or transportation of such resources or by-products.
>
> **Unregulated fishing** activities are those carried out over aquatic animals, in respect of which no management measures are applied and that are not in accordance to the responsibilities of the State in terms of conservation of marine resources in light of international law; and those activities carried out by vessels without nationality or those operating in an area subject to the application of management measures by a Regional Organisation, which flag a flag of a State that is not part of such organisation.
>
> **Unreported fishing** means such fishing activities that have not been reported on a timely basis or that have been reported in an imprecise manner to the competent authority, in violation of domestic and international regulations.[17]

In terms of measures adopted to address IUU fishing, most NPOA-IUU only either list or describe measures available in existing fisheries regulations. The NPOA-IUU of Ghana, Seychelles, Tonga, Spain, New Zealand, Namibia, and Zambia also focus on recommended actions that would need to be taken in order

---

Korea, Ministry of Maritime Affairs and Fisheries, *National Plan of Action to Prevent, Deter, and Eliminate Illegal, Unreported and Unregulated Fishing*, no date; Australia, Department of Agriculture, Fisheries and Forestry, *National Plan of Action to Prevent, Deter, and Eliminate Illegal, Unreported and Unregulated Fishing*, July 2005; Republic of Namibia, *National Plan of Action to Prevent, Deter, and Eliminate Illegal, Unreported and Unregulated Fishing*, January 2007; The Republic of Zambia, Ministry of Agriculture and Cooperatives, Fisheries Department, *National Plan of Action to Prevent, Deter, and Eliminate Illegal, Unreported and Unregulated Fishing*, Lusaka, Zambia, 09 October 2007.

[16] *New Zealand NPOA-IUU*, sec. 1.3; *Republic of Korea NPOA-IUU*, sec. 3; *Ghana NPOA-IUU*, at 7.

[17] *Chile NPOA-IUU*, para. 3.

to address gaps in the existing policy and legal framework on fisheries. These recommendations largely conform to the provisions of the IPOA-IUU.

Other States take a broader strategic approach in addressing IUU fishing. Canada, for example, in its NPOA-IUU provides for the review and strengthening of its fisheries policies through efforts to streamline fisheries programmes, modernising the governance of fisheries, amendment of its *Fisheries Act,* and implementation of the Atlantic Fisheries Policy Review.[18] Canada's NPOA-IUU also includes measures to improve MCS operations and dockside monitoring programme, and implementation of market related measures, as well as proposals to strengthen RFMO mechanisms to address IUU fishing. Similarly, Australia has highlighted specific measures and actions taken by its states and territory governments to ensure responsible fisheries management, particularly with respect to addressing illegal and unreported fishing for particular species such as abalone, rock lobster and finfish.[19] Furthermore, Australia has included in its NPOA-IUU the cooperative efforts it is undertaking with Papua New Guinea in the Torres Strait region and the support it provides in the Pacific to address IUU fishing.[20]

4.3.3 *Regional Plans of Action*

The IPOA-IUU does not specifically call for the development of regional plans of action to combat IUU fishing; however this is implied from several references to cooperation between and among States and the role of RFMOs to combat IUU fishing.[21] In regional areas outside the framework of RFMOs, some States have entered into cooperative arrangements to address common problems of IUU fishing. The EU is one of the first regional organisations to adopt a Community action plan in 2002 for the eradication of IUU fishing.[22] The Community Action Plan contained 15 actions related to the control of nationals, identification and monitoring of IUU vessels, identifying and quantifying illegal catches, certification and documentation of catch, improvement of information on fishing vessels, international cooperation, and provision of assistance for developing States to control illegal fishing. Four years later, the EU adopted a Strategy to Combat IUU Fishing which shifted the focus from flag State implementation to the control of IUU-caught fish and fishery products from third countries which enter the

[18] Canada, *National Plan of Action to Prevent, Deter, and Eliminate Illegal, Unreported and Unregulated Fishing*, March 2005.

[19] *Australian NPOA-IUU*, Annex 4 with Attachment 1.

[20] *Australian NPOA-IUU*, Annex 5.

[21] *IPOA-IUU*, paras. 9.1, 28, and 78–84.

[22] Commission of the European Communities, Communication from the Commission, Community Action Plan for the Eradication of Illegal, Unreported and Unregulated Fishing, *COM(2002) 180 final*, Brussels, 28 May 2002.

EU market.[23] This strategy aims to fill the gap in the enforcement of fisheries regulations to cover fishing vessels entering EU waters and fish and fishery products that enter EU territories. Despite a number of criticisms on the strategy,[24] the EU Strategy to Combat IUU Fishing has become the basis for the adoption of the EU Regulation on IUU Fishing which is discussed in Chapter 10.

In the Southeast Asian region, the Regional Plan of Action (RPOA) to Promote Responsible Fishing, including Combating IUU Fishing in the Region was adopted in 2007.[25] The RPOA draws on core principles from already established international fisheries instruments for promoting responsible fishing practices, including the IPOA-Sharks, IPOA-Capacity, IPOA-IUU, IPOA-Seabirds, and the FAO Model Scheme on Port State Measures to Combat IUU Fishing. The objective of the RPOA is to enhance and strengthen the overall level of fisheries management in the region. The actions adopted under the RPOA cover conservation of fisheries resources and their environment, managing fishing capacity, and combating IUU fishing in three areas: the South China Sea, the Arafura-Timor Seas, and the Sulu-Sulawesi Seas (Celebes Sea).

The range of actions adopted under the RPOA include the compilation of information on artisanal and industrial fishing, assessment of the current status of fish stocks, trade flows and markets; improvement of data collection systems; sharing of information about vessels, fishing effort, catch levels, fish landings and sales of fish and fish products; development of a regional approach to identify, compile and exchange information on vessels; introduction of management measures to prevent overfishing and reduction of overcapacity; cooperation to manage straddling stocks; and provision of assistance to traditional, artisanal and small scale fisheries. The RPOA also calls on its participants to standardise catch and landing documentation and implement trade certification schemes for high value products,

---

[23] Communication from the Commission to the European Parliament, the Council, the European Economic and Social Committee and the Committee of the Regions, On the new strategy for the Community to prevent, deter, and eliminate Illegal, Unreported and Unregulated Fishing, Brussels, 17.10.2007, *COM(2007) 601 final.*

[24] Examples of these criticisms include the possible exclusion of products from developing States from the EU markets due to stricter controls and the lack of capability of States in implementing certification systems for fishery products. *See* Consultation on the Elaboration of a new Strategy against IUU fishing by the European Community, Response Document Resulting from a Stakeholder Consultation Meeting, Brussels, 20 February 2007, http://ec.europa.eu/fisheries/cfp/external_relations/illegal_fishing_en.htm, 16 March 2007, at 4.

[25] The RPOA was approved by Indonesia, Australia, Brunei Darussalam, Cambodia, Malaysia, Papua New Guinea, The Philippines, Singapore, Thailand, Timor-Leste and Vietnam. *See* Joint Ministerial Statement, *Regional Ministerial Meeting on Promoting Responsible Fishing Practices including Combating Illegal, Unreported, Unregulated (IUU) Fishing in the Region*, Bali, Indonesia, 04 May 2007; Regional Plan of Action (RPOA) to Promote Responsible Fishing Practices including Combating Illegal, Unreported, Unregulated (IUU) Fishing in the Region, Bali, Indonesia, 04 May 2007.

monitor and control the transshipment of fisheries resources, strengthen MCS measures, and implement flag and port State measures.

A Coordination Committee comprising officials from participating States has been established in order to provide strategic advice to the RPOA members. Regular meetings are held to discuss and monitor the implementation of the RPOA. At its initial meeting, five strategic priority areas were identified by RPOA participating countries for the purpose of strengthening and implementing further measures. These areas are MCS, coastal State responsibilities, regional capacity building, evaluation of the current resource and management situation in the region, and port State measures.[26] Although the implementation of the RPOA is at its early stages, RPOA participating countries are cooperating and coordinating their activities to implement the regional plan of action. Some of the RPOA members have adopted domestic measures consistent with the RPOA actions. Brunei for example, has developed a national plan of action to combat IUU fishing based on the RPOA.

The Southern African Development Community (SADC) is another regional organisation that has undertaken actions to combat IUU fishing. In July 2008, a Statement of Commitment by SADC Ministers Responsible for Marine Fisheries on Illegal, Unreported and Unregulated Fishing was adopted. Four priority areas have been identified by SADC: improving regional and inter-regional cooperation; strengthening of fisheries governance and legal frameworks to eliminate illegal fishing; development of a regional MCS strategy; and strengthening of MCS capacity.[27] The SADC Statement of Commitment also pledged to implement no later than 2010, measures consistent with the IPOA-IUU and the FAO Model Scheme on Port State Measures, such as vessel monitoring system, progressive ban on transshipment at sea in the SADC region, and prohibition of port access to IUU vessels.[28] In order to achieve the goals set under its Statement of Commitment, SADC directed its officials to work towards developing a plan of action by June 2009.

Other regional economic and political organisations have recognised the problem of IUU fishing and encourage cooperation among their members to address this concern. The Asia Pacific Economic Cooperation (APEC) for example, has

---

[26] See reports of the Workshop on Monitoring, Control and Surveillance for the Implementation of the Regional Plan of Action to Promote Responsible Fishing Practices including Combating IUU Fishing in the Region, 4–6 March 2008, Bali, Indonesia; Pre-meeting of the Regional Plan of Action to Promote Responsible Fishing Practices including Combating IUU Fishing in the Region Coordination Committee on Monitoring, Control and Surveillance for the implementation of the RPOA, 28 April 2008, Manila, Philippines.

[27] Southern African Development Community, *Statement of Commitment by SADC Ministers Responsible for Marine Fisheries on Illegal, Unreported and Unregulated Fishing*, 04 July 2008, Windhoek, Namibia, Preamble.

[28] *SADC Statement of Commitment*, para. 15.

called on its member economies to eliminate IUU fishing in the region,[29] particularly through the use of at-sea, port State, and trade related measures in accordance with international law.[30] APEC has also endeavoured to assess the extent and impacts of IUU fishing in the region.[31] More recently, the Association of Southeast Asian Nations (ASEAN) adopted a Political-Security Community Blueprint which called for close cooperation among ASEAN member States to address non-traditional security issues such as IUU fishing, including through the implementation of the IPOA-IUU and the establishment of an ASEAN Fisheries Consultative Forum.[32] The Organisation for Economic Co-operation and Development (OECD) has also adopted a programme of work for its Committee for Fisheries to study the environmental, economic and social effects of IUU fishing.[33]

A less formal example of cooperation at a multilateral level is the creation of a Ministerially led Task Force on Illegal, Unreported and Unregulated Fishing on the High Seas in 2003 by a group of fisheries ministers and international non-government organisations.[34] The High Seas Task Force developed an action plan designed to combat IUU fishing on the high seas and identified nine practical initiatives to expose and deter IUU fishing activities. These activities include the commitment to the International MCS Network, development of a global information system on high seas fishing vessels, adoption of a preliminary set of guidelines on flag State performance, improvement of port State controls, and addressing the specific needs of developing States.[35] The members of the High

---

[29] Asia Pacific Economic Cooperation (APEC), *Seoul Oceans Declaration*, Adopted at the First APEC Ocean-related Ministerial Meeting, 25–26 April 2002, Seoul, Korea, para. 6.

[30] APEC Joint Ministerial Statement, *Bali Plan of Action: Towards Healthy Oceans and Coasts for the Sustainable Growth and Prosperity of the Asia-Pacific Community*, Adopted at the 2nd APEC Ocean-related Ministerial Meeting, Second Ministerial Meeting, Bali, Indonesia, 16–17 September 2005, secs. I.c.iii and I.c.v.

[31] *See* Frank Meere and Mary Lack, *Assessment of Impacts of Illegal, Unreported and Unregulated (IUU) Fishing in the Asia Pacific* (Singapore: APEC Secretariat, 2008); Mary Ann Palma and Martin Tsamenyi, *Case Study on the Impacts of Illegal, Unreported and Unregulated Fishing in the Sulawesi Sea*, APEC Fisheries Working Group, FWG 02/2007 (Singapore: APEC Secretariat, April 2008); Sea Resources Management Sdn Bhd, *Case Study on Illegal, Unreported and Unregulated (IUU) Fishing off the East Coast of Peninsular Malaysia: Final Report*, APEC#208-FS-01.4 (Singapore: APEC Secretariat, 2008).

[32] Association of Southeast Asian Nations (ASEAN), *ASEAN Political-Security Community Blueprint*, Adopted at the 14th ASEAN Summit, 26 February-01 March 2009, Cha-am, Phetchaburi, Thailand, para. B.4.1.xiv.

[33] Organisation for Economic Co-operation and Development (OECD), Trade and Agriculture Directorate, *Past Work Programmes*, www.oecd.org. Accessed on 15 March 2009.

[34] The Task Force members included ministers from Australia, Canada, Chile, Namibia, New Zealand, and the UK.

[35] *See* High Seas Task Force, *Closing the Net: Stopping Illegal Fishing on the High Seas*, Governments of Australia, Canada, Chile, Namibia, New Zealand, and the United Kingdom, WWF, IUCN and the Earth Institute at Columbia University (London: DEFRA, 2006).

Seas Task Force continue to conduct activities to achieve these proposed initiatives. The United Kingdom has adopted an action plan to implement the recommendations of the High Seas Task Force on IUU Fishing.[36]

4.4 State Control over Nationals

According to the FAO, "(o)ne of the reasons why IUU fishing has been such a persistent problem is that many States have not been successful in controlling the fishing activities of their nationals".[37] Accordingly, the IPOA-IUU identifies State control over their nationals as one of the primary measures to be addressed by "All States" to combat IUU fishing.[38] The implementation of the control of nationals requirements under the IPOA-IUU is to be achieved in accordance with the relevant provisions of the LOSC and "without prejudice to the primary responsibility of the flag State on the high seas."[39]

The LOSC, which permits States to draw on general international law,[40] gives exclusive jurisdiction over events on board a vessel on the high seas to the flag State.[41] One of the recognised bases of criminal jurisdiction in international law is jurisdiction based on the nationality of the perpetrator of a criminal act. This is often referred to as the active nationality principle. There is therefore no fundamental principle which would prevent a State from enacting a law that punishes its nationals for taking part in IUU fishing operations, even if that national is on board a vessel flying the flag of another State.

In the context of the IPOA-IUU, an important preliminary question is the definition of "nationals". This issue is important because under international law, nationals can be defined to include individuals, companies and vessels. The starting point for a discussion of the scope and content of the control of national requirement is the LOSC. Article 116 of the LOSC grants all States "the right for their respective nationals to engage in fishing on the high seas" subject to specific qualifications.[42] Article 117 of the LOSC then goes on to provide that: "All States have the duty to take, or to cooperate with other States in taking such measures

---

[36] United Kingdom, *The UK Action Plan on Implementing the Recommendations of the High Seas Task Force on Illegal, Unregulated and Unreported (IUU) Fishing*, May 2006.

[37] *FAO Technical Guidelines for Responsible Fisheries No. 9*, *supra* note 3, at 12

[38] *IPOA-IUU*, para. 18.

[39] *IPOA-IUU*, para. 18.

[40] For example, the final preambular paragraph of the LOSC provides that: "Affirming that matters not regulated by this Convention continue to be governed by rules and principles of international law".

[41] *LOSC*, Art. 92.

[42] These qualifications, as outlined in the Article 116 of the LOSC, are treaty obligations, the rights and duties as well as the interests of coastal States provided in Article 63 paragraph 2, and Articles 64 to 67, and the provisions of Part VII Section 2 of the LOSC.

for their respective nationals as may be necessary for the conservation of the living resources of the high seas." Article 118 of the LOSC further requires States whose nationals fish on the high seas to cooperate in conserving and managing living resources on the high seas. Finally, Article 119(3) of the LOSC requires States whose nationals fish on the high seas to cooperate in conserving the living resources on the high seas, and in so doing, "shall ensure that conservation measures and their implementation do not discriminate in form or in fact against fishermen of any State."

Significantly, all these articles use the term "national" or "fishermen," thus suggesting that the term "nationals" takes its ordinary meaning of individuals of the relevant nationality. The narrow approach taken by the LOSC is in contrast to the preceding 1958 Convention on Fishing and Conservation of the Living Resources of the High Seas, where it is apparent that the term "nationals" is used in a different way. It is defined in Article 14 of the 1958 High Seas Fishing Convention retained the traditional meaning of "nationals" as "fishing boats or craft of any size having the nationality of the State concerned, according to the law of that State, irrespective of the nationality of the members of their crews". However, this definition was dropped from the LOSC.

The initiative to develop a global plan of action to address IUU fishing has taken up the matter of control of nationals. In this regard, the IPOA-IUU provides:

> 18. In the light of Article 117 of the 1982 UN Convention, and without prejudice to the primary responsibility of the flag State on the high seas, each State should, to the greatest extent possible, take measures or cooperate to ensure that nationals subject to their jurisdiction do not support or engage in IUU fishing. All States should cooperate to identify those nationals who are operators or beneficial owners of vessels involved in IUU fishing.
>
> 19. States should discourage their nationals from flagging fishing vessels under the jurisdiction of a State that does not meet its flag State responsibilities.

What is interesting about these provisions is that the term "nationals" has been used with a deliberate ambiguity. On the one hand, it was merely a call to use existing jurisdiction over nationals on the basis of the so called active nationality principle of jurisdiction in international law. On the other hand, it was also intended to leave untouched the ambiguous reference in Articles 116 to 119 of the LOSC about the use of that term.[43]

---

[43] For a discussion of this term in the context of the high seas provisions of the LOSC, *see* William Edeson, 'Tools to AIUU Fishing: the Current Legal Situation', in FAO, Report of and Papers Presented at the Expert Consultation on Illegal, Unreported and Unregulated Fishing, Sydney, Australia, 15–19 May 2000, *FAO Fisheries Report No. 666* (Rome: FAO, 2001), at 145–158.

In the context of combating IUU fishing, using jurisdiction over nationals may be considered as the last throw of the dice—in other words, if all else fails, a State may urge the State whose nationality an individual bears to take action. While the right of a State to exercise jurisdiction over its nationals wherever they are is not in dispute, there are important practical problems associated with this type of jurisdiction. A critical problem arising out of jurisdiction over nationals is a practical one: how does the State effectively enforce such jurisdiction in its law?

Apart from the issues of determining nationality and having dual nationality, there are likely to be evidential problems in cases dealing with nationals outside the jurisdiction of one State, for example witnesses to an event might not be readily accessible or documentary evidence may not be easily obtained. There may also be difficulties in obtaining custody of the alleged offender if the events occurred on board a foreign fishing vessel. Thus, while the control of nationals provision in the IPOA-IUU is an important addition to the weaponry for tackling IUU fishing at a theoretical level, there may not be many opportunities at the practical level to use it.

The possibility of a person being subject to more than one jurisdiction is another real issue granting that many people have dual nationality these days. The line of defence against second prosecutions could be to put in a clause to the effect that, where one State has already exercised jurisdiction over a national in respect of the same offence, whether that led to a conviction or not, then another State should not exercise jurisdiction over that offence. Another approach is to rely on the fact that most States have fundamental human rights provisions which will or at least should protect a person from *autrefois convict* and *autrefois acquit*. In other words, if a person has already been convicted or acquitted in respect of the same offence, then that person should not be punished a second time. The unknown element here is whether the human rights provisions would operate in respect of an offence committed elsewhere under the laws of another State.

A particularly strong example of the implementation of the control of nationals requirements is found in s 133E of the New Zealand *Fisheries Act 1996* which give effect to the UN Fish Stocks Agreement:

> (1) No New Zealand national may use a vessel that is not registered under the Ship Registration Act 1992, or a tender of that vessel, to take (by any method) on the high seas any fish, aquatic life, or seaweed for sale, or to transport any fish, aquatic life, or seaweed taken on the high seas, except in accordance with an authorisation issued by a state specified in subsection (2).

The authorisations referred to in subsection 2 may be issued by a Party to either the UN Fish Stocks Agreement or the FAO Compliance Agreement, or by a State that is Party to or has accepted the obligations of a global regional or subregional organisation or arrangement to which the authorisation relates.[44] The provision relates further to an authorisation by a State that "(i) is a signatory to the UN Fish Stocks Agreement; and (ii) has legislative and administrative mechanisms to control its vessels on the high seas in accordance with that agreement."[45] Any person who contravenes this provision commits an offence and is liable to a penalty. Section 113F of the New Zealand *Fisheries Act 1996* also provided a number of exemptions from the application of s 113E.[46] The importance of this provision is that it does not leave action solely to the flag State to control unauthorised fishing on the high seas. However, one important safeguard is written into the New Zealand legislation, namely that the consent of the Attorney-General is required before proceedings can be instituted under these provisions.[47] These provisions are mechanisms intended to ensure that the primacy of the jurisdiction of the flag State is protected, as well as providing a means of avoiding the risk of double jeopardy, jurisdictional conflicts, and other difficulties which might arise.

The application of the New Zealand legislation was given its judicial test in 2007 in the case of *Ministry of Fisheries v Tukunga*.[48] The defendants in this case, Sekope Tukunga, a Tongan citizen who has New Zealand permanent residence and Glen John Boyes, a New Zealand citizen, were charged with offences pursuant to s 113A of the New Zealand *Fisheries Act 1996*, namely that they took or transported fish in the national fisheries jurisdiction of Australia other than in accordance with the laws of that jurisdiction. The key question before the Court was whether New Zealand had jurisdiction to prosecute the defandants under the control of nationals provisions of the *Fisheries Act 1996* and whether a person who holds a foreign passport but who resides in New Zealand and holds permanent New Zealand residency is a "New Zealand national" for the purposes of s 113A. The court examined all the complexities associated with implementing the control of nationals requirements under relevant international fisheries instruments such as the LOSC, UN Fish Stocks Agreement, and the IPOA-IUU and convicted the defendants accordingly.

---

[44] New Zealand, *Fisheries Act 1996 Amendment Act (No. 2) 1999*, sec. 113(2).

[45] New Zealand, *Fisheries Act 1996 Amendment Act (No. 2) 1999*, sec. 113(2)(d).

[46] Examples of such exemptions include when the applicant is a citizen of another country which has jurisdiction over the applicant's proposed fishing activities on the high seas and if New Zealand is not a participant in, or a member of, or has not accepted the obligations of a global, regional, or subregional fisheries organisation or arrangement that covers the area of the high seas in which the applicant proposes to conduct fishing activities.

[47] New Zealand, *Fisheries Act 1996 Amendment Act (No. 2) 1999*, sec. 113ZE. Similar provisions can be found in the Australian *Fisheries Management Act 1991 (Cth)*, sec. 105.

[48] New Zealand, District Court at Wellington, *Ministry of Fisheries v Tukunga*, CRI-2005-085-6992 & 6989, Judgment 30 March 2007.

The Court justified the conviction of Boyes, a New Zealand citizen on the grounds that assertion of such jurisdiction is justified by the LOSC, UN Fish Stocks Agreement and the IPOA-IUU. The court decided that "(n)o authority is needed for the proposition that, apart altogether from treaty law, international law recognises the right of states to legislate in respect of the conduct of their own nationals wherever such conduct may have occurred."[49] Additionally, "the international instruments collectively demonstrate the purpose of the international community to control illegal, unregulated and unreported fishing, by mandating states parties to legislate in respect of their own nationals and ships flying their own flags, wherever such nationals and such ships might be found."[50] In relation to the first defendant, the Court held that Mr Tukunga was born in Tonga, is not a New Zealand citizen, and does not hold a New Zealand passport. However, the definition of a "New Zealand national" in s 2 of the *Fisheries Act 1996* is wider than that, and relevantly includes "a person who is ordinarily resident in New Zealand". It follows that, in principle, a foreign passport holder who is ordinarily resident in New Zealand can be a "New Zealand national" for the purposes of s 113A".[51]

4.5 Elimination of Economic Incentives

The IPOA-IUU requires States to avoid providing subsidies and economic support to companies, vessels or persons that are involved in IUU fishing.[52] The FAO provides that there is no justification for States to continue assisting those who carry out IUU fishing.[53]

Provision of subsidies occurs in different fishing regions of the world. In the Caribbean region, subsidised fuel, duty concessions on the purchase of fishing gear, equipment, fishing boats and engines, and subsidised loans to fishers are provided.[54] In the South Pacific, although governments no longer provide significant subsidies to their fisheries, many of the fisheries in this region were developed

---

[49] *Ibid.*, para. 40.

[50] *Ibid.*, para. 42.

[51] *Ibid.*, para. 54. The court decided that should counsel determine that an issue remains as to whether Mr Tukunga is a "New Zealand national" for the purposes of s 113A, they will agree on a set of facts upon which to argue the issue. See para. 139 of the *Ministry of Fisheries v Tukunga* (2007).

[52] *IPOA-IUU*, para. 23.

[53] *FAO Technical Guidelines No. 9*, *supra* note 3, at 17.

[54] FAO, Report of the Third Ad Hoc Meeting of Intergovernmental Organisations on Work Programmes Related to Subsidies in Fisheries, Rome, Italy, 23–25 July 2003, *FAO Fisheries Report No. 719* (Rome: FAO), at 3.

with the support of subsidies.[55] The EU fisheries support reached nearly USD600 million in 1990, excluding the support provided by individual States.[56]

As discussed in Chapter 3, the WTO Agreement on Subsidies and Countervailing Measures provides a concrete definition of subsidies to include grants, loans, equity infusion, loan guarantees, tax credits, provision of goods and services other than general infrastructure, and government support of prices and incomes.[57] In the case of fisheries, the FAO categorises different forms of subsidies such as direct government payments to or on behalf of the industry; tax waivers and deferrals; government fishery loans; loan guarantees and insurance; implicit payment to, or charges against, the fisheries industry; and other general programmes that affect fisheries.[58] More specifically, fisheries subsidies under these categories include grants to purchase new or old vessels; vessel decommissioning payments; compensation for damage to fish stocks; transport subsidies; fuel tax exemption; and grants to small fisheries and direct aid to participants in specific fisheries.[59] If paragraph 23 of the IPOA-IUU on the elimination of subsidies and economic support is read consistent with the WTO Agreement on Subsidies and Countervailing Measures, it would mean than these fisheries subsidies should be denied from those involved in IUU fishing. It also implies that a mechanism would need to be in place in order to determine whether a vessel owner or company is entitled to receive subsidies based on a record of compliance with fisheries regulations.

Subsidies in fisheries are associated with overcapacity in the fishing fleet and inadvertently cause IUU fishing. Some economic incentives also create negative environmental effects as they tend to reinforce overfishing and overinvestment in fisheries[60] not only in the EEZ but also on the high seas.[61] In such cases, the elimination of economic incentives on the basis of the IPOA-Capacity applies.

---

[55] *Ibid.*, at 4.

[56] William E. Schrank, 'Subsidies for Fisheries: A Review of Concepts', Paper Presented at the Expert Consultation on Economic Incentives and Responsible Fisheries, Rome, Italy, 28 November-01 December 2000, *FAO Fisheries Report No. 638* Supplement (Rome: FAO, 2000), at 28.

[57] WTO Agreement on Subsidies and Countervailing Measures, concluded on 15 April 1994, 1867 UNTS 14. *Hereinafter referred to as* SCM Agreement, Art. 1.1(a)(1).

[58] Schrank, *supra* note 56, at 33–34.

[59] *Ibid.,* at 33.

[60] J.C. Seijo, 'Estimating Fisheries Subsidies and Their Impact on Sustainability of Fish Resources: Dealing with Dynamics, Risk and Uncertainty,' Paper Presented at the Expert Consultation on Economic Incentives and Responsible Fisheries, Rome, Italy, 28 November-01 December 2000, *FAO Fisheries Report No. 638 Supplement* (Rome: FAO, 2000), at 41.

[61] Dominique F. Gréboval, 'The International Plan of Action for the Management of Fishing Capacity and Selected Issues Pertaining Illegal, Unreported and Unregulated Fishing,' in FAO, Report of and Papers Presented at the Expert Consultation on Illegal, Unreported and Unregulated Fishing, Sydney, Australia, 15–19 May 2000, *FAO Fisheries Report No. 666* (Rome: FAO, 2001), at 234.

Some of the proposed methods for the reduction of excess capacity include fleet reduction programmes, including buy back schemes,[62] which is referred to as conservation subsidies and one of the incentive blocking measures to control fishing fleet capacity.[63] Examples of the States which have operated buy back programmes include Japan, the U.S., Canada, Norway, Australia and some EU member States.[64]

---

[62] A buy back programme buys and removes vessels, licences or vessel capacity units from a fleet to decrease capacity. *See* J.M. Ward, J.E. Kirkley, R. Metzner, and S. Pascoe, Measuring and Assessing Capacity in Fisheries. 1. Basic Concepts and Management Options, *FAO Fisheries Technical Paper No. 433/1* (Rome: FAO, 2004) at 33.

[63] FAO, Report of the FAO Technical Working Group on the Management of Fishing Capacity, La Jolla, United States of America, 15–18 April 1998 *FAO Fisheries Report No. 586* (Rome: FAO, 1998), at para. 59.

[64] *FAO Fisheries Technical Paper No. 433/1*, *supra* note 62, at 33.

Chapter Five

Flag State Measures

5.1. Freedom of the High Seas and Flag State Jurisdiction

The freedom of the seas, as propounded by Hugo Grotius in the 1600s, has evolved into one of the principles of customary international law known as the freedom of the high seas. It began as an attempt to contest the monopoly of Spain and Portugal over trade and navigation to what was called the 'New World'.[1] In his dissertation, Grotius stated that "the sea is common to all, because it is so limitless that it cannot become a possession of any one, and because it is adapted for the use of all, whether we consider it from the point of view of navigation or of fisheries."[2] Since then, this doctrine has been widely proclaimed and supported[3] and has been the legal basis for the development of flag State jurisdiction.

The freedom of the high seas was codified in the 1958 Convention on the High Seas and the LOSC. Under the 1958 Convention on the High Seas, the freedom of the high seas includes the freedom of navigation, freedom of fishing, freedom to lay submarine cables and pipelines, and freedom to fly over the high seas.[4] Other types of high seas freedom were later on added under the LOSC, to include the freedom to construct artificial islands and other installations and the freedom of scientific research.[5] Related to these freedoms is the right of every State to sail ships flying its flag on the high seas,[6] signifying the primacy of the authority of

---

[1] Hugo Grotius, *The Freedom of the Seas or The Right which Belongs to the Dutch to Take Part in the East Indian Trade* (New York: Oxford University Press, 1916) at viii; *See also* C.G. Roelofsen, 'The Sources of Mare Liberum; The Contested Origins of the Doctrine of the Freedom of the Seas,' in Wyobo P. Heere, (ed.), *International Law and its Sources: Liber Amirocum Maarten Bos* (Hague: T.M.C. Asser Institut, 1989), at 93–124.

The doctrines in question were The Inter Caetera Papal bull of 4 May 1493 by Alexander VI and the Treaty of Tordesillas of 7 June 1494, which demarcated the world from pole to pole with an imaginary line and granted possession of unclaimed territories to Spain and Portugal.

[2] Grotius, *supra* note 1, at 28.

[3] *See* D.P. O'Connell, *The International Law of the Sea*, Vol. 1 (New York: Oxford University Press, 1983); *See also* Lawrence Juda, *International Law and Ocean Use Management: the Evolution of Ocean Governance* (New York: Routledge, 1996), at 8–17.

[4] *Convention on the High Seas Fishing*, Geneva, Switzerland, concluded on 29 April 1958, in force 30 September 1962, 13 UST 2312; 450 UNTS 11, Art. 2.

[5] United Nations Convention on the Law of the Sea, Montego Bay, Jamaica, concluded on 10 December 1982, in force 16 November 1994, 1833 UNTS 3; 21 ILM 1261 (1982). *Hereinafter referred to as* LOSC. Art. 87(1).

[6] *LOSC*, Art. 90.

the flag State of the vessels. This right, however, is subject to a number of conditions. Article 87 of the LOSC provides that these freedoms are to be exercised under the conditions laid down by the LOSC and by other rules of international law, with due regard for interests of other States in their exercise of the freedom of the high seas, and also with due regard for rights with respect to the activities in the Area.[7] Similarly, as part of the freedom of fishing, all States have the right for their nationals to engage in fishing on the high seas, subject to the provisions of the LOSC, treaty obligations, and rights and duties as well as interests of coastal States with respect to shared and straddling stocks.[8]

As will be explained in this chapter, flag State duties with respect to fisheries conservation and management are embodied in the LOSC, the FAO Compliance Agreement, the UN Fish Stocks Agreement, and the FAO Code of Conduct for Responsible Fisheries. Under international law, the flag State has the duty to control the fishing activities of its vessels, no matter where they operate. If the vessel is fishing in the waters of the flag State, the jurisdiction of the flag State is exclusive. If the vessel is fishing on the high seas, the flag State has the exclusive responsibility for controlling the activities of the vessel. However, as will be discussed in later chapters, a number of international agreements have given States other than the flag State, certain rights to take action with respect to fishing vessels on the high seas in order to ensure that such vessels do not undermine international conservation and management measures. Vessels fishing in waters under the sovereignty and sovereign rights of coastal States are also subject to the enforcement jurisdiction of that coastal State.

The IPOA-IUU emphasises the need for a State to ensure that vessels flying its flag do not engage in IUU fishing. International fisheries instruments such as the UN Fish Stocks Agreement and FAO Compliance Agreement further provide the requirement for a flag State to authorise vessels to fish only when the State is able to exercise effectively its responsibilities with respect to the activities of its vessels.[9] The fact that IUU fishing activities continue to be a global fisheries problem

---

[7] The 'Area' is defined in Article 1 of the LOSC as the seabed and ocean floor and subsoil thereof, beyond the limits of national jurisdiction.

[8] Convention on Fishing and Conservation of Living Resources of the High Seas, Geneva, Switzerland, concluded on 29 April 1958, in force 20 March 1966, 17 UST 138; 559 UNTS 285. *Hereinafter referred to as High Seas Fishing Convention.* Art. 1; *LOSC*, Arts. 63, 64, 67, and 116.

[9] FAO, *International Plan of Action to Prevent, Deter, and Eliminate Illegal, Unreported, and Unregulated Fishing*, adopted on 23 June 2001 at the 120th Session of the FAO Council. *Hereinafter referred to as* IPOA-IUU, para. 35; Agreement for the Implementation of the Provisions of the United Nations Convention on the Law of the Sea of 10 December 1982 relating to the Conservation and Management of Straddling Fish Stocks and Highly Migratory Fish Stocks, concluded on 4 September 1995, in force 11 December 2001, 34 ILM 1542 (1995); 2167 UNTS 88. *Hereinafter referred to as* UN Fish Stocks Agreement, Art. 18(2); FAO, Agreement to Promote Compliance with International Conservation and Management Measures by fishing Vessels on the High Seas, Rome, Italy, concluded on 24 November 1993, entered into force 24 April 2003, 33 ILM 968. *Hereinafter referred to as* the FAO Compliance Agreement, Art. III(3).

suggests the failure of flag States to exercise effective control over their vessels.[10] In order to combat IUU fishing and discharge their duties, flag States are required under the IPOA-IUU and other international instruments to adopt four general measures to effectively address IUU fishing: fishing vessel registration, authorisation to fish, record of fishing vessels, and flag State enforcement actions.

5.2. Fishing Vessel Registration

The LOSC provides the right of every State to sail ships flying its flag on the high seas.[11] This right is balanced with the obligation to fix the conditions for the grant of its nationality to ships, for the registration of ships in its territory, and for the right to fly its flag.[12] Article 91(1) of the LOSC also provides that "there must exist a genuine link between the State and the ship". On the basis of these conditions, ships acquire the nationality of the State whose flag they are entitled to fly and are required to sail under the flag of that State only.[13] The LOSC further provides that a ship may not change its flag during a voyage or while in a port of call, except in the case of a real transfer of ownership or change of registry.[14] Apart from these general provisions, the LOSC does not elaborate on the conditions that may be imposed by flag States to register vessels.

There are two aspects of vessel registration. The first pertains to the public law function of registration which includes the allocation of a vessel to a specific State and subjection of the vessel to the jurisdiction of that State. It also involves the conferment of the right to fly the flag, the right to diplomatic and naval protection, and the right to engage in activities such as fishing.[15] The second pertains to the private law functions of registration including the protection of the title of the registered owner and the preservation of priorities between persons holding security interests over the vessel such as mortgages, liens, and encumbrances.[16] A vessel that is registered and given the right to fly the flag of a State is entered into a public record or a Register, and sometimes in the case of fishing vessels, a Register of Fishing Vessels.

---

[10] *See* Kevin Bray, 'A Global Review of Illegal, Unreported, and Unregulated (IUU) Fishing', in FAO, Report of and Papers Presented at the Expert Consultation on Illegal, Unreported and Unregulated Fishing, Sydney, Australia, 15–19 May 2000, *FAO Fisheries Report No. 666* (Rome: FAO, 2001), at 88–134.

[11] *LOSC*, Art. 90.

[12] *LOSC*, Art. 91(1).

[13] *LOSC*, Art. 92(1).

[14] *LOSC*, Art. 92(1).

[15] John Fitzpatrick, 'Measures to Enhance the Capability of a Flag State to Exercise Effective Control Over a Fishing Vessel', in FAO, Report of and Papers Presented at the Expert Consultation on Illegal, Unreported and Unregulated Fishing, Sydney, Australia, 15–19 May 2000, *FAO Fisheries Report No. 666* (Rome: FAO, 2001), at 209.

[16] *Ibid.*

The FAO Code of Conduct and the FAO Technical Guidelines for Responsible Fisheries No. 1 provide some guidance for States in allocating their flags to fishing vessels. According to the FAO Code of Conduct and the FAO Technical Guidelines, a flag State needs to ensure that vessels to which it has allocated its flag carry onboard the original Certificate of Registry or a document that would attest to the nationality of the fishing vessels.[17] In the case of a newly built or newly sold fishing vessel en-route to the flag State, the FAO Technical Guidelines provide that a flag State may issue a document in which it calls on all other States to recognise that the vessel is sailing under its protection.[18] On arrival in the new flag State, a Certificate of Registry or a Provisional Certificate of Registry would normally be issued.[19] In the case of closure of registry, a flag State may require information such as the reason for closure (i.e. decommissioning, scrapping, or sale), and if applicable the names(s) and nationality or nationalities of the new owners.[20] The details of the closure must then be submitted to the appropriate international organisations and to State Parties to any relevant international agreement.[21]

International law imposes obligations on a flag State to effectively exercise its jurisdiction and control in administrative, technical, and social matters over ships flying its flag.[22] Effective control and jurisdiction over fishing vessels are therefore not only limited to the registration of fishing vessels but also to generally accepted international regulations on the construction, equipment, seaworthiness of ships, safety at sea, and labour standards, which are embodied in IMO and ILO conventions, regulations, and standards. The LOSC provides the requirement for a flag State to adopt measures necessary to ensure:

- That each ship, before registration and thereafter, is surveyed by a qualified surveyor of ships, and has on board navigational equipment and nautical publications appropriate for the safe navigation of ships;
- That each ship is in the charge of a master and officers who possess appropriate qualifications; and
- That the master, officers, and to the extent appropriate, the crew are fully conversant with and required to observe international regulations concerning the safety of life at sea, the prevention of collisions, the prevention, reduction

[17] FAO, *Code of Conduct for Responsible Fisheries*, Adopted at the 28th Session of the FAO Conference, Rome, Italy, 31 October 1995. *Hereinafter referred to as* FAO Code of Conduct. Para. 8.2.2; *See also* FAO, 'Fishing Operations,' *FAO Technical Guidelines for Responsible Fisheries No. 1* (Rome: FAO, 1996), at para. 26.

[18] *Ibid*., para. 27.

[19] *Ibid*.

[20] *Ibid*., para. 28.

[21] *Ibid*., para. 28.

[22] *LOSC*, Art. 94(1).

and control of marine pollution, and the maintenance of communications by radio.[23]

Consistent with the general flag State duty to exercise jurisdiction and control in administrative, technical, and social matters over ships, the safety and labour standards for fishing vessels are covered under the 1993 Torremolinos Protocol for the Safety of Fishing Vessels,[24] the 1995 International Convention on Training, Certification and Watchkeeping for Fishing Vessels (STCW-F),[25] as well as a number of fishing vessel safety codes and voluntary guidelines.

Similar to the FAO Compliance Agreement, the 1993 Torremolinos Protocol for the Safety of Fishing Vessels applies to fishing vessels of 24 metres in length and over.[26] The Annexes to the 1993 Torremolinos Protocol provide regulations for the construction and equipment of fishing vessels, which include the requirement to conduct an initial survey, and periodical surveys and inspections at appropriate intervals, on the vessel to ensure their safety.[27] An International Fishing Vessel Safety Certificate is issued if a vessel complies with the requirements of these regulations. In terms of certification of fishing vessel crew, the 1995 STCW-F provides the specific minimum requirements for the certification of personnel on board various types of fishing vessels, particularly open and decked fishing vessels of less than 12 metres, decked fishing vessels of 12 metres in length and over but less than 24 metres or fishing vessel with less than 750 kW propulsion power, and fishing vessels of 24 metres in length and over or with 750 kW propulsion power and more. While these regulations promote the safety of fishing vessels, they are yet to be fully implemented because of the lack of ratification of the 1993 Torremolinos Protocol and STCW-F by fishing States.

The uniform standards formulated jointly by FAO, ILO, and IMO on the Code of Safety for Fishermen and Fishing Vessels (Fishing Vessel Code) also provide some guidelines on promoting the improvement of safety and health on board fishing vessels. The Fishing Vessel Code provides information on the use of navigational equipment, mechanical equipment, and safety on deck.[28] The Code also covers the safety of fishing operations, particularly trawling, purse seining, Danish seining, longline fishing, tuna pole and line fishing, and fish and ice

---

23 *LOSC*, Arts. 94(2) and 94(3).

24 International Maritime Organization (IMO), *Torremolinos Protocol of 1993 relating to the Torremolinos International Convention for the Safety of Fishing Vessels 1977*, Torremolinos, Spain, 2 April 1993, www.imo.org. *Hereinafter referred to as* 1993 Torremolinos Protocol.

25 IMO, *International Convention on Standards of Training, Certification and Watchkeeping for Fishing Vessel Personnel*, concluded on 7 July 1995, in force 1 February 1997. *Hereinafter referred to as* STCW-F.

26 *1993 Torremolinos Protocol*, Art. 3(4).

27 *1993 Torremolinos Protocol*, Annex, Reg. 6.

28 IMO, *Code of Safety for Fishermen and Fishing Vessels, Part A Safety and Health Practice for Skippers and Crews* (London: IMO, 2005).

handling, which are not discussed in any other international guidelines or codes. The FAO, ILO, and IMO have also formulated guidelines on the construction and design of smaller fishing vessels from 12 metres to 24 metres, as well as measures to protect and accommodate crew on these vessels.[29] The additional focus on smaller fishing vessels is necessary if proper flag State jurisdiction is to be exercised, since such vessels comprise the majority of the world's fishing fleet and are distributed across coastal waters. The structure and technological capacities and training needs of personnel in such vessels vary greatly from large vessels.

Effective jurisdiction and control of States in social matters over ships flying their flags further involves the adherence to maritime labour standards. These standards are related to the minimum age, medical examination, accommodation, articles of agreement, competency certificates, vocational training, and hours of work, which are included in the comprehensive Maritime Labour Convention of 2006, and the most recently adopted Work in Fishing Convention 2007 (No. 188) and Work in Fishing Recommendation 2007 (No. 199).[30] These conventions provide the overall responsibility of flag States to ensure the rights of fishers in relation to their service on board fishing vessels, as well as to adopt laws and regulations that will ensure fishing vessel owners are responsible for making available to fishers agreements that will address their living and working conditions.

In addition, the Work in Fishing Convention 2007 specifically provides that "(E)ach member shall effectively exercise its jurisdiction and control over vessels that fly its flag by establishing a system for ensuring compliance with the requirements of this Convention including, as appropriate, inspections, reporting, monitoring, complaint procedures, appropriate penalties and corrective measures, in accordance with national laws or regulations."[31] As part of compliance and enforcement mechanisms under the Convention, State parties are required to establish an effective system of inspection and issuance of documents concerning living and working conditions onboard fishing vessels.[32] As a safeguard, each State party is required to apply the Work in Fishing Convention in such a way as to ensure that fishing vessels flying the flag of any State that has not ratified this Convention do not receive more favourable treatment than fishing vessels that fly the flag of any Member that has ratified it.[33]

---

[29] IMO, *FAO/ILO/IMCO Voluntary Guidelines for the Design, Construction and Equipment of Small Fishing Vessels* (London: IMO, 2005).

[30] ILO, *Maritime Labour Convention,* 23 February 2006, 45 I.L.M. 792 (2006); ILO, Report I (1B), International Labour Conference, 94th (Maritime) Sess. (2006); *Work in Fishing Convention 2007 (No. 188)* and *Work in Fishing Recommendation 2007 (No. 199)*, adopted on 14 June 2007, www.ilo.org. *Hereinafter referred to as* Work in Fishing Recommendation.

[31] *The Work in Fishing Convention*, Art. 40.

[32] *The Work in Fishing Convention*, Art. 41.

[33] *The Work in Fishing Convention*, Art. 44.

Another flag State duty is to establish requirements for the marking of fishing vessels in accordance with the FAO Standard Specifications for the Marking and Identification of Fishing Vessels.[34] These standard specifications are based on the International Telecommunication Union (ITU) Radio Call Signs (IRCS) system which is an established international system from which the identity and nationality of vessels can be readily determined. Under these standard specifications, vessels should be marked with their IRCS.[35] Vessels to which an IRCS has not been assigned shall be marked with characters allocated by the ITU to the flag State, followed by the licence or registration number assigned by the flag State.[36] According to the FAO Standard Specification for the Marking and Identification of Fishing Vessels, apart from the name of the vessel or identification mark and the port of registry required by international practice or national legislation, the marking system shall be the only other vessel identification mark consisting of letters and numbers to be painted on the hull or superstructure.[37] Technical specifications are also provided on the use of letters and number and display of the marking on the fishing vessels.[38]

Flag States are also required to mark fishing gears in accordance with uniform and internationally recognisable vessel gear marking systems.[39] The FAO has proposed legal and technical measures for the marking of fishing gears. The FAO Technical Guidelines for Responsible Fisheries provide that national legislation should contain a requirement for the marking of fishing gear and fishing implements, including nets, lines and fish aggregating devices (FADs), in order to identify the owner of the gear.[40] Furthermore, the FAO has developed proposals that call for the adoption of a system for the marking of fishing gears that will reflect the special requirements of vessels fishing on the high seas, vessels fishing in waters of States other than those of the flag State, and vessels of a coastal State fishing in its own waters.[41]

The IPOA-IUU enumerates other measures that a flag State needs to take into consideration when registering fishing vessels. These include requirements for the State to ensure that fishing vessels flying its flags, including chartered vessels,

---

[34] *FAO Compliance Agreement*, Art. III(6); *UN Fish Stocks Agreement*, Art. 18(3)(d); *FAO Code of Conduct*, paras. 8.2.3 and 8.2.4; FAO, *The Standard Specifications for the Marking and Identification of Fishing Vessels* (Rome: FAO, 1989).

[35] *FAO Standard Specifications for the Marking and Identification of Fishing Vessels*, para. 2.1.2.

[36] *Ibid*., para. 2.1.3.

[37] *Ibid*., para. 2.1.5.

[38] *Ibid*., para. 3.

[39] *UN Fish Stocks Agreement*, Art. 18(3)(d); *FAO Code of Conduct*, Art. 8.2.4; *See also* FAO, Report of the Expert Consultation on the Marking of Fishing Gear, Victoria, British Columbia, Canada, 14–19 July 1991, *FAO Fisheries Report No. 485* (Rome: FAO, 1993).

[40] *FAO Technical Guidelines for Responsible Fisheries No. 1*, *supra* note 17, para. 36.

[41] FAO, *Proposed System for the Marking of Fishing Gear*, Annex III to the FAO Technical Guidelines for Responsible Fisheries No. 1: Fishing Operations, (Rome: FAO, 1996), at para. 2.3.

do not engage in IUU fishing,[42] the requirement to avoid flagging vessels with a history of non-compliance,[43] and the requirement to deter vessels from reflagging or flag-hopping for the purposes of non-compliance with conservation and management measures.[44] Under the IPOA-IUU, a flag State is required to avoid flagging vessels with a history of non-compliance except under two conditions. The first condition is where the ownership of the vessel has subsequently changed and the new owner has provided sufficient evidence demonstrating that the previous owner or operator has no further legal, beneficial or financial interest in, or control of, the vessel.[45] The second condition is when the flag State determines that flagging the vessel would not result in IUU fishing.[46] Similar measures are provided in other international instruments.[47]

Two major interrelated issues emerge from the flag State duty to register and exercise effective jurisdiction and control over vessels flying its flag. These issues relate to the concept of a genuine link and controlling the activities of vessels flying the flags of non-parties to regional fisheries management agreements. Article 91 of the LOSC provides the right of every State to register and grant its nationality to ships. The same article states that "(t)here must exist a genuine link between the State and the ship." The acquisition by a vessel of the nationality of a State generates legal rights and obligations for both the vessel and the flag State.[48] The genuine link requirement would also allow the flag State to regulate the operations of its vessels, ensure that such vessels comply with fisheries conservation and management measures, and consequently ensure that those vessels do not engage in IUU fishing. Although the genuine link concept has been discussed since the First United Nations Conference on the Law of the Sea (UNCLOS I);[49] its scope and content are yet to be elaborated clearly in international law.

The United Nations Conference on Trade and Development (UNCTAD) considers the concept of a genuine link as an economic link.[50] UNCTAD maintains

---

[42] *IPOA-IUU*, paras. 34 and 37.

[43] *IPOA-IUU*, para. 36.

[44] *IPOA-IUU*, para. 38. According to IPOA-IUU, flag-hopping is the practice of repeated and rapid changes of a vessel's flag for the purposes of circumventing fisheries conservation and management measures or provisions adopted at a national, regional or global level or of facilitating non-compliance with such measures or provisions. *See* IPOA-IUU, para. 39; Agenda 21, par. 17.52; UN Fish Stocks Agreement, Preamble; FAO Compliance Agreement, Preamble.

[45] *IPOA-IUU*, para. 36.1.

[46] *IPOA-IUU*, para. 36.2.

[47] *See* LOSC, Art. 91 and 94; UN Fish Stocks Agreement, Art. 18; FAO Compliance Agreement, Art. III.

[48] Annick Van Houtte, 'Flag State Responsibility and the Contribution of Recent International Instruments in Preventing, Deterring, and Eliminating IUU Fishing,' Appendix E in Report of the Expert Consultation on Fishing Vessels Operating Under Open Registries and their Impact on Illegal, Unreported and Unregulated Fishing, Miami, Florida, USA, 23–25 September 2003, *FAO Fisheries Report No. 722* (Rome: FAO, 2003) at 48.

[49] *See* Convention on the High Seas, Art. 5(1).

[50] G.P. Pamborides, *International Shipping Law: Legislation and Enforcement* (Athens: Ant. N. Sakkoulas Publishers, 1999), at 16.

that several elements are normally relevant in determining whether a genuine link exists. These elements include: the contribution of the merchant fleet to the national economy of the country; the treatment of revenues and expenditure of shipping in the national balance of payment accounts; the employment of nationals on vessels; and the beneficial ownership of vessels.[51] Hence, it can be said that a vessel which flies the flag of an open registry State is one that has no real economic connection with the country whose flag it flies.[52] In this case, the exercise of effective jurisdiction and control by a flag State, whether it is a traditional register or open register State, is a separate issue.

The International Tribunal on the Law for the Sea (ITLOS), in the *M/V Saiga* case, decided that the purpose of the LOSC on the need for a genuine link between a ship and a flag State is "to secure more effective implementation of the duties of the flag State and not to establish criteria by reference to which the validity of the registration of ships in a flag State may be challenged by another State."[53] The meeting of the Joint FAO/IMO Ad Hoc Working Group on IUU Fishing in 2001 suggested that there is little benefit in defining the concept of genuine link; but rather that States focus on what might constitute effective flag State control over fishing vessels.[54]

The Organisation for Economic Co-operation and Development (OECD) suggests that the problem with the identification of the beneficial owners of vessels not only lies with the ship registration procedures of a flag State but also with the general corporate instruments that provide the principal measures to effectively cloak the identity of vessel owners.[55] The ability of corporations to be registered as owners of vessels provide mechanisms whereby nominee shareholders, nominee directors, corporate shares, or intermediaries may act on the behalf of owners.[56] Similarly, there are institutional means that require the minimal involvement of beneficial owners.[57] Although more common in open registries, these

[51] Awni Benham, 'Ending Flag State Control,' in Andree Kirchner, (ed.), *International Marine Environment Law: Institutions, Implementation and Innovations* (Hague: Kluwer Law International, 2003), at 126.

[52] *Ibid.*

[53] International Tribunal on the Law of the Sea (ITLOS), *M/V Saiga No.2*, Judgment, 1999, www.itlos.org. Accessed on 15 January 2004, para. 83.

[54] FAO, Committee on Fisheries, Twenty-fourth Session, Rome, Italy, 26 February-02 March 2001, Summary of the Report of the Joint FAO/IMO Ad Hoc Working Group on Illegal, Unreported and Unregulated Fishing and Related Matters, Rome, Italy, 09–11 October 2000, *COFI/2001/Inf.10*, (Rome: FAO, 2001), at para. 1; *See also* Judith Swan, Fishing Vessels Operating Under Open Registers and the Exercise of Flag State Responsibilities, *FAO Fisheries Circular No. 980* (Rome: FAO, 2002), at 5.

[55] OECD Maritime Transport Committee, *Ownership and Control of Ships*, March 2003, www.oecd.org. Accessed on 02 May 2008, para. 28. This report also provides a table of registration requirements of some 32 flag States to illustrate how such requirements facilitate the anonymity of beneficial owners of vessels.

[56] *Ibid.*, paras. 32–44.

[57] *Ibid.*, paras. 45–60.

mechanisms are also available to fishing vessels flagged to States with traditional registers. In such cases, a flag State would need to strengthen its corporate governance framework that includes at the very least, mechanisms that would enable authorities to maintain and obtain information on beneficial ownership and share such information with other regulators and enforcement authorities, both domestically and internationally, taking into account confidentiality of information.[58]

The vagueness of the concept of genuine link is also usually associated with two issues related to IUU fishing. The first issue is the increasing number of fishing vessels flying the flags of open register States while the second is the continuous reflagging of vessels, or flag-hopping, for the purposes of avoiding compliance with conservation and management measures. However, while vessels flying the flags of open register States are more likely to engage in IUU fishing, there is no guarantee that vessels flying the flag of traditional register States do not engage in IUU fishing. All flag States share a common problem when it comes to IUU fishing, which is the lack of effective control over fishing vessels.

A number of case studies illustrate the good level of jurisdiction and control exercised by open register States over their fishing vessels. For example, Belize has been reported to have taken 17 enforcement actions against fishing vessels on its registry from 1997 to 2001.[59] Such enforcement actions included the imposition of fines as well as the deletion of vessels from its registry.[60] Cyprus and Panama have taken similar actions, including the imprisonment of owners or operators of the vessels.[61] These examples do not suggest, however, that open register States are more effective in addressing IUU fishing nor imply that establishing a genuine link between the vessel and the flag State is not ideal. They merely show that even with the absence of a real connection between the vessel's owners and the flag State, mechanisms are available to all flag States to exercise effective control over fishing vessels to combat IUU fishing.

Compared to other international agreements, the 1986 Convention on the Registration of Ships (Ship Registration Convention),[62] although not yet in force, sets out more detailed requirements for the registration of ships. The Ship

---

[58] OECD, Steering Group on Corporate Governance, *Options for Obtaining Beneficial Ownership and Control Information*, September 2002, www.oecd.org. Accessed on 02 May 2008, para. 9.

[59] Matthew Gianni, 'IUU Fishing and the Cost to Flag of Convenience Countries,' in OECD, *Fish Piracy: Combating Illegal, Unreported and Unregulated Fishing* (Paris: OECD, 2004), at 285.

[60] *Ibid.*

[61] *See* Gabriel P. Gabrielides, 'Cyprus-A Case Study,' and Arnulfo Franco, 'Panama-Case Study,' in Report of the Expert Consultation on Fishing Vessels Operating Under Open Registries and their Impact on Illegal, Unreported and Unregulated Fishing, Miami, Florida, USA, 23–25 September 2003, *FAO Fisheries Report No. 722* (Rome: FAO, 2003), at Appendices I and J.

[62] *United Nations Convention on the Conditions for the Registration of Ships*, Geneva, Switzerland, concluded on 07 February 1986, UN Doc. TD/RS/CONF/19/Add.1; 26 I.L.M. 1229 (1987). This Convention has not yet entered into force and does not apply to fishing vessels.

Registration Convention aims to strengthen the genuine link between a flag State and vessels entitled to fly its flag, including bareboat chartered vessels, by providing conditions that would determine the legitimate interest of owners or operators in the management and operation of their vessels. These conditions include the identification of the owner and operator of the vessels,[63] the participation of the nationals of the flag State in the operation of the vessels,[64] manning of ships,[65] and establishment of the principal place of business of the shipowning company in the territory of the flag State.[66] Although the Ship Registration Convention is not in force, it can potentially form the basis for any future initiative by the FAO to provide some guidelines with respect to the registration and establishing the genuine link of fishing vessels.[67]

There is no internationally agreed registration system exclusively for fishing vessels. Similarly, most national vessel registration processes do not distinguish fishing vessels from other vessels, hence contributing to the failure to address specific issues related to the registration of fishing vessels. A good benchmark practice is EU procedures for communicating information about the characteristics and identification features of EU vessels. These procedures include the requirement for EU member States to collect and transmit vessel information to the European Commission for inclusion on its fishing vessel register, including: the internal number;[68] updating indicator;[69] country of registration; flag; registration number; name of vessel; port of registration, international radio call sign; external marking; type(s) of fishing gear; length overall; length between perpendiculars; tonnage; engine power; hull material; date of entry into service; year of construction; importing/exporting country; name of agent; agent's address; name of owner; owner's address; and place of construction.[70] These data requirement seem numerous, however, they do not include information that would disclose the history of ownership and nationality of a fishing vessel.

---

[63] *UN Convention on the Conditions for the Registration of Ships*, Art. 6.

[64] *UN Convention on the Conditions for the Registration of Ships*, Art. 7.

[65] *UN Convention on the Conditions for the Registration of Ships*, Art. 9.

[66] *UN Convention on the Conditions for the Registration of Ships*, Art. 10(1).

[67] *FAO Fisheries Circular No. 980*, *supra* note 54, at 10.

[68] The internal number is allocated definitively to a fishing vessel when it is first registered in a national register. This number is neither modified nor reallocated even if the vessel is transferred from or to another EU member State, destroyed, withdrawn from fishing or reassigned. *See* European Community, Commission Regulation (EC) No. 2090/98 of 30 September 1998 concerning the fishing vessel register of the Community, *OJ L 266* 1.10.1998, p. 27, Art. 5a.

[69] Updating Indicator is a form of classification for an EU-flagged vessel based on whether or not the vessel is of new construction, has changed activity; modified, withdrawn from fishing by exportation, cessation, or destruction, *etc. See* Table 1 of EC No. 2090/98.

[70] EU, EC No. 2090/98 as amended by EC No. 839/2002 of 21 May 2002 amending Regulation (EC) No 2090/98 concerning the fishing vessel register of the Community, *OJ L 134*, 22/05/2002, pp. 5–8, Annex.

Another shortcoming of the international ship registration requirements is that they do not adequately respond to the issue of reflagging of fishing vessels to avoid compliance with fisheries laws and regulations. Measures related to the jurisdiction and control in the administrative, technical, and social matters of vessels do very little to address concerns about flag hopping. While the collection of information that would clearly identify owners and operators of vessels is important, it is argued that unless the activities of such vessels are closely monitored, the frequency of reflagging of a particular vessel would be difficult to trace.

A practical approach to avoid the continuous reflagging of fishing vessels for the purpose of avoiding compliance with fisheries laws and regulations is to require vessel owners seeking to register fishing vessels to provide full information on previous owners and flag States, and explain any frequent changes in the registration of the vessel.[71] Should a pattern of possible flag-hopping emerge from such information, it may be presumed that the vessel has been used in IUU fishing[72] and be denied the right to fly the flag of that State. This approach would also enable a flag State to identify persons or entities with legitimate interest in the fishing vessel and those who may be held responsible for the activities of the vessel.

Flag States would also need to address the issue of the registration of chartered fishing vessels to ensure that such vessels do not engage in IUU fishing. To this end, paragraph 37 of the IPOA-IUU provides that "States involved in a chartering arrangement, including flag States and other States that accept such arrangements, should, within the limits of their respective jurisdictions, take measures to ensure that chartered vessels do not engage in IUU fishing". To achieve this, transparency in the chartering arrangement would need to be required. For example, a flag State would need to require the foreign register to provide details of the legal owners, mortgages, liens, and other encumbrances.[73] The underlying registration would also need to be cancelled for the period of the charter.[74]

In order to effectively address IUU fishing, there is further need for States to register all types and categories of fishing vessels.[75] This would enable States to assess the history of compliance of fishing vessels before they are granted the right to fly their flag.[76] Maintaining a register of all types of fishing vessels, for example small scale and commercial fishing vessels, would also facilitate the

---

[71] FAO, Fisheries Department, Implementation of the International Plan of Action to Prevent, Deter and Eliminate Illegal, Unreported and Unregulated Fishing, *FAO Technical Guidelines for Responsible Fisheries No. 9* (Rome: FAO, 2002), at para. 4.1.

[72] *Ibid.*

[73] *Fitzpatrick*, *supra* note 15, at 212.

[74] *Ibid.*

[75] *FAO Technical Guidelines No. 9*, *supra* note 71, para. 4.1.

[76] In particular, fishing vessels of less than 24 metres in length. *See* FAO Compliance Agreement, Art. II(2).

monitoring of the activities of such vessels within national jurisdiction and on the high seas. Admittedly, the implementation of these measures may pose a challenge particularly for developing States considering the significant number of small and traditional fishing vessels that operate in their coastal waters.

5.3. Authorisation to Fish

A flag State can exercise effective control over fishing vessels not only through vessel registration but also through the issuance of licences or authorisations to fish. The IPOA-IUU provides the link between the two measures. Paragraph 40 of the IPOA-IUU provides that flag States would need to consider conducting the separate functions of registration and licensing of fishing vessels in a manner that ensures each gives appropriate consideration to the other. For example, a flag State should consider making its decision to register a fishing vessel conditional upon it being prepared to provide an authorisation to fish to that vessel.[77] Furthermore, according to the FAO Compliance Agreement, if a fishing vessel that has been authorised to be used for fishing on the high seas by a Party to the Agreement ceases to be entitled to fly the flag of that Party, the authorisation to fish on the high seas shall be deemed to have been cancelled.[78] Since the functions of fishing vessel registration and licensing often fall under the jurisdiction of different authorities, cooperation and information sharing between responsible agencies at the national level will be critical.[79]

A flag State has the duty to issue licences to fishing vessels on the high seas, including in areas managed by RFMOs to which the flag State is a member. Similarly, a flag State needs to provide licences to vessels fishing in areas under the jurisdiction of a coastal State where an access agreement or arrangement exists. Under international fisheries instruments, there is a requirement for a flag State to ensure that only vessels with authorisations to fish conduct operations in areas under the jurisdiction of RFMOs or on the high seas.[80] Article III(5)(a) of the FAO Compliance Agreement specifically provides that no Party shall authorise any fishing vessel previously registered in the territory of another Party that has undermined the effectiveness of international conservation and management measures to be used for fishing on the high seas, unless it is satisfied that two conditions have been met. The first condition is where any period of suspension by another Party of an authorisation to fish on the high seas has expired.

---

[77] *IPOA-IUU*, para. 41.

[78] *FAO Compliance Agreement*, Art. III(4).

[79] *IPOA-IUU*, para. 40.

[80] *IPOA-IUU*, para. 44; *UN Fish Stocks Agreement*, Art. 18(3); *FAO Compliance Agreement*, Art. III(2); *FAO Code of Conduct*, Art. 8.2.2.

The second condition is where no authorisation to fish on the high seas for such fishing vessel has been withdrawn by another Party within the last three years. Article III(5)(b) extends to fishing vessels previously registered in the territory of a State which is not a Party to the FAO Compliance Agreement, provided that sufficient information is available on the circumstances in which the authorisation to fish was suspended or withdrawn. A flag State Party to the Agreement may issue an authorisation to fish to a vessel if the ownership of that vessel has subsequently changed, and the new owner has provided sufficient evidence demonstrating that the previous owner or operator has no further legal, beneficial or financial interest in, or control of the fishing vessel.[81] A flag State may also issue an authorisation to fish if, after having taken into account all relevant facts, that flag State has determined that to grant an authorisation to fish on the high seas to the vessel would not undermine the object and purpose of the FAO Compliance Agreement.[82]

The effectiveness of a fishing vessel licensing system to combat IUU fishing does not solely depend on the issuance of a valid authorisation to fish, but more specifically on the enforcement of the terms and conditions of a fishing licence. The UN Fish Stocks Agreement provides that one of the duties of a flag State is to establish regulations for the application of certain terms and conditions on a fishing licence;[83] however, it does not provide any details on these terms and conditions.

The IPOA-IUU enumerates some of the conditions under which a fishing licence may be issued. A fishing licence must contain basic information such as the name of the vessel, and where appropriate, the natural or legal person authorised to fish, as well as the areas, scope and duration of the authorisation, and authorised species and fishing gear and other applicable management measures.[84] Additional conditions that may be attached to fishing licences, include:

- vessel monitoring systems;
- catch reporting conditions;
- conditions related to transshipment, if permitted;
- observer coverage;
- maintenance of fishing and related logbooks;
- navigational equipment;
- marking of fishing vessels and gears according to international standards;
- use of internationally recognised fishing vessel identification number; and

---

[81] *FAO Compliance Agreement*, Art. III(5)(c).
[82] *FAO Compliance Agreement*, Art. III(5)(d).
[83] *UN Fish Stocks Agreement*, Art. 18(3)(b)(i).
[84] *IPOA-IUU*, para. 46.

- compliance with measures related to maritime safety, protection of the marine environment, and other conservation and management measures.[85]

One of the conditions that would need to be included in every authorisation to fish is the requirement for the recording and timely reporting of vessel position, catch of target and non-target species, fishing effort and other relevant fisheries data in accordance with sub-regional, regional and global standards for the collection of data.[86] Fisheries data that would need to be reported include: time series of catch and effort statistics; total catch in number, nominal weight, or both, by species (both target and non-target); discard statistics reported as number or nominal weight by species; effort statistics for each fishing method; and fishing location, date and time fished and other statistics on fishing operations.[87] Means to verify the accuracy of reported data also need to be developed by the flag State, such as routine inspections at port and use of independent observers on board vessels.

The marking of fishing vessels is also an important condition of an authorisation to fish. Although such a condition may vary in detail in particular circumstances, an authorisation to fish would need to include a requirement to provide information relating to the gear type, principal target species,[88] the reporting of fishing gear lost, abandoned or otherwise discarded, as well as the disposal of old and unwanted fishing gears.[89] The authorisation to fish would also need to include conditions relating to the deployment of fish aggregating devices (FADs), type of FADs to be used, their location, and the fishing activities permitted around FADs.[90]

Authorisation to fish must also ensure that transport and support vessels do not support or engage in IUU fishing. To this end, flag States are required under the IPOA-IUU to ensure that all fishing, transport and support vessels involved in transshipment at sea have a prior authorisation to transship.[91] As such, a fishing licence may include requirements relating to the date and location of all their transshipments of fish at sea, the weight by species and area of the catch transshipped, information related to the identification of the vessels involved in transshipment, and the port of landing of the transhipped catch.[92]

The conditions of a fishing licence also need to take into account the regulatory regime present on the high seas where a fishing vessel is entitled to conduct its

[85] *IPOA-IUU*, para. 47.
[86] *UN Fish Stocks Agreement*, Art. 18(3)(e).
[87] *UN Fish Stocks Agreement*, Annex I, Art. 3(1); *IPOA-IUU*, para. 47.2.
[88] *FAO Proposed System for the Marking of Fishing Gear*, *supra* note 41, para. 2.2.
[89] *Ibid.*, para. 2.6.
[90] *Ibid.*, para. 6.1.
[91] *IPOA-IUU*, para. 49.
[92] *IPOA-IUU*, para. 49.

operations. These conditions include the obligations under conservation and management measures of an RFMO to which the flag State is a member and any other internationally agreed fisheries management measures. Similarly, if a State has entered into an agreement or an arrangement with another State under which fishing vessels flying its flag are entitled to fish, the flag State may also impose conditions on the authorisation to fish based on the provisions of the agreement.

The mere imposition of these conditions would not result in control of flag States of the activities of their fishing vessels. Flag State control would not be effective without the capacity to monitor the compliance of fishing vessels with the terms and conditions of their licences. Furthermore, since breach of any of the conditions of a fishing licence is illegal fishing, the measures that a flag State adopts to address such violations would determine how effective a licensing system is to combat IUU fishing.

5.4. Record of Fishing Vessels

Related to the obligations of registering and licensing fishing vessels is the requirement under international fisheries instruments for States to maintain a record of fishing vessels.[93] A record of fishing vessels is a record of vessels entitled to fly the flag of a State which are authorised to fish on the high seas. Article VI(1) of the FAO Compliance Agreement enumerates the information that is required to be entered in a record of fishing vessels, including the name of fishing vessel, registration number, previous names (if known), and port of registry; previous flag (if any); name and address of owner or owners; where and when built; type of vessel; and length. Non-mandatory information that may be entered in the record of fishing vessels and provided to the FAO under the FAO Compliance Agreement include the name and address of operator(s) or manager(s), types of fishing method(s), moulded depth, beam, gross register tonnage, and power of main engine or engines.[94] Article 4 of Annex I of the UN Fish Stocks Agreement also provides for the collection of similar vessel related data, although it does not specifically require Parties to include such data in their record of fishing vessels.

Article VI of the FAO Compliance Agreement provides the basis for the establishment of the High Seas Vessels Authorization Record (HSVAR), a database available on the FAO website which contains distinctive and descriptive elements of vessels fishing on the high seas, as well as information on their identification, registration, authorisation status, infringements, and other related

---

[93] *UN Fish Stocks Agreement*, Art. 18(3)(c); *FAO Compliance Agreement*, Arts. III(6), IV, VI(1) and VI(2); *FAO Code of Conduct*, Art. 8.2.1.

[94] *FAO Compliance Agreement*, Art. IV(2).

information.[95] For the purpose of HSVAR, fishing vessels include any vessels used or intended for use for the commercial exploitation of living marine resources, including mother ships and any other vessels directly engaged in such fishing operations.[96]

The HSVAR was developed in 1995 when FAO requested States which have accepted the FAO Compliance Agreement to provide data on vessel authorisation to facilitate the testing of the database. The U.S., Canada, Norway, Japan, and the EU initially agreed to provide the requested information which resulted in the inclusion of 5,672 fishing vessels in the database.[97] In 2003, the FAO reminded States which have accepted the Agreement to comply with their obligations under Article VI, particularly on the exchange of information on vessels authorised to fish on the high seas. There are now 5,943 fishing vessels from 12 States on the HSVAR database.[98]

In addition to the information provided under Article VI of the FAO Compliance Agreement, the IPOA-IUU also identified the following information that may be included in a record of fishing vessels: the previous names of the fishing vessel, if any and if known; name, address and nationality of the natural or legal person in whose name the vessel is registered; name, street address, mailing address and nationality of the natural or legal persons responsible for managing the operations of the vessel; name, street address, mailing address and nationality of natural or legal persons with beneficial ownership of the vessel; name and ownership history of the vessel, and, where this is known, the history of non-compliance by that vessel, in accordance with national laws, with conservation and management measures or provisions adopted at a national, regional or global level; and vessel dimensions, and where appropriate, a photograph, taken at the time of registration or at the conclusion of any more recent structural alterations, showing a side profile view of the vessel.[99]

This information not only pertains to the basic characteristics of fishing vessels but also determines the history of registration and beneficial ownership of vessels, as well as the history of non-compliance with conservation and management measures adopted at the national, regional, or global levels. The IPOA-IUU also provides that flag States may require the inclusion of this information in their record of fishing vessels on fishing vessels that are not authorised to fish on the

[95] FAO Website, *High Seas Vessels Authorization Record*, www.fao.org. Accessed on 13 May 2008.

[96] *FAO Compliance Agreement*, Art. I(a).

[97] FAO Website, *High Seas Vessels Authorization Record, supra* note 95.

[98] Parties to the FAO Compliance Agreement which have provided information on their fishing vessels to be entered in HSVAR are Belize, Benin, Canada, Ghana, Japan, Madagascar, Morocco, Namibia, New Zealand, Norway, Syrian Arab Republic, U.S., and the EU.

[99] *IPOA-IUU*, para. 42.

high seas.[100] Such information will enable States to identify fishing vessels which have engaged in IUU fishing.

The international obligation to establish and maintain a national record of fishing vessels is intended to enable a flag State to effectively exercise control over its vessels. Chapter 9 analyses the measures adopted by RFMOs that require flag States to submit records of vessels authorised to fish in the areas of competence of RFMOs. Together, national and RFMO records of fishing vessels are effective means not only to establish the reporting obligations of fishing vessels but also to determine the history of their flags and ownership. However, for the purpose of ascertaining that fishing vessels are not engaged in IUU fishing, the information provided in the records of fishing vessels may not suffice because such records do not necessarily include details on the actual operations of a fishing vessel, for example data on catches, transshipment, landings, or violations of national and regional fisheries regulations. Therefore, there is need for the record of fishing vessels to be integrated with relevant databases on quota allocations, catch and landing statistics, post-harvest activities, refuelling and resupply activities, number and type of violations, and other fisheries related information in order to allow States to establish whether a fishing vessel has engaged in or supported IUU fishing. Records of fishing vessels would also need to be linked to registers of fishing vessels.

Similar challenges exist with respect to the maintenance of an international record of fishing vessels such as HSVAR. Apart from the lack of apparent connection between the general details of fishing vessels and their fishing activities, there are other issues related to the maintenance of HSVAR. HSVAR is limited to the scope of application of the FAO Compliance Agreement; hence, the database only includes vessels of 24 metres in length and over which are authorised to fish on the high seas. HSVAR is known to contain poor quality and inadequate quantity of data.[101] For example, there is inadequate information on the beneficial owners of vessels because such information is not frequently provided to FAO by flag States.[102] This is linked with the third problem with the implementation of HSVAR which is that not all States provide their vessel information to be included in the database.

In order to address these issues, FAO convened an Expert Consultation on the Development of a Comprehensive Global Record of Fishing Vessels, Refrigerated Vessels, and Fishing Support Vessels in February 2008 to determine the feasibility of a global record and how such an initiative can be

---

[100] *IPOA-IUU*, para. 43.

[101] Gail Lugten, 'Current Legal Developments: Food and Agriculture Organization, The FAO Global Record of Fishing Vessels, Refrigerated Vessels and Fishing Support Vessels', 23 *International Journal of Marine and Coastal Law* (2008), at 763.

[102] *Ibid.*

progressed.[103] The Expert Consultation considered four options. The first option is to rename HSVAR the FAO Global Record. The second option is to develop or extend HSVAR into a new and extensive vessel database. This could be done by combining the existing HSVAR data with data from Lloyds Register Fairplay, RFMOs registries and State registries, to produce a comprehensive Global Record of vessels. The third option is to amend the HSVAR-related provisions of the FAO Compliance Agreement and the contents of the database in order to establish the Global Record. The new regime is considered to correct the shortcomings in the existing regime. The fourth option is to attach the Global Record to a new legally binding instrument, such as the FAO Port State Measures Agreement. The Expert Consultation recommended that the FAO Committee on Fisheries (COFI) consider progressing the Global Record through both mandatory and market driven approaches such as the use of international binding or non-binding instruments and any other means such as catch certification, trade documentation and port State measures.[104]

Since the FAO Global Record aims to include all types and sizes of vessels, including refrigerated transport vessels and supply vessels, the Expert Consultation proposed a phased approach starting with vessels of 100GT and above, followed by vessels of 55GT and above but less than 100GT, and then finally by vessels of 10GT and above but less than 55GT.[105] It was further agreed that the Global Record would need to use a system of unique vessel identifiers which would not change, even if a vessel changed flag, owner or name.[106]

A number of legal and operational issues must be resolved in establishing a Global Record of Fishing Vessels. There is a need to determine the legal basis for the submission of fishing vessel information to the global record. The operational requirements for the creation and maintenance of a global record and its technical and financial implications for developing States would need to be considered. The proposed extensive coverage of the global record may impose a burden on some States to collect data from a large number of fishing vessels. Another issue that needs to be considered is the determination of the main purpose of the Global Record of Fishing Vessels and how States may use the information included in the record. It is essential to determine if the global record will only be a database of vessel information or if it will provide any legal basis for determining whether a fishing vessel has engaged in IUU fishing. Although a Global Record of Fishing Vessels will provide a complete picture of the fishing capacity in the world, this

---

[103] *See* FAO, Report of the Expert Consultation on the Development of a Comprehensive Global Record of Fishing Vessels, Rome, 25–28 February 2008, *FAO Fisheries Report No. 865* (Rome: FAO, 2008).

[104] *Ibid.*, at 9.

[105] *Ibid.*, at 4.

[106] *Ibid.*

initiative is quite ambitious and may require considerable amount of effort and resources to establish and maintain. Without clear and specific objectives and proper implementation, the Global Record of Fishing Vessels may not be an effective global tool to combat IUU fishing.

5.5. Flag State Enforcement

Article 94(5) of the LOSC provides that in exercising effective jurisdiction and control over its vessels, the flag State is required to take any steps which may be necessary to secure observance of generally accepted international regulations, procedures and practices. Article 217 of the LOSC specifically provides the enforcement powers of a flag State with respect to the control, reduction, and prevention of marine pollution. Article 217(1) provides the right and obligation of flag States to provide effective enforcement of relevant rules, standards, laws and regulations, irrespective of where a violation occurs. A flag State is required to investigate any matter, cooperate with other States, institute proceedings, and take other actions with respect to any alleged violation involving any maritime casualty, incident of navigation, serious injury to nationals of another State, or damage to ships or the marine environment.[107] A flag State is also required to promptly inform other relevant States and international organisations of the outcome of any investigation or action taken against the vessel.[108]

The UN Fish Stocks Agreement lays down the obligations of a State to ensure that fishing vessels flying its flag comply with subregional and regional conservation and management measures for straddling and highly migratory fish stocks. Article 19 of the UN Fish Stocks Agreement provides that the duties of a flag State are to:

- enforce measures irrespective of where the violations occur;
- investigate immediately and fully any alleged violation of subregional or regional conservation and management measures, which may include the physical inspection of the vessels concerned, and report promptly to the State alleging the violation on the progress and outcome of the investigation;
- require any vessel flying its flag to give information to the investigating authority regarding vessel position, catches, fishing gear, fishing operations and related activities in the area of the alleged violation;
- if satisfied that sufficient evidence is available in respect of an alleged violation, refer the case to its authorities with a view to instituting proceedings without delay in accordance with its laws, and detain the vessel if appropriate;

---

[107] *LOSC*, Arts. 94(6), 94(7), 217(4), 217(5), and 217(6).
[108] *LOSC*, Art. 217(7).

- ensure that the vessel does not engage in fishing operations on the high seas until such time as all outstanding sanctions imposed by the flag State in respect of the violation has been complied with; and
- carry out investigations and judicial proceedings expeditiously.

The UN Fish Stocks Agreement also provides the obligation of a flag State to cooperate with other States in enforcing subregional and regional conservation and management measures. A flag State conducting an investigation of an alleged violation of conservation and management measures for straddling and highly migratory fish stocks may request the assistance of any other State whose cooperation may be useful in the conduct of that investigation.[109] A flag State has a further duty to cooperate with the coastal State where one of its fishing vessels has conducted unauthorised fishing to take appropriate enforcement action.[110] In cases where a flag State is a Party to a subregional or regional fisheries management agreement or arrangement, it may take action in accordance with established procedures under that agreement or arrangement.[111]

In general therefore, a flag State is required to take enforcement measures with respect to its vessels which act in contravention of internationally agreed conservation measures, conservation and management measures adopted by RFMOs, and bilateral agreements or arrangements. Flag State enforcement action is necessary to combat IUU fishing. The effectiveness of flag State enforcement action, however, will depend on a sound framework for fisheries sanctions, which include making the contravention of international conservation measures an offence under national legislation.[112] Similarly, non-compliance with measures adopted under bilateral, subregional, or regional agreements or arrangements would need to be considered an offence under national laws.

The UN Fish Stocks Agreement and the FAO Compliance Agreement provide that sanctions for violating high seas fisheries conservation and management measures may involve the withdrawal or suspension of the authorisation to fish of a vessel by the flag State.[113] The FAO Compliance Agreement provides that if a fishing vessel, previously registered in another State, has undermined the effectiveness of international conservation and management measures, the State Party to the Agreement may refuse to issue an authorisation to fish.[114] A fishing licence may also be withdrawn, suspended, or cancelled if a fishing vessel has contravened the provisions of the FAO Compliance Agreement.[115] Under the UN Fish

---

[109] *UN Fish Stocks Agreement*, Art. 20(2).
[110] *UN Fish Stocks Agreement*, Art. 20(6).
[111] *UN Fish Stocks Agreement*, Art. 20(7).
[112] *FAO Compliance Agreement*, Art. III(8); *FAO Code of Conduct*, para. 8.2.7.
[113] *UN Fish Stocks Agreement*, Art. 19(2); *FAO Compliance Agreement*, Art. III(8); *FAO Code of Conduct*, para. 7.7.2.
[114] *FAO Compliance Agreement*, Art. III(5)(b).
[115] *FAO Compliance Agreement*, Art. III(8).

Stocks Agreement the authorisation to fish of a vessel may be cancelled if the master of the vessel refuses, upon the direction of the flag State, to submit to high seas boarding and inspection procedures.[116] A vessel which has engaged in a serious fisheries violation may not engage in fishing on the high seas until all outstanding sanctions imposed by the flag State have been met.[117]

With respect to areas of the high seas managed by RFMOs, members of regional fisheries and economic organisations have the primary responsibility as flag States to take enforcement actions against their vessels which are believed to have carried out IUU fishing activities. In addition to the power of the flag States to institute sanctions against vessels flying their flags, RFMOs can also adopt trade related sanctions such as the prohibition of imports or landings to effectively address IUU fishing activities. These measures are further discussed in Chapter 9.

116 *UN Fish Stocks Agreement*, Art. 22(4).
117 *UN Fish Stocks Agreement*, Art. 19(1)(e).

Chapter Six

Coastal State Measures

6.1 General Framework for Coastal State Measures

The declaration of fisheries zones and exclusive economic zones (EEZ) by coastal States in the 1970s and 1980s resulted in the vast areas of the most lucrative fishing grounds coming under coastal State jurisdiction. The subsequent adoption of the LOSC displayed the old freedom of fishing for most parts of the oceans and provided the legal basis for coastal States to manage, control, regulate and obtain economic rent from distant water fishing nations seeking access to their EEZs.

The establishment of fisheries zones and EEZs created favourable conditions for IUU fishing in a number of ways. The sheer size of these zones placed enormous burden on coastal States in terms of effective management and enforcement of jurisdiction, thus providing fertile ground for IUU fishing. A classic illustration of this problem is in the South Pacific islands region where EEZ claims have resulted in large areas of ocean space coming under the fisheries jurisdiction of Pacific Island States. For example, Kiribati has an EEZ of 3.1 million square km and a land area of only 690 square km, resulting in a land to sea ratio of 1:6000.[1] Additionally, problems such as marginal high seas fishing, underreporting and misreporting of fish catches, the use of efficient and destructive fishing technology and the widespread reflagging of fishing vessels, result in the increase in IUU fishing activities in the waters of Pacific Island States.

Recognising the constraints facing many coastal States in policing their EEZs, the IPOA-IUU identifies a number of tools that coastal States can use to prevent, deter and eliminate IUU fishing. Paragraph 51 of the IPOA-IUU provides that "(i)n the exercise of the sovereign rights of coastal States for exploring and exploiting, conserving and managing the living marine resources under their jurisdiction, in conformity with the LOSC and international law, each coastal State should implement measures to prevent, deter and eliminate IUU fishing in the exclusive economic zone."[2] The IPOA-IUU then goes on to enumerate eight specific measures, including effective monitoring, control and surveillance

---

[1] *See* N. Douglas (ed), *Pacific Islands Yearbook*, 17th Edition, (Suva: Fiji Times Ltd, 1994).

[2] FAO, *International Plan of Action to Prevent, Deter, and Eliminate Illegal, Unreported, and Unregulated Fishing*, adopted on 23 June 2001 at the 120th Session of the FAO Council. *Hereinafter referred to as* IPOA-IUU, para. 51.

(MCS), cooperation and exchange of information with other States, issuance of valid authorisations to fish, maintenance of records of fishing vessels, maintenance of logbooks, control of at-sea transhipment, regulation of fishing access and avoiding granting licences to vessels with known history of IUU fishing.[3] It is worth noting that the various coastal State measures identified in paragraph 51 of the IPOA-IUU are not new, but simply reiterate the existing powers available to coastal States under the LOSC.

In practical terms, many of the coastal State measures outlined in IPOA-IUU overlap with the actions that may be taken by all States, flag States and port States to combat IUU fishing. For example, in cases where IUU fishing activities are conducted by vessels flying the flag of the coastal State itself, the coastal State is also the flag State. As such, the coastal State would also need to discharge its duties as a flag State with respect to its vessels operating in the EEZs of other States and on the high seas. The coastal State will also be acting under the umbrella of "All States" in implementing some of the measures identified by the IPOA-IUU in respect of All State Responsibilities discussed in Chapter 4. In addition, the coastal State will be acting as a port State if it takes action to prohibit IUU fishing vessels from entering into its ports. Because of these overlaps, it is important for coastal States to develop comprehensive measures across all areas of action identified by the IPOA-IUU to effectively combat IUU fishing. This chapter focuses on a range of measures which are particularly unique to coastal States.

6.2 Coastal State Fisheries Jurisdictional Framework

The LOSC provides the primary legal framework for coastal State measures to combat IUU fishing. As discussed in Chapter 3, the LOSC recognises two broad maritime zones that come under the jurisdiction of coastal States. These are zones under the sovereignty of coastal States (comprising the internal waters, archipelagic waters and territorial seas); and zones under the sovereign rights of coastal States (encompassing the EEZ and the continental shelf). For fisheries purposes, it is important to note that the contiguous zone is part of the EEZ.

In terms of the sanctions imposed by the coastal State on IUU fishers, coastal State measures in maritime zones under sovereignty would need to be distinguished from coastal State measures in maritime zones under sovereign rights. This is precisely because the LOSC only constraints the exercise of fisheries powers by coastal States in maritime zones under sovereign rights, leaving the coastal State with more flexibility in its maritime zones under sovereignty.

---

[3] *IPOA-IUU*, para. 51.

6.2.1 *Coastal State Measures to Combat IUU Fishing in Maritime Zones under Sovereignty*

The internal waters, archipelagic waters and territorial sea fall under the full sovereignty of coastal States. Sovereignty under international law is absolute and is limited only by the expressed international obligations of States.[4] The LOSC does not contain specific provisions on the management of fisheries resources in these zones.[5]

Consistent with coastal State sovereignty, the navigational freedoms permitted by the LOSC also place some limitations on foreign fishing vessels traversing maritime zones under the sovereignty of coastal States. For example the laws and regulations that the coastal State is permitted to implement relating to innocent passage of foreign vessels through its territorial sea include the conservation of the living resources of the sea.[6] It is a violation of the right of innocent passage for a foreign vessel to engage in any fishing activities while passing through the territorial sea of the coastal State.[7] Similarly, in straits used for international navigation, most of which fall within territorial seas, the coastal State has the power to regulate fishing vessels, including the stowage of fishing gear.[8] Some States have specifically adopted provisions in their national legislation on maritime zones in this regard.[9]

For various policy reasons, including food security and protection of artisanal fisheries, many coastal States prohibit the licensing of foreign fishing vessels in their territorial seas and archipelagic waters.[10] Unlike the EEZ, the LOSC also does not constrain the enforcement powers of the coastal State for fisheries violations by foreign fishing vessels in maritime zones under its sovereignty. The coastal State measures proposed by the IPOA-IUU to combat IUU fishing only apply in the EEZ.[11] This is more a recognition of the sovereignty of coastal States

[4] The extent of sovereignty is provided in the LOSC as follows: "(T)he sovereignty of a coastal State extends, beyond its land territory and internal waters, and in the case of an archipelagic State, its archipelagic waters, to an adjacent belt of seas described as the territorial sea." *See* United Nations Convention on the Law of the Sea, Montego Bay, Jamaica, concluded on 10 December 1982, in force 16 November 1994, 1833 UNTS 3; 21 ILM 1261 (1982). *Hereinafter referred to as* LOSC. Art. 87(1). Art. 2(1).

[5] *LOSC*, Arts. 2 and 49(1).

[6] *LOSC*, Art. 21(1)(d).

[7] *LOSC*, Art. 19(2)(i).

[8] *LOSC*, Art. 42(1)(c).

[9] See for example Bahamas, *Chapter 282 on Archipelagic Waters and Maritime Jurisdiction, Act No. 37 of 1993*, Art. 5.

[10] See for example Republic of the Philippines, Republic Act (RA) 8550, *An Act Providing for the Development, Management and Conservation of the Fisheries and Aquatic Resources, Integrating all Laws Pertinent thereto, and for other Purposes, also known as the Philippine Fisheries Code of 1998*, 25 February 1998, Sec. 2(d); Tokelau, *Territorial Sea and Exclusive Economic Zone Act 1977*, Art. 6.

[11] *IPOA-IUU*, para. 51.

in the maritime zones under their sovereignty to take any actions they deem fit, and not a suggestion that IUU fishing does not take place in the internal waters, archipelagic waters and territorial sea.

Some coastal States draw a distinction between penalties for fisheries violations in maritime zones under their sovereignty and violations in the EEZ. For example, under the Australian *Fisheries Management Act 1991* (Cth), the penalty for using a foreign fishing vessel in the territorial sea within the Australian Fisheries Zone[12] include both fine and imprisonment, while only fines apply for a similar offence in the Australian Fisheries Zone outside the territorial sea.[13] Strict liability also applies to fisheries offences in the territorial sea.[14] Similarly, under the *Territorial Waters Act* of Grenada, any foreign vessel found guilty of fishing in the territorial sea is liable on conviction to a fine not exceeding $10,000 or imprisonment for a term not exceeding five years or both; or on summary conviction to a fine not exceeding $5,000 or imprisonment for a term not exceeding 2 years of both.[15]

Other States, however, do not distinguish between zones under sovereignty and EEZs in applying sanctions. For example, in Indonesia, any person who intentionally conducts prohibited fishing activities within the fisheries management areas (which includes archipelagic waters, the territorial sea and the EEZ) is liable to a term of imprisonment not exceeding ten years or a fine of up to two billion rupiahs.[16] The lack of distinction between penalties for fisheries violations in zones under sovereignty and zones under sovereign rights is the legacy of the 1970s practice of fisheries zone declarations extending from the baseline to 200 nautical miles. The key issue for coastal States in this category is the danger of contravening the fisheries enforcement limitations under Article 73 of the LOSC in respect of violations by foreign fishing vessels in the EEZ. The enforcement power of coastal States and its limitations are discussed in the succeeding sections.

6.2.2 *Coastal State Measures to Combat IUU Fishing in Maritime Zones under Sovereign Rights*

The two maritime zones of jurisdiction subject to the sovereign rights of the coastal State are the EEZ and the continental shelf. The jurisdictional competence

[12] The Australian fishing zone is defined as (a) the waters adjacent to Australia within the outer limits of the exclusive economic zone adjacent to the coast of Australia; and (b) the waters adjacent to each external territory within the outer limits of the EEZ adjacent to the coast of the external Territory; but does not include coastal waters of, or waters within the limits of, a State or internal Territory; or waters that are excepted waters. Australia, *Fisheries Management Act 1991* (Cth), Sec. 4.

[13] Australia, *Fisheries Management Act 1991* (Cth), Secs. 100A and 100B.

[14] Australia, *Fisheries Management Act 1991* (Cth), Sec. 100B(1A).

[15] Grenada, *Territorial Waters Act, No. 17 of 1978*, Sec. 12(2).

[16] See Indonesia, *Law of the Republic of Indonesia No. 31 of 2004 Concerning Fisheries*, Chapter XV.

of the coastal State in the EEZ and on the continental shelf is characterised by the LOSC in terms of "sovereign rights" as opposed to "sovereignty". The concept of sovereign rights means that the coastal State only has a right of jurisdiction that is related to specific purposes identified in the LOSC.[17] Beyond the specified purposes, the coastal State has no jurisdiction over foreign vessels and their activities. The sovereign rights of coastal States extend to the exploration, exploitation, conservation, and management of natural resources of the seabed, subsoil and superjacent waters.[18]

The coastal State's sovereign rights on the continental shelf apply largely to non-living resources, but also to sedentary species, defined as "organisms which, at the harvestable stage, either are immobile on or under the seabed or are unable to move except in constant physical contact with the seabed or the subsoil."[19] Globally, IUU fishing on the continental shelf is not yet a significant issue largely because of the physical constraints on fishing on the continental shelf.[20] In practice, coastal States exercise continental shelf fisheries enforcement powers through EEZ fisheries powers.

Articles 62(4) and 73 of the LOSC provide the scope and content of coastal State measures to combat IUU fishing in the EEZ which are enumerated in paragraph 51 of the IPOA-IUU. Article 62(4) which deals largely with licensing of foreign fishing vessels requires nationals of other States fishing in the EEZ to comply with the conservation measures and with the other terms and conditions established in the laws and regulations of the coastal State. Article 73(1) of the LOSC reinforces Article 62(4) by providing the scope of the enforcement powers of the coastal State in respect of both licensed and unlicensed fishing vessels. The enforcement powers of the coastal State under Article 73(1) include boarding,

---

[17] C.A. Fleischer, 'The Exclusive Economic Zone under the Convention Regime and in State Practice', 17 *Law of the Sea Institute Proceedings* (1984), at 253. See also the International Law Commission's discussion in the Report of the International Law Commission covering the work of its Eighth Session, 23 April–4 July 1956, *UN General Assembly Official Records*, 11th Session, supp. No. 9 (A/3159), at 42–43.

[18] *LOSC*, Art. 56.

[19] *LOSC*, Art 77(4).

[20] However, in the case of Australia and Indonesia, there had been problems relating to illegal fishing in the delimited areas of the EEZ and continental shelf around the Arafura and Timor Seas. The maritime boundary agreements between the two States created an area where the Australian continental shelf extends under the Indonesian water column. Among other factors, this has resulted in a difficulty to enforce the sovereign rights of each State in their respective maritime zones, according to the agreed treaties, without affecting the rights of the other. This issue is being resolved through continuing bilateral cooperation between the two States. For the relevant treaties, *see* Agreement between the Government of the Commonwealth of Australia and the Government of the Republic of Indonesia establishing Certain Seabed Boundaries, concluded on Canberra, Australia, concluded on 18 May 1971, in force 8 November 1973, ATS 1973 No. 31; Agreement between the Government of the Commonwealth of Australia and the Government of the Republic of Indonesia establishing certain seabed boundaries in the area of the Timor and Arafura Seas, Supplementary to the Agreement of 18 May 1971, Jakarta, Indonesia, concluded on 9 October 1972, in force 8 November 1973, ATS 1973 No. 32.

inspection, arrest and judicial proceedings, as may be necessary to ensure compliance with the laws and regulations adopted by the coastal State.

Article 62(4) of the LOSC is primarily directed at licensed fishing vessels while Article 73(1) addresses both licensed and unlicensed fishing vessels. In theory, the two Articles combined provide the framework for coastal States to combat IUU fishing. However, the critical challenge facing many coastal States is their capacity to develop and effectively implement the necessary tools to give effect to these provisions of the LOSC.

6.3 Monitoring, Control and Surveillance

The implementation of measures under Articles 62(4) and 73(1) of the LOSC to combat IUU fishing requires effective MCS measures, which have been identified as the "best hope for preventing, deterring, and eliminating IUU fishing."[21] Accordingly, the IPOA-IUU calls on States to apply specific MCS related measures from the commencement of the fishing activity to the final destination of caught fish.[22] These measures include the issuance of fishing licences; maintenance of a record of fishing vessels; implementation of vessel monitoring system (VMS); observer programmes; boarding and inspection measures; acquisition, storage, and dissemination of MCS data; training and education of all persons involved in MCS operations; and promoting understanding of MCS in national judicial systems.[23]

Following the definitions outlined by FAO, MCS is generally defined as having three distinct, but interrelated components:

> Monitoring – the continuous requirement for the measurement of fishing effort characteristics and resource yield;
>
> Control – the regulatory conditions under which the exploitation of the resources may be conducted; and
>
> Surveillance – the degree and types of observations required to maintain compliance with the regulatory controls imposed on fishing activities.[24]

The Southern African Development Community (SADC) has adopted a similar definition of MCS. According to the SADC Protocol on Fisheries, monitoring is

---

21 *FAO Technical Guidelines No. 9, supra* note 2, para. 3.2.5.

22 *IPOA-IUU*, para. 24.

23 *IPOA-IUU*, para. 24; *See also* Agreement for the Implementation of the Provisions of the United Nations Convention on the Law of the Sea of 10 December 1982 relating to the Conservation and Management of Straddling Fish Stocks and Highly Migratory Fish Stocks, concluded on 4 September 1995, in force 11 December 2001, 34 ILM 1542 (1995); 2167 UNTS 88. *Hereinafter referred to as* UN Fish Stocks Agreement, 18(3)(g).

24 *See* Peter Flewwelling, An Introduction to Monitoring, Control and Surveillance Systems for Capture Fisheries, *FAO Technical Paper No. 338* (Rome: FAO, 2004), at 3.

not only limited to the requirement to measure fishing effort but also involves the collection and analysis of fisheries and all fisheries-related activities.[25] SADC defines control to include the enforcement of legal and administrative measures,[26] and considers surveillance as not limited to "observations" but also the supervision of fishing activities to ensure compliance with control measures.[27]

Traditionally the term MCS has been associated with deterrence, principally law enforcement, through boarding and inspections at sea. Currently however, MCS is viewed differently and has two aspects, namely, preventative MCS and deterrent MCS.[28] Both aspects of MCS provide information to support the design and implement fisheries management plans at national and regional levels, as well as enforce such plans.

There are three spatial components of MCS: land, sea, and air.[29] The land component of MCS serves as the base of operations and involves the coordination of MCS activities and deployment of available resources. It also includes port inspections and monitoring of transshipment in port, as well as the trade in fish. The sea component of MCS includes the surveillance of maritime zones through the use of several tools such as vessel platforms. The air component includes air surveillance, which is usually the first level of response against IUU fishing activities. It also includes the use of satellite based technology to identify possible fisheries violations.[30] States use different approaches to implementing these spatial components of MCS, depending on their financial capability, fisheries management risks and priorities and organisational structure. MCS tools at the national level also include appropriate participatory management plans developed with stakeholder input, enforceable legislation and control mechanisms such as licensing, data collection systems such as dockside monitoring, deployment of observers, sea and port inspections, supporting communication systems, and linked land based monitoring.[31]

In summary, an MCS system is the totality of coastal State measures designed to compel obedience to fisheries laws and regulations. It requires and has increasingly come to involve an elaborate network of institutions, rules, procedures and specialised personnel. While some aspects of an MCS system are located

---

25 Southern African Development Community (SADC), *Protocol on Fisheries*, 26 August 2002, Art. 1.

26 *SADC Protocol on Fisheries*, Art. 1.

27 *SADC Protocol on Fisheries*, Art. 1. In 1981, the Committee for the Eastern Central Atlantic Fisheries has adopted a definition of MCS similar to those of SADC and FAO.

28 *FAO Technical Paper No. 338*, *supra* note 24, at 13.

29 For complete discussion of the spatial components and tools of MCS, *see* Peter Flewwelling, Cormac Cullinan, David Balton, Raymond P. Sautter and J. Eric Reynolds, Guide to Monitoring, Control, and Surveillance Systems for Coastal and Offshore Capture Fisheries, *FAO Fisheries Technical Paper. No. 415* (Rome: FAO, 2003).

30 *FAO Fisheries Technical Paper No. 338*, *supra* note 24, at 13–14.

31 *Ibid.*

within the fisheries administration, other aspects function outside the scope of fisheries. The large number of agencies required to ensure an effective MCS system creates the potential for administrative complexity which requires coordination and careful management. An effective MCS response against IUU fishing requires a co-ordinated approach at the national level in which the fisheries MCS task becomes part of a total system concerned with maritime MCS as a whole. This approach is particularly effective where IUU fishing is intertwined with other illegal activities, such as drug trafficking and armed robbery. The inter-agency mechanism adopted by States must not only maximise the efficiency of operational MCS but would also need to be able to strategically respond to the changing nature of maritime enforcement threats and issues. Additionally, establishing an effective coordinating mechanism for all national maritime agencies can reduce duplication of effort and jurisdictional conflicts, as well as facilitate exchange of information required for MCS operations.

The sections below examine some specific examples of MCS tools that coastal States can employ to combat IUU fishing in their maritime zones of jurisdiction.

6.3.1 *Licensing of Fishing Vessels in the EEZ*

At the heart of coastal State MCS measures to combat IUU fishing is effective and transparent licensing system for fishing vessels. The significance of the role of the licensing system in terms of combating IUU fishing is underscored by the fact that six out of the eight examples of coastal State measures enumerated in paragraph 51 of the IPOA-IUU relate to licensing. These include: authorisation to fish; establishment of records of fishing vessels; maintenance of logbooks for licensed vessels; authorisation and control of transshipment activities; regulating of fishing access generally to the EEZ; and denying licences to vessels with known history of IUU fishing.

The LOSC provides sufficient legal basis for the licensing authority of the coastal State. The LOSC requires the coastal State to determine its capacity to harvest the living resources in the EEZ, and where it does not have the capacity to harvest the total allowable catch, allow other States access to the surplus resources through agreements.[32] It is therefore the sovereign right of the coastal State to prescribe laws and regulations pertaining to access to fisheries resources in its EEZ by foreign fishing vessels.[33]

Among the terms and conditions the coastal State can impose on foreign fishing vessels under Article 62(4) of the LOSC include:

- licensing of fishermen, fishing vessels and equipment, including payment of fees and other forms of remuneration, which, in the case of developing coastal

[32] *LOSC*, Art. 62(2).
[33] *LOSC*, Art. 62(4).

States, may consist of adequate compensation in the field of financing, equipment and technology relating to the fishing industry;

- determining the species which may be caught, and fixing quotas of catch, whether in relation to particular stocks or groups of stocks or catch per vessel over a period of time or to the catch by nationals of any State during a specified period;
- regulating seasons and areas of fishing, the types, sizes and amount of gear, and the types, sizes and number of fishing vessels that may be used;
- fixing the age and size of fish and other species that may be caught;
- specifying information required of fishing vessels, including catch and effort statistics and vessel position reports;
- requiring, under the authorisation and control of the coastal State, the conduct of specified fisheries research programmes and regulating the conduct of such research, including the sampling of catches, disposition of samples and reporting of associated scientific data;
- the placing of observers or trainees on board such vessels by the coastal State;
- the landing of all or any part of the catch by such vessels in the ports of the coastal State;
- terms and conditions relating to joint ventures or other cooperative arrangements;
- requirements for the training of personnel and the transfer of fisheries technology, including enhancement of the coastal State's capability of undertaking fisheries research; and
- enforcement procedures.

The regulatory competence of the coastal State under Article 62(4), which is not exhaustive, can be grouped into three categories.[34] The first category includes provisions which are directed at regulating fishing effort and the areas of fishing. The second category covers the collection and verification of data, while the third category involves enforcement mechanisms.[35] Although the licensing requirements under Article 62(4) are directed at foreign fishing vessels, the point needs to be made that most of the requirements listed under the Article are equally applicable to domestic fishing vessels. Implementation of these terms and conditions across the entire fisheries will ensure that the conservation and management objectives of the coastal State are met to help prevent, deter, and eliminate IUU fishing in the EEZ.

In a study conducted by the World Wildlife Fund for Nature (WWF), several elements have been identified as essential in any access agreement in order to

---

[34] Stuart M. Kaye, *International Fisheries Management* (Hague: Kluwer Law International, 2001), at 108–109.

[35] *Ibid.*

prevent the depletion of fisheries resources in the EEZs of coastal States.[36] These elements include access authorisation, catch and effort restrictions, closed areas and seasonal restrictions, requirement to comply with laws of the coastal State, flag State responsibilities, licensing and fees, vessel reporting requirements, observers, transshipment, discards, waste and non-target species, other MCS measures, including enforcement and penalties, port access, and other restrictions on fishing operations. A coastal State may also include general provisions in an access agreement, such as cooperation on scientific research, dispute resolution, transparency and confidentiality, as well as review, entry into force, and termination of the agreement.[37] These elements represent a comprehensive list of requirements that are imposed by coastal States around the world on foreign fishing vessels and fishers accessing their EEZs.

State practice in the Pacific Island region provide the international benchmark of the range of terms and conditions that coastal States may impose on foreign fishing access to their EEZs. The Pacific Islands Forum Fisheries Agency (FFA) Harmonised Minimum Terms and Conditions for Foreign Fishing Vessel Access require, as a condition of licence, that all foreign fishing vessels comply with a range of measures, including conditions relating to: the maintenance and submission of catch logs in EEZs and high seas; vessel reporting; port to port VMS reporting; carriage of observers; appointment of agents; port State requirements; and acceptance of flag State or fishermen's associations responsibilities.[38] In 2008, the Pacific Island States which are Parties to the Nauru Agreement (PNA) agreed to impose additional conditions relating to full catch retention, prohibition on the use of fish aggregating devices during specified periods and, more significantly, prohibition on fishing in certain high seas pockets by foreign fishing vessels licensed to fish in the waters of the PNA countries.[39]

---

[36] *See* Will Martin, Michael Lodge, John Caddy, and Kwame Mfodwo, *A Handbook for Negotiating Fishing Access Agreements* (Washington: WWF, 2001). This study presented model fisheries access agreements between a fishing State and a coastal State, a coastal State and national fisheries association of the accessing State, and a coastal State and fishing company.

[37] *See ibid.*, at Section 3. Similar initiative has been previously undertaken by the Asian-African Legal Consultative Committee (AALC). The AALC drafted guidelines for model legislation and model bilateral agreements allowing foreign fishing in the EEZ, with provisions for assistance to be rendered to coastal States to develop their national fishing industries. *See* International Law Commission, 1606th Meeting, Co-operation with other bodies, Item 10 of the agenda, A/CN.4/SR.1606, 1 *Yearbook of the International Law Commission* (1980), at para. 6.

[38] Forum Fisheries Agency (FFA), *The Harmonised Minimum Terms and Conditions for Foreign Fishing Vessel Access*, As Amended by FFC67, May 2008. www.ffa.int. Accessed on 13 August 2009.

[39] The Third Arrangement Implementing the Nauru Agreement Setting Forth Additional Terms and Conditions of Access to the Fisheries Zones of the Parties, Koror, Republic of Palau, concluded on 16 May 2008. The members of the Parties to the Nauru Agreement (PNA) are: Federated States of Micronesia, Kiribati, Marshall Islands, Nauru, Palau, Papua New Guinea, Solomon Islands and Tuvalu.

As part of their licensing systems, coastal States may also establish records of foreign fishing vessels which are granted access in their EEZ. Vessels with known history of violations, may be excluded from the register, and are consequently denied licences. A good illustration is the FFA Regional Register of Foreign Fishing Vessels under which all foreign fishing vessels wishing to obtain a national fishing licence from any FFA Member State must apply for good standing on the FFA vessel register.[40] The acquisition of a good standing status is based on data obtained from fishing vessel registration and the VMS.[41] The FFA Vessel Register also creates an additional "layer" for verifying the validity of vessel information contained in national fishing vessel registers of FFA Members.

6.3.2 *Observer Programmes*

In order to ensure that fishing operations are documented and that fishing vessels comply with conservation and management measures, States are encouraged to establish observer programmes.[42] Similar to the implementation of VMS discussed in the next section, the LOSC and the UN Fish Stocks Agreement provide the legal basis for the implementation of observer programmes. Under the LOSC, States have the right to place observers on board vessels in exercising their sovereign rights over marine resources in their EEZs.[43] There is also a duty under the UN Fish Stocks Agreement to implement national observer programmes, participate in subregional or regional observer programmes, and permit observers of other States to carry out functions agreed under such programmes.[44] These international instruments, however, do not provide the specific functions of and the process involved in conducting observer programmes.

Observer programmes are usually implemented in order to generate data for fishery science and compliance purposes.[45] Implementation of an observer programme for fishery science involves the estimation of total catch and effort,

---

[40] FFA, *Guide to Application and Renewal of Good Standing on the FFA Vessel Register*, www.ffa.int, Accessed on 13 August 2009, at 2.

[41] The application form for the acquisition of a good standing status requires detailed information on the vessel, vessel's previous details vessel contact details and owners and/or charterers, vessel type and specifications, and mobile transceiver unit (MTU). Aside from filling out the application form, foreign fishing vessels are also required to submit supporting documents such as the vessel's flag State registration certificate, company ownership details and proof of corporate registration, photo of the vessel, vessel's schematic stowage plan, and MTU inspection report. *See* FFA, *Application for Registration*, 30 March 2009, www.ffa.int. Accessed on 13 August 2009.

[42] *IPOA-IUU*, paras. 24.4; 47.4; and 80.9; FAO, *Code of Conduct for Responsible Fisheries,* Adopted at the 28th Session of the FAO Conference, Rome, Italy, 31 October 1995. *Hereinafter referred to as* FAO Code of Conduct. Para. 8.4.3.

[43] *LOSC*, Art. 62(4)(g); See also *FAO Code of Conduct*, para. 7.7.3.

[44] *UN Fish Stocks Agreement*, Art. 18(3)(g)(ii).

[45] Sandy I. Davies; J. Eric Reynolds, (eds.), Guidelines for Developing an At-Sea Fishery Observer Programme, *FAO Fisheries Technical Paper No. 414* (Rome: FAO, 2002), at 5.

including by-catch and discards, and biological sampling of catches.[46] To ensure compliance with fisheries laws and regulations, observers may be given the power to validate logbooks and inspect other relevant documents, visit fishing vessels, and collect catch data.[47] The information obtained from both scientific and compliance observer programmes is necessary for effective fisheries management.[48] However, the implementation of observer programmes to collect information for compliance purposes may be considered more relevant in effectively addressing IUU fishing as such a programme directly assists in the monitoring of illegal and unreported fishing activities.

A number of States have implemented national observer programmes. Papua New Guinea has an observer unit within the MCS Group of the National Fisheries Authority, which is responsible for placing trained observers on board licensed fishing vessels.[49] The national observer programme of Papua New Guinea gathers information for both fishery science and compliance purposes.[50] In Canada, a corps of observers is deployed for about 22,000 days of at sea monitoring a year.[51] The objectives of the Canadian observer programme vary, although in general, it is to monitor compliance with regulations and gather data for stock assessment to provide input into opening and closing of fisheries.[52] The Namibian observer programme has similar objectives.[53]

The primary advantage of establishing an observer programme is that it collects data required for determining the status of living marine resources and the consequences of commercial fishing operations.[54] Two of the most developed national observer programs in terms of scope and length of implementation and use in fisheries management are those of the U.S. and Australia. The U.S. started using observers to collect fisheries data from 1972 for a range of conservation and management issues. Through the National Marine Fisheries Service, the U.S. deploys over 60,000 observers at sea annually to monitor 42 different fisheries and collect catch and bycatch data from U.S. commercial fishing and processing

---

[46] *Ibid.*

[47] Karl Laubstein, 'Fisheries Observers on Fishing Vessels', in CP Nolan, (ed.), *Proceedings of the International Conference on Integrated Fisheries Monitoring,* Sydney, Australia, 1–5 February 1999 (Rome: FAO, 1999), at 166–167.

[48] *FAO Fisheries Technical Paper No. 414*, *supra* note 45, at 5.

[49] Papua New Guinea (PNG), National Fisheries Authority Website, *Control Surveillance*. www .fisheries.gov.pg. Accessed on 17 August 2006.

[50] *Ibid.*

[51] William A. Karp and Howard McElderry, 'Catch Monitoring by Fisheries Observers in the United States and Canada', in CP Nolan, ed., *Proceedings of the International Conference on Integrated Fisheries Monitoring,* Sydney, Australia, 1–5 February 1999 (Rome: FAO, 1999), at 269.

[52] *Ibid.*

[53] *See* National Marine Fisheries Service, *Proceedings of the Third International Fisheries Observer Conference*, New Orleans, Louisiana, USA, 18–21 November 2002, US Department of Co mmerce NOAA Technical Memorandum NMFS-SPO-64 (Maryland: NMFS, 2004), at Appendix 2.

[54] Karp and McElderry, *supra* note 51, at 283.

vessels.[55] Australia started its observer programme in 1991 by employing observers to accompany foreign fishing vessels to monitor compliance with licensing and access arrangement, collect data on all species caught, gears, and other information. The number of observers deployed at sea has significantly increased to cover 18 fisheries in 2007. This increase was driven mainly by the ongoing need to enhance monitoring of fishing activities, as well as to address gaps highlighted in fisheries strategic assessments, by-catch action plans, and ecological risk assessments.[56]

As a component of MCS, an observer programme allows for the verification of reported fisheries data, such as information recorded by fishing vessels in their logbooks, which is an effective means to detect unreported fishing. Observer records and information may also be required as evidence in the prosecution of a violation by a vessel, owner or company.[57] Such evidence will be more admissible if the observer programme has developed standardised formats, methods and protocols for recording and handling compliance related issues.[58] For the purpose of using observer reports in establishing a fisheries violation, there is a need for the competence of observers to be established, particularly in the event of litigation, by means of standard training.[59] Aside from the proper execution of his or her rights and responsibilities, there is also a need for an observer to have a common understanding of the interpretation of the provisions of relevant legal instruments or agreements.[60]

6.3.3 *Vessel Monitoring Systems*

The IPOA-IUU encourages States to implement, where appropriate, a VMS, in accordance with relevant national, regional or international standards, including the requirement for vessels under their jurisdiction to carry VMS on board.[61] The implementation of a VMS has its legal basis in the LOSC and the UN Fish Stocks Agreement. The LOSC empowers coastal States to require vessels of other States fishing in their EEZs to provide certain information, such as vessel position reports.[62] Similarly, the UN Fish Stocks Agreement provides that one of the duties of flag States is the development and implementation of VMS in accordance with

---

[55] NOAA Fisheries, Office of Science and Technology, *National Observer Programme*, www.st.nmfs.gov. Accessed on 17 May 2008.

[56] Australian Fisheries Management Authority, *The Observer Program*, www.afma.gov.au. Accessed on 9 June 2008.

[57] *FAO Fisheries Technical Paper 414*, *supra* note 45, at 41.

[58] *Ibid.*

[59] Laubstein, *supra* note 47, at 165.

[60] *Ibid.*

[61] *IPOA-IUU*, para. 24.3.

[62] *LOSC*, Art. 62(4)(e).

regional, subregional or global programmes.[63] Therefore VMS can either be a compliance tool by the coastal State to monitor the activities of licensed fishing vessels in its maritime zone of jurisdiction, or it can be a tool by the flag State to monitor the activities of vessels flying its flag fishing on the high seas or in the jurisdiction of other States. The analysis here focuses on the use of the VMS as a compliance tool by the coastal State.

VMS responds to the international requirement of collecting and verifying fishing catch and effort, and other fishing activities for more effective fisheries management. There are different types of VMS. The more conventional type of VMS relies on vessel movement report through radio, aerial or surface surveillance, land based radar, sea based sonar, observer programmes or incidental reports by other fishing vessels or airplanes.[64] This type of VMS is used to monitor areas in the immediate vicinity and is therefore more local in coverage. The other type of VMS is satellite based. This type of VMS depends on satellite systems such as Inmarsat, Argos, and Euteltracs which may be used for fisheries monitoring and other purposes.

Satellite based VMS consists of several components, namely a transmitter or receiver, an automated reporting system, and a communications system. The transmitter or transreceiver is carried by each fishing vessel participating in the system and is capable of fixing a position and calculating the speed and course of the vessel. The Global Positioning System (GPS) receiver is used to transmit vessel data because of its high level of accuracy, availability, and relatively low equipment cost. The automated reporting system then controls the transmission of the position data and other data through a communications system. It is the communication system that conveys the data to a monitoring agency.[65]

An increasing number of States are now implementing satellite based VMS.[66] In general, VMS provides monitoring agencies with accurate locations, at periodic time intervals, of fishing vessels participating in the VMS. In the South Pacific, the FFA recognises the principal role of VMS as support to existing surveillance assets such as patrol vessels, surveillance flights, and observer programmes that currently operate in the Western and Central Pacific Ocean. One of the conditions for the operation of the FFA VMS is the installation, maintenance,

---

[63] *UN Fish Stocks Agreement*, Art. 18(3)(g)(iii). *See also* UN Fish Stocks Agreement, Art. 5(j) and 18(3)(e) and FAO Code of Conduct, Art. 7.7.3.

[64] FAO, Fishing Operations: Vessel Monitoring Systems', *FAO Technical Guidelines for Responsible Fisheries No. 1 Suppl. 1* (Rome: FAO, 1998), at 10.

[65] *Ibid*., at 8.

[66] Philip Marshall, 'Electronic Monitoring', in CP Nolan, (ed.), *Proceedings of the International Conference on Integrated Fisheries Monitoring, Sydney*, Australia, 01–05 February 1999 (Rome: FAO, 1999), at 304–306. *See also* FAO, Report of the Sub-regional Fisheries Commission Workshop on Vessel Monitoring Systems, Saly, Senegal, 14–17 October 2002, *FAO Fisheries Report No. 696* (Rome: FAO, 2002).

and operation of a registered mobile transceiver unit (MTU) on board foreign fishing vessels in accordance with the specifications and operating instructions of manufacturers and FFA standards.[67] Pacific Island States have also adopted and implemented VMS standards in their domestic legislation consistent with FFA regulations.[68]

There are a number of advantages in implementing a VMS. First, the use of VMS has revealed its utility in stock assessment, management and conservation.[69] Satellite imagery can give a good indication of vessel densities in specific regions over specific periods of time.[70] Second, VMS positions can provide timely access to the positions of all licensed vessels which gives the authorities a considerable advantage in detecting unlicensed vessels.[71] Third, the awareness by vessel operators that they are being tracked encourages compliance with fisheries regulations.[72]

In terms of IUU fishing, while VMS may detect illegal fishing activities of vessels in areas where fishing is prohibited, it does not necessarily provide information on other IUU fishing activities which are not directly associated with the location of a vessel in a fisheries management area, such as fishing beyond allowable quota or other catch allocation, use of an illegal fishing method, or restriction of fishing activities on the basis of nationality or licence status. It is therefore imperative that the implementation of VMS be supplemented by other traditional MCS tools such as fishing patrols, at sea boarding and port inspections and other technologies.

An effective legal framework is required to make the VMS an effective MCS tool. A number of legal considerations will need to be taken into consideration in implementing VMS. The legal issues will differ depending on whether a particular VMS is intended to regulate national fishing vessels exclusively, foreign fishing vessels, or meant to discharge flag State duties on the high seas. Legislation implementing VMS should be unambiguous, regardless of whether or not it is acceptable by those it seeks to regulate and conducive to effective and easy enforcement. Penalties should be adequate in severity to deter violations of VMS regulations. More specifically related to satellite-based VMS, the use of logbooks

---

[67] FFA, The Harmonised Minimum Terms and Conditions for Foreign Fishing Vessel Access, *supra* note 38, sec. 14.

[68] See for example, Fiji, *Marine Spaces (Foreign Fishing Vessels) Amendment Regulations,* 2002; Republic of Palau, *Vessel Monitoring Regulations*, 2000; Samoa, *Fisheries (Vessel Monitoring System) Regulations*, 1999.

[69] N. Kourti, I. Shepherd, H. Greidanus, *et al.,* 'Integrating Remote Sensing in Fisheries Control', 12 *Fisheries Management and Ecology* (2005), at 303.

[70] *Ibid.*

[71] Navigs s.a.r.l., 'Fishing Vessel Monitoring Systems: Past, Present and Future', *Prepared for the High Seas Task Force*, OECD, 17 October 2005. www.high-seas.org. Accessed on 12 February 2006, at 10.

[72] *Ibid.*

and automatic location communicators (ALCs) should be underpinned by a number of requirements aimed at avoiding obstruction of regulatory efforts.

Although the legal requirements in every State with regard to the VMS will differ, there are certain core issues which must be covered in the regulatory framework for any VMS. The key components of an effective VMS legislation include clear identification of types and length of vessels to have the VMS installed, specification of technical requirements such as hardware specification, communication specification, installation and service activation requirements and operational requirements, clear provisions on communication service providers, procedures governing the installation and operation of approved ALCs, provisions on confidentiality of VMS data and information, ownership of VMS data and information, evidentiary issues such as admissibility of VMS data in court proceedings, and penalties for breach of VMS requirements.

6.3.4 *Requiring Entry and Exit Reports in the EEZ*

Article 58(1) of the LOSC grants the freedom of navigation to foreign fishing vessels through EEZs of coastal States. The exercise of such freedom is subject to the sovereign rights of the coastal State in the EEZ. Given the large sizes of EEZs under the jurisdiction of coastal States, some of which are adjacent to areas of the high seas, and the limited surveillance and enforcement assets by many coastal States, Article 58 has the potential to provide a loophole for IUU fishing by unlicensed foreign fishing vessels in the EEZ.

Consistent with its sovereign rights under the LOSC, the coastal State can legitimately regulate the activities of unlicensed foreign fishing vessels exercising their freedom of navigation through the EEZ of the coastal State, provided freedom of navigation is not denied. Foreign vessels exercising their freedom of navigation through the EEZ shall have due regard to the rights and duties of the coastal State and shall comply with the laws and regulations adopted by the coastal State in accordance with the provisions of the LOSC.[73]

A number of coastal States have taken legislative and enforcement measures to ensure that freedom of navigation through their EEZs by foreign fishing vessels is not used as a cover for IUU fishing. Foreign fishing vessels wishing to navigate through the EEZ of a coastal State without conducting fishing activities are normally required to notify relevant authorities their entry into and exit from the EEZ within a certain period.[74] They are also required to abide by the applicable laws of the State in the EEZ, such as the stowing of fishing gears while on transit,

---

[73] *LOSC*, Art. 58(3).

[74] See for example Malaysia, Act No. 317, *Fisheries Act 1985*, Sec. 16; Seychelles, *Fisheries Act 1986*, Sec. 8.

and providing information required by the State such as the name of the vessel, flag State, and location.

6.4 Enforcement of Fisheries Laws and Regulations

As highlighted earlier in this chapter, Articles 62(4) and 73(1) of the LOSC provide the international legal basis for coastal State enforcement measures to combat IUU fishing. Under Article 62(4), the laws and regulations coastal States may implement pursuant to their sovereign rights in the EEZ include enforcement procedures.[75] This provision is reinforced by Article 73(1) which provides that the enforcement powers of the coastal State in the EEZ include "boarding, inspection, arrest and judicial proceedings, as may be necessary to ensure compliance with the laws and regulations adopted by it in conformity with this Convention."

For analytical purposes the fisheries enforcement powers of the coastal State can be categorised into two, namely: the procedures or mechanisms which are "active" in character in the sense that they involve a police function as their principal characteristic; and those requirements which are relatively "passive". The active procedures include surveillance over fishing vessel activity, inspection of fishing vessels both at sea and in port, and hot pursuit of alleged offenders. The passive enforcement actions available to coastal States which should supplement the active enforcement process include intelligence gathering and sharing and use of VMS data. The succeeding discussions focus on the "active" fisheries enforcement powers of coastal States.

6.4.1 *Boarding, Inspection, and Arrest of Vessels*

In general boarding and inspection involves five key steps: detection, approach, boarding, inspection, and disembarkation.[76] It may involve a routine boarding and inspection or boarding with suspected violation. In case of the latter, the designated authority of a coastal State which is responsible for conducting boarding and inspection would need to establish the violation of the fishing vessel and take into account the safety of the vessel in its approach to, boarding of, and disembarkation from the vessel. The actual at sea inspection of fishing vessels focuses on data gathering primarily to determine compliance with the terms of the licence and fisheries regulations. The verification of the logbooks should

---

[75] *LOSC*, Art. 62(4)(k).

[76] C. Mees, 'Points to Remember in Connection with Armed Boardings', Technical Papers Presented at the Regional Workshop on Fisheries Monitoring, Control, and Surveillance, Kuala Lumpur and Kuala Terengganu, Malaysia, 29 June-3 July 1998, *GCP/INT/648/NOR1* (Rome: FAO, 2000), at 147–150.

enable a fisheries officer to reconstruct the fishing activities of the vessel since its entry into the EEZ of the coastal State.[77]

However, an effective legal framework for boarding and inspection in the EEZ does not stop with the acts of boarding and inspection.[78] It also involves arrest of vessels and seizure of fishing gears and other fishing implements, and documents and other records, as well as the prosecution of fisheries offenders and application of sanctions and penalties.

6.4.2 *Hot Pursuit*

Another enforcement power granted to a coastal State, but which is not directly provided in the IPOA-IUU, is the right of hot pursuit. Hot pursuit was based on Article 23 of the 1958 Geneva Convention on the High Seas, which provided for this right in the internal waters, territorial sea, and contiguous zone of a coastal State. The LOSC has extended the application of the right of hot pursuit to archipelagic waters, EEZ, and the continental shelf.[79] Hence, the right of hot pursuit is conferred on a coastal State to pursue and arrest on the high seas a ship which has violated its laws and regulations within the limits of national jurisdiction, including those pertaining to the conservation and management of fisheries resources.

The LOSC attaches a number of conditions to the exercise of the right of hot pursuit by coastal States. Hot pursuit may only be commenced while a foreign fishing vessel is within the relevant zone of jurisdiction where the alleged violation has taken place and only after a visual or auditory signal to stop which enables it to be seen or heard by the foreign ship.[80] Furthermore, the pursuit should not be interrupted and should cease as soon as the ship pursued enters the territorial sea of its own State or of a third State. The LOSC also provides that the right of hot pursuit may be exercised only by warships or military aircraft, or other ships or aircraft clearly marked and identifiable as being on government service and authorised to serve such purpose.[81] In case of an aircraft carrying out the hot pursuit, the aircraft giving the order to stop must itself pursue the ship until a ship or another aircraft of the coastal State, summoned by the aircraft, arrives to take over the pursuit.[82] It is not sufficient to justify an arrest outside the territorial sea that the ship was merely sighted by the aircraft as an offender or suspected

---

[77] *FAO Fisheries Technical Paper No. 415*, *supra* note 29, at 91.

[78] *See* Erik Franckx, Fisheries Enforcement: Related Legal and Institutional Issues: National, Sub regional or Regional Perspectives, *FAO Legislative Studies, No. 71* (Rome: FAO, 2001), at Part II.

[79] *LOSC*, Art. 111(2).

[80] *LOSC*, Art. 111(4).

[81] *LOSC*, Art. 111(5).

[82] *LOSC*, Art. 111(6).

offender, if it was not both ordered to stop and pursued by the aircraft itself or other aircraft or ships which continued the pursuit without interruption.[83]

Poulantzas (2004) has not only analysed extensively the right of hot pursuit under international law but has also documented numerous State legislation and practices on hot pursuit around the world.[84] According to him, coastal States have incorporated similar procedures on hot pursuit in domestic legislation even before the codification of international law of the sea. Various coastal States exercise the right of hot pursuit for the purpose of protecting other coastal rights, including customs, fiscal, security, sanitary, and fishing.[85] With proper procedure and effective implementation, hot pursuit can be an effective coastal State enforcement measure to prevent and deter IUU activities of foreign fishing vessels.

6.4.3 *Application of Sanctions*

International fisheries instruments such as the IPOA-IUU, UN Fish Stocks Agreement, FAO Compliance Agreement, and FAO Code of Conduct all require States to develop and apply consistent and transparent sanctions with sufficient severity to effectively prevent IUU fishing and to deprive offenders from accruing benefits from such activities.[86]

There are two basic requirements for the application of sanctions against fishing violations. The first relates to the application of sanctions in a consistent and transparent manner while the second is instituting sanctions of sufficient severity. The purpose of applying consistent and transparent sanctions is to ensure that the measures are fair and non-discriminatory.[87] The need for sanctions to be of sufficient severity arises from the fact that IUU fishing provides avenue for potential illegal gain from the activity;[88] therefore, to deter IUU fishing, the penalties imposed would need to outweigh the benefits derived from conducting any IUU fishing activity. However, no guidelines have been provided in the IPOA-IUU as to what may constitute transparent and consistent sanctions which are severe enough to combat IUU fishing. Apart from the general criteria in applying fisheries related sanctions, the IPOA-IUU does not clearly stipulate the different forms of penalties that may be imposed on fishing vessels or persons engaged in IUU

---

[83] For more discussions on the right of hot pursuit by an aircraft, *see* Nicholas Grief, *Public International Law in the Airspace of the High Seas* (Dordrecht: Martinus Nijhoff Publishers, 1994).

[84] *See* Nicholas M. Poulantzas, *The Right of Hot Pursuit in International Law* (Hague: Martinus Nijhoff Publishers, 2004), at 93–117.

[85] *Ibid.*, at 94.

[86] *IPOA-IUU*, para. 21; *UN Fish Stocks Agreement*, Art. 19(2); *FAO Compliance Agreement*, Art. III(8); *FAO Code of Conduct*, para. 8.2.7.

[87] *See* IPOA-IUU, paras. 52 and 65. Sanctions may be in a form of denial of landing and transshipment of fish or market restrictions.

[88] K. Kuperan and Jon G. Sutinen, 'Blue Water Crime: Deterrence, Legitimacy, and Compliance in Fisheries', 32 *Law and Society Review* (1998), at 313.

fishing. This is a matter left to each coastal State to address in accordance with national policy. Some States treat fishing violations by national and foreign fishing vessels, particularly in their territorial seas as criminal actions. States such as Tanzania, Micronesia, Malaysia, Barbados, and Nigeria provide for the imprisonment of masters, owners, and charterers of foreign fishing vessels for up to two years.[89] Grenada provides for the punishment of two to five years imprisonment of foreign fisheries offenders[90] while Iceland limits imprisonment of skippers to 6 months.[91] Other States such as Indonesia and Malawi provide for the imprisonment of foreign offenders for up to 10 years.[92]

Unlike in maritime zones under sovereignty, the application of sanctions against fisheries violations in the EEZ is, however, constrained by Article 73 of the LOSC. Pursuant to its sovereign rights in the EEZ, a coastal State may take measures such as the arrest of foreign fishing vessels and subjecting them to judicial proceedings.[93] Arrested vessels and their crews are to be promptly released upon posting of a reasonable bond or other security. The LOSC also provides that sanctions for fisheries violation in the EEZ by foreign fishing vessels may not include imprisonment, unless provided in agreements between States concerned, or any form of corporal punishment.[94]

The issue of the "reasonableness" of a bond has been a major point of contention. There is no minimum or maximum amount fixed for a bond or financial security to be considered reasonable. Different elements are also taken into account in determining the reasonableness of a bond. In its judgment in the *M/V Saiga* case,[95] the International Tribunal for the Law of the Sea (ITLOS) stated that the criterion of reasonableness encompasses the amount, nature, and form of

---

[89] Tanzania, *Territorial Sea and EEZ Act*, 1989, Sec. 17(f); Federated States of Micronesia, *Chapter 3 of Title 18 of the Code of Federated States of Micronesia-Regulation of Foreign Vessels*, Sec. 317; Malaysia, *Fisheries Act 1985 (Act 317),* Sec. 25; Barbados, *Fisheries Act, Cap. 391 Fisheries (Management) Regulations*, 1998, Sec. 15; Nigeria, *Sea Fisheries Decree 1992 (No. 71 of 1992)*, Art. 10.

[90] Grenada, *Territorial Waters Act, No. 17 of 1978*, Sec. 12(2).

[91] Iceland, *Act Regarding Fisheries Within the Fisheries Jurisdiction of Iceland, 1976*, Art. 20.

[92] Indonesia, *Law of the Republic of Indonesia No. 31 of 2004 Concerning Fisheries*, Art. 20; Malawi, *Fisheries Conservation and Management Act, 1997 (No. 25 of 1997)*, Sec. 12 and 13.

[93] *LOSC*, Art. 73(1).

[94] *LOSC*, Art. 73(3); *See also* LOSC, Art. 292. These restrictions on the enforcement only apply in the EEZ and not in other areas of national jurisdiction such as the internal waters, archipelagic waters, and territorial sea where coastal States may exercise full sovereignty. *See* LOSC, Art. 2.

[95] *M/V Saiga* was an oil tanker registered under the Saint Vincent and the Grenadines. On 28 October 1997, in one of its bunkering operations, it was boarded and arrested by Guinean authorities in the EEZ. The master of the vessel was charged for crimes of contraband, fraud and tax evasion. Its flag State submitted to ITLOS a request for the prompt release of the *M/V Saiga* and its crew under Article 292 of the LOSC. The Judgment ordered that Guinea promptly release the *M/V Saiga* and its crew upon posting of a reasonable bond or security by Saint Vincent and the Grenadines. The security consisted of the gas oil discharged from the *Saiga* by the authorities of Guinea plus an amount of $400,000.

the bond or financial security.[96] In the *Camuoco* case,[97] ITLOS considered other factors in the assessment of the reasonableness of bonds and other financial security, such as the gravity of the alleged offences, the penalties imposed or imposable under the laws of the detaining State, the value of the detained vessel and of the seized cargo, and the amount of the bond imposed by the detaining State and its form.[98] The Tribunal also clarified that these criteria do not represent a complete list of factors that are relevant in assessing the reasonableness of a bond or financial security, nor does ITLOS intend to lay down rigid rules as to the exact weight to be attached to each of the factors.[99] ITLOS jurisprudence also suggests that a reasonable bond or other security in Article 73(2) of the LOSC refers to bond or security of a financial nature; hence, non-financial conditions cannot be considered as components of a valid bond or financial security with respect to an alleged fisheries violation of a foreign fishing vessel in the EEZ of a coastal State.[100] According to ITLOS:

> The object and purpose of article 73, paragraph 2, read in conjunction with article 292 of the Convention, is to provide the flag State with a mechanism for obtaining the prompt release of a vessel and crew arrested for alleged fisheries violations by posting a security of a financial nature whose reasonableness can be assessed in financial terms. The inclusion of additional non-financial conditions in such a security would defeat this object and purpose.[101]

Some States utilise administrative sanctions to punish fishing violations in their EEZs. Administrative sanctions refer to "sanctions imposed by an administrative agency or an independent institution for a breach of a regulation or rule established by that agency or institution or enacted by parliament without intervention by a court."[102] There are certain advantages in instituting administrative penalties for fisheries violations. Administrative procedures are alternatives to criminal

---

[96] ITLOS, *The M/V Saiga Case (Saint Vincent and the Grenadines v. Guinea) (Prompt Release)*, Case No. 1, Judgment of 04 December 1997, para. 82.

[97] *Camuoco* is a Panamanian flagged fishing vessel engaged in longline fishing. On 28 September 1999, the *Camuoco* was boarded by French authorities in the EEZ of the Crozet Island for unlawful fishing. The French confirmed the arrest of the *Camouco* and ordered that the release of the arrested vessel would be subject to the condition that prior payment be made of a bond in the amount of FF20 million. The vessel owner and the master of the *Camuoco* filed a summons before the court of first instance in order to secure prompt release of seized items and seek a reduction of the bond. The Tribunal decided that the bond imposed was not reasonable and that France must promptly release the *Camuoco* and its master upon posting of a bond or other financial security in the amount of FF8 million.

[98] ITLOS, *The "Camuoco" Case (Panama v. France) (Prompt Release)*, Case No. 5, Judgment of 07 February 2000, para. 67.

[99] ITLOS, *The "Monte Confurco" Case (Seychelles v. France) (Prompt Release)*, Case No. 6, Judgment of 18 December 2000, para. 76.

[100] ITLOS, *The "Volga" Case (Russian Federation v. Australia) (Prompt Release)*, Case No. 11, Judgement on 23 December 2002, para. 77.

[101] *Ibid.*

[102] P. Cacaud, M. Kuruc, and M. Spreij, Administrative Sanctions in Fisheries Law, *FAO Legislative Study No. 82* (Rome: FAO, 2003), at 2.

proceedings and entail a lesser burden of proof and swift and economic settlement of violations, including negotiated settlements.[103] The use of administrative sanctions avoids lengthy processes involved in criminal proceedings.[104] It also represents a less expensive way of enforcing fisheries prohibitions, compared to judicial proceedings.[105]

Even though international and regional fisheries instruments stipulate the requirement to impose adequate sanctions against fisheries violations, these instruments do not provide clear guidelines on how severe the penalties should be to effectively address IUU fishing. This issue has been raised by Tonga when a fishing vessel *FV Chu Huai No 638* flying the flag of Taiwan (Chinese Taipei) was caught fishing illegally in Tongan waters in January 2008. Taiwan took action against the vessel according to its national fisheries regulations by suspending the fishing licence of the vessel for three months, as well as the licence of the captain of the vessel for six months in addition to the training that he needs to undertake in his home country.[106] Tonga upheld that the measures taken by Taiwan towards *FV Chu Huai No 638* were not severe enough for the offences that it has committed under Tongan legislation, and required the vessel to pay a penalty of $500,000 pan'ga (or about USD238,000).[107]

Another issue which requires further development in relation to the application of sanctions is the need for penalties that reflect all economic and environmental costs of an IUU fishing activity. The economic and environmental costs to be considered include, at the very least, the value of the resource affected by unsustainable fishing practices, the effect of the loss of that resource to the ecosystem, and the impact of the activity on the environment.

6.5 Cooperation with Other States

In combating IUU fishing, a coastal State is not only required to cooperate with a flag State in respect of the licensing of fishing vessels accessing the EEZ, but is also required to cooperate with other coastal States with regard to the management of shared fisheries resources. A coastal State, under international law, is required to cooperate with other States either directly or through subregional or regional management organisations to ensure the conservation and development

---

[103] Blaise Kuemlangan, 'National Legislative Options to Combat IUU Fishing', in FAO, Report of and Papers Presented at the Expert Consultation on Illegal, Unreported and Unregulated Fishing, Sydney, Australia, 15–19 May 2000, *FAO Fisheries Report No. 666* (Rome: FAO, 2001), at 163.

[104] Cacaud, Kuruc, and. Spreij, *supra* note 102, at 54.

[105] *Ibid.*

[106] Pacnews, 'Taiwan tuna boat pays $500,000 pan'ga penalty for illegal fishing', 18 December 2008. www.islandbusiness.com. Accessed on 01 February 2009.

[107] *Ibid.*

of straddling stocks,[108] highly migratory species,[109] anadromous stocks,[110] and catadromous species.[111] In the case of straddling and highly migratory fish stocks, a coastal State also has the obligation to cooperate with States fishing on the high seas to ensure that compatible measures are applied in respect of those stocks in the EEZ and on the high seas.[112] Thus, in granting access to its EEZ, the coastal State would also need to take into account conservation and management measures established by RFMOs. This would require the coastal State to monitor the compliance of foreign fishing vessels with relevant regional fisheries conservation and management measures. The coastal State may also require compliance with these regional measures as a condition for access to its EEZ. States bordering an enclosed or semi-enclosed sea also have the obligation to cooperate and coordinate the management, conservation, exploration, and exploitation of the living resources of the sea.[113]

Given the various limitations on the ability of coastal States to individually enforce their fisheries jurisdictions, cooperation between and among coastal States can be a powerful tool to overcome some of these shortcomings. Inter-coastal State cooperation can take a variety of form, including intelligence sharing, mutual criminal assistance, sharing of assets, logistics support, joint patrols and coordinated patrols.

The Niue Treaty on Cooperation in Fisheries Surveillance and Law Enforcement in the South Pacific Region (Niue Treaty)[114] provides one of the best global examples of the benefits of bilateral and regional cooperation among coastal States to combat IUU fishing. Article III of the Niue Treaty provides a general obligation for the parties to cooperate in the enforcement of their fisheries laws and regulations and to develop regionally agreed procedures for the conduct of fisheries surveillance and law enforcement. Other provisions of the Niue Treaty deal with cooperation with regard to exchange of information,[115] prosecution[116] and enforcement of penalties.[117] Article VI of the Niue Treaty on the development and

---

[108] *LOSC,* Art. 63.

[109] *LOSC,* Art. 64.

[110] *LOSC,* Art. 66.

[111] *LOSC,* Art. 66.

[112] *UN Fish Stocks Agreement,* Art. 7(2).

[113] *LOSC,* Art. 123.

[114] Niue Treaty on Cooperation in Fisheries Surveillance and Law Enforcement in the South Pacific Region, Honiara, Solomon Islands, concluded on 9 July 1992, in force 20 Mary 1993, ATS 1993 No. 31. *Hereinafter referred to as* the Niue Treaty.

[115] *Niue Treaty,* Art. V.

[116] *Niue Treaty,* Art. VII.

[117] Art. VIII of the Niue Treaty provides that "(S)hould two or more Parties wish to provide that a penalty imposed by one party under its fisheries laws be enforced by another Party, they may, by way of provisions in a Subsidiary Agreement or otherwise, agree on procedures for that purpose consistent with their national laws."

implementation of subsidiary agreements, is perhaps one of the important cornerstones of cooperation under the treaty. Article VI(1) and (2) provide:

> 1. A Party may, by way of provisions in a Subsidiary Agreement or otherwise, permit another Party to extend its fisheries surveillance and law enforcement activities to the territorial sea and archipelagic waters of that Party. In such circumstances, the conditions and method of stopping, inspecting, detaining, directing to port and seizing vessels shall be governed by the national laws and regulations applicable in the State in whose territorial sea or archipelagic waters the fisheries surveillance or law enforcement activity was carried out.
>
> 2. Vessels seized by another Party pursuant to an agreement under paragraph 1 of this Article in the territorial sea or archipelagic waters of a Party shall, together with the persons on board, be handed over as soon as possible to the authorities of that Party.

These provisions provided the avenue for some of the parties to the Niue Treaty to cooperate effectively on fisheries enforcement matters in their EEZs.[118]

The 2003 Treaty[119] and the 2007 Agreement[120] between Australia and France provide another illustration of bilateral MCS cooperation. The 2003 Australia-France Treaty covers cooperative surveillance of fishing vessels within the Area of Cooperation,[121] the exchange of information on the details of fishing vessels, including their licensing, cooperative surveillance activities and prosecution,[122] and logistical support for the exercise of hot pursuit.[123] The 2007 Australia-France Agreement gives effect to some aspects of the 2003 Treaty by making provisions for bilateral cooperative enforcement activities and hot pursuit.[124]

The provisions of this bilateral agreement on hot pursuit, although still subject to interpretation with respect to the LOSC, provide an important development in State practice in response to IUU fishing.[125] The 2007 Australia-France Agreement goes further than the state practice demonstrated in the pursuits of the *South*

---

[118] Examples of these subsidiary agreements are those entered into between Tonga and Tuvalu, Federated States of Micronesia and the Republic of the Marshall Islands, and Australia and Papua New Guidea.

[119] Treaty between the Government of Australia and the Government of the French Republic on cooperation in the maritime areas adjacent to the French Southern and Antarctic Territories (TAAF), Heard Island and the McDonald Islands, Canberra, Australia, adopted 24 November 2003, in force 1 February 2005, ATS [2005] ATS 6. *Hereinafter referred to as* Australia-France Treaty.

[120] Agreement on Cooperative Enforcement of Fisheries Laws between the Government of Australia and the Government of the French Republic in the Maritime Areas Adjacent to the French Southern and Antarctic Territories, Heard Island and the McDonald Islands, Paris, France, adopted 8 January 2007, not yet in force, [2007] ATNIF 1. *Hereinafter referred to as* Australia-France Agreement on Cooperative Enforcement.

[121] *Australia-France Treaty*, Art. 3 and Annex I.

[122] *Australia-France Treaty*, Art. 5.

[123] *Australia-France Treaty*, Art. 4.

[124] *Australia-France Agreement on Cooperative Enforcement*, Arts. 3 and 4.

[125] Warwick Gullett and Clive Schofield, 'Pushing the Limits of the Law of the Sea Convention: Australian and French Cooperative Surveillance and Enforcement in the Southern Ocean', 22 *The International Journal of Marine and Coastal Law* (2007), at 573.

Tomi[126] and *Viarsa 1*[127] in the Southern Oceans not only by providing that an authorised vessel of either Party may commence a hot pursuit, but also by specifically providing that an authorised vessel of one State may "take over" the pursuit commenced by an authorised vessel of the other State.[128] However, what *South Tomi* and *Viarsa 1* demonstrated is that cooperative arrangements for hot pursuit need not necessarily be through formalised treaty provisions but can be undertaken on an ad hoc basis. Molenaar (2004) explains how "multilateral" hot pursuit can be interpreted as consistent with Article 111 of the LOSC, as well as the IPOA-IUU.[129] The author maintains that should an international court or tribunal be asked to rule on the legality of hot pursuit cases involving government ships and aircrafts of several nationalities, it may take into account State practice, including bilateral and regional agreements on surveillance and enforcement.

Another example of coastal State cooperation to combat IUU fishing is the International Network for the Cooperation of Fisheries-related Monitoring, Control and Surveillance Activities (International MCS Network). The International MCS Network was created in 2000 to improve the efficiency and effectiveness of fisheries related MCS activities through enhanced cooperation, coordination, information collection and exchange among national organisations and institutions responsible for fisheries related MCS.[130] The Network aims to strengthen MCS capabilities in participating countries through coordination and cooperation with the goal of deterring, reducing, and eliminating IUU fishing and other harmful activities to living marine resources in the world's oceans. It is designed to support States in their attempts to meet their obligations arising from international agreements and their national responsibilities in performing MCS functions.

---

[126] In March 2001, Australian fishing authorities pursued a fishing vessel, *South Tomi*, flying the flag of Togo which was sighted and suspected for conducting illegal fishing for toothfish off the west coast of Australia. The pursuit of *South Tomi* lasted 14 days (3,300 nautical miles) and was assisted by South African and French authorities.

[127] In August 2003, the *Viarsa 1* flagged under Uruguay, was suspected of illegal fishing for toothfish in the Australian territorial sea near Heard Island. Australian authorities pursued the vessel when it refused to stop. The hot pursuit, which continued for 21 days (3,900 nautical miles), was assisted by South African authorities.

[128] Gullett and Schofield, *supra* note 125, at 573.

[129] *See* Erik Jaap Molenaar, 'Multilateral Hot Pursuit and Illegal Fishing in the Southern Ocean: The Pursuits of the Viarsa 1 and the South Tomi,' 19 *The International Journal of Marine and Coastal Law* (2004), at 30–37.

[130] International Network for the Cooperation and Coordination of Fisheries Related Monitoring, Control and Surveillance Activities Website. www.imcsnet.org. Accessed on 17 February 2008.

Chapter Seven

Port State Measures

7.1 The Importance of Port States in Addressing IUU Fishing

Fishing vessels seek port access for many reasons, including refuelling, reprovisioning, landing catch, transshipment of fish, effecting repairs, and in emergencies. In recognition of the significance of port access for fishing vessels, port State control is now considered one of the most important tools to prevent, deter and eliminate IUU fishing. The growing importance of port State measures in the global fight against IUU fishing also stems from the failure by flag States to exercise effective control and jurisdiction over vessels flying their flags. Since not all flag States are capable or willing to discharge their flag State duties to combat IUU fishing, the adoption and implementation of port State measures addresses this shortcoming to a certain extent.

As provided in Chapter 3, international law provides for the exclusive jurisdiction of States over their ports and the discretion to determine conditions for entry by foreign vessels.[1] Therefore it follows that a port State can adopt measures that would supplement the prescriptive and enforcement actions taken by both flag States against their own vessels and coastal States against fishing vessels that fish illegally in their waters. In this respect, port State measures would ensure that IUU vessels which escape detection by flag and coastal States and enter or are about to enter ports are made accountable for their actions and reported to the concerned States. The application of port State measures also limits the extent to which some port States provide shelter to vessels engaged in IUU fishing by allowing the transit of fish derived from IUU fishing.

A port State can play a critical role in preventing IUU caught fish from entering the market and international trade, hence reducing the financial incentives that may be gained from such activities. Port State control can also act as a disincentive to IUU operators by increasing the cost of their operations, for example by forcing them to call into more remote ports with less stringent port State measures.[2] Port State measures are therefore effective means to make the movement of IUU fish from the shore to the consumer more difficult.

---

[1] United Nations Convention on the Law of the Sea, Montego Bay, Jamaica, concluded on 10 December 1982, in force 16 November 1994, 1833 UNTS 3; 21 ILM 1261 (1982). *Hereinafter referred to as* LOSC. Arts. 25 and 218.

[2] High Seas Task Force, *Promoting Responsible Ports*, HSTF/06, (Paris: OECD, 2005). www .high-seas.org. Accessed on 12 February 2006, at 1.

IUU fishing is not only caused by ineffective flag State control and inadequate coastal State enforcement. A major gap that contributes to the proliferation of IUU fishing is the existence of "ports of convenience". In the maritime world, "ports of convenience" is a term used to describe ports and terminals around the world which apply substandard shipping and labour standards. In the fisheries context, "ports of convenience" also refer to ports where authorities have no established rules and procedures to ensure that only legally caught fish are landed or transhipped in their ports. Such ports often conduct inadequate inspection of fishing vessels, gears, and fish catch, or provide little to no investigation into the origin of the fish, compliance with fisheries regulations, and the authenticity of relevant documents.[3] Ports of convenience provide easy avenue for the laundering of fish and are deemed to "undermine, frustrate and neutralise efforts to prevent, deter and eliminate IUU fishing".[4]

An associated issue is the lack of capacity of most port States, particularly in developing States, to conduct adequate port inspection and take enforcement action against IUU vessels. The general lack of port State mechanisms and capabilities to monitor the activities of fishing vessels that land or transship their fish, create opportunities for vessels conducting IUU fishing to take advantage of ports with more lenient regulations or those with inadequate capacities to monitor fishing activities.

The IMO process has addressed the lack of effective flag State control and inadequate domestic procedures for port inspection through the establishment of regional port State control regimes, starting from the Paris Memorandum of Understanding on Port State Control adopted in 1982.[5] Regional port State control mechanisms involve the exchange of information and harmonisation of port measures and are considered the basis for preventing substandard shipping, improving safety of shipping, and preventing and controlling marine pollution.[6] On the basis of these regional mechanisms, IMO has developed a

---

[3] An example of a port of convenience reported by the Environmental Justice Foundation (EJF) as providing services to IUU fleets operating off the coast of West Africa is Las Palmas de Gran Canaria in Spain. This port is also known as a gateway through which illegally caught fish can enter the EU market without proper inspection as to its origin. *See* EJF Website, *Las Palmas: Port of Convenience and Gateway to Europe*, www.ejf.org. Accessed on 20 November 2008.

[4] FAO, Committee on Fisheries, Eleventh Session, Bremen, Germany, 2–6 June 2008, Trade Related Measures for Sustainability: Progress on a Binding Instrument on Port State Measures, *COFI:FT/XI/2008/6*, para. 12.

[5] The existing port State control MOUs are: Europe and the north Atlantic (Paris MOU); Asia and the Pacific (Tokyo MOU); Latin America (Acuerdo de Viña del Mar); Caribbean (Caribbean MOU); West and Central Africa (Abuja MOU); the Black Sea region (Black Sea MOU); the Mediterranean (Mediterranean MOU); the Indian Ocean (Indian Ocean MOU); and the Arab States of the Gulf (GCC MoU (Riyadh MoU)). See also Terje Lobach, Port State Control of Foreign Fishing Vessels, *FAO Fisheries Circular No 987* (Rome: FAO, 2003), at 2.

[6] Joseph E Vorbach, 'The Vital Role of Non-Flag State Actors in the Pursuit of Safer Shipping', 32 *Ocean Development and International Law* (2001), at 34.

global strategy for port State control which includes inspection procedures and training and qualification of control officers.[7]

As discussed in Chapter 3, the FAO recognises the need to enhance port State measures as part of the international efforts to combat IUU fishing through the adoption of the FAO Model Scheme to Combat Illegal, Unreported and Unregulated Fishing developed in 2005[8] and the FAO Port State Measures Agreement to Prevent, Deter and Eliminate Illegal, Unreported and Unregulated in 2009.[9] The FAO Model Scheme reinforces the relevant provisions of the IPOA-IUU and provides international best practice and guidance on a range of issues, including prior notification requirements, inspection procedure, types of information to be reported, training of inspectors, and exchange of information Most these measures have been adopted under the FAO Port State Measures Agreement.

The UN General Assembly, in its resolution in 2005 on sustainable fisheries, urged States to cooperate, in particular at the regional level, and through RFMOs to apply the FAO Model Scheme and to promote its application.[10] Similarly, in 2006, the UN General Assembly further recognised the need for enhanced port State controls to combat IUU fishing, and urges States to cooperate to adopt all necessary port measures, consistent with international law, taking into account Article 23 of the UN Fish Stocks Agreement, and to promote the development and application of minimum standards at the regional level.[11]

7.2 Application of Port State Measures

Because under international law States have full sovereignty over their ports and to control their nationals, discussions among States on port State measures have mostly concentrated on foreign vessels. Accordingly, the IPOA-IUU and the

---

[7] Lobach, *supra* note 5, at 3.

[8] FAO, *Model Scheme on Port State Measures to Combat Illegal, Unreported and Unregulated Fishing* (Rome: FAO, 2007).

[9] FAO, Agreement on Port State Measures to Prevent, Deter and Eliminate Illegal, Unreported and Unregulated Fishing, Adopted in November 2009, Appendix V of the FAO Council, Hundred and Thirty-seventh Session, Rome, 28 September-02 October 2009, Report of the 88th Session of the Committee on Constitutional and Legal Matters (CCLM), 23–25 September 2009, *CL 137/5*, September 2009. *Hereinafter referred to as* the FAO Port State Measures Agreement.

[10] United Nations General Assembly (UNGA), Fifty-ninth session, Agenda Item 49(b), Resolution adopted by the General Assembly, Sustainable fisheries, including through the 1995 Agreement for the Implementation of the Provisions of the United Nations Convention on the Law of the Sea of 10 December 1982 relating to the Conservation and Management of Straddling Fish Stocks and Highly Migratory Fish Stocks, and related instruments, *A/RES/59/25*, 17 January 2005, para. 42.

[11] UNGA, Sixtieth session, Agenda Item 75(b), Resolution adopted by the General Assembly, Sustainable fisheries, including through the 1995 Agreement for the Implementation of the Provisions of the United Nations Convention on the Law of the Sea of 10 December 1982 relating to the Conservation and Management of Straddling Fish Stocks and Highly Migratory Fish Stocks, and related instruments, 10 March 2006, *A/RES/60/31*, para. 42.

FAO Model Scheme apply to foreign fishing vessels[12] seeking entry into a port within the jurisdiction of a State. The FAO Port State Measures Agreement follows this trend, with two exceptions. The first exception includes "vessels of a neighbouring State that are engaged in artisanal fishing for subsistence provided that the port State and the flag State cooperate to ensure that those vessels do not engage and/or support illegal, unreported and unregulated fishing."[13] The second exception applies to "container vessels that are not carrying fish or, if carrying fish, only fish that have been previously landed, provided that there are no clear grounds for suspecting that those vessels have engaged in illegal, unreported and unregulated fishing related activities."[14] In these circumstances, the port State would need to ensure that such vessels are subject to effective measures that prevent IUU fishing.

The FAO Port State Measures Agreement takes the application of the IPOA-IUU and FAO Model Scheme further and covers not only the actual taking or harvesting of fish but also fishing related activities, which includes "any operation in support of, or in preparation for, fishing, including the processing, transshipment or transport of fish that have not been previously landed and offloaded at a port, as well as the provision of personnel, fuel, gear and other supplies at sea."[15] The definition of fishing related activities excludes the activities of reefer vessels.

The FAO Port State Measures Agreement requires a party to the agreement to take necessary measures against IUU fishing and activities in support of IUU fishing as laid out in the agreement. Similar to the FAO Model Scheme, the FAO Port State Measures Agreement sets the minimum requirements for entry into port, inspection, and enforcement actions on foreign fishing vessels accessing ports and allows a port State the flexibility to adopt more rigorous requirements as they deem appropriate under national law. In the exercise of sovereignty over their ports, port States may adopt more stringent measures in accordance with international law.[16] The FAO Port State Measures Agreement also recognises relevant measures adopted under RFMOs to combat IUU fishing. However, the agreement provides that a party does not become bound by measures or decisions of any RFMO of which it is not a member.[17]

---

[12] Foreign fishing vessels referred to in the FAO Model Scheme includes any vessel used or intended for use for the purpose of fishing, including support ships, carrier vessels and any other vessels directly involved in fishing operations. See *FAO Model Scheme*, para. 1.2; FAO, *International Plan of Action to Prevent, Deter, and Eliminate Illegal, Unreported, and Unregulated Fishing*, adopted on 23 June 2001 at the 120th Session of the FAO Council. *Hereinafter referred to as* IPOA-IUU. Para. 53; *FAO Port State Measures Agreement*, Art. 3.

[13] *FAO Port State Measures Agreement* Art. 3(1)(a).

[14] *FAO Port State Measures Agreement*, Art. 3(1)(b).

[15] *FAO Port State Measures Agreement*, Art. 1(d).

[16] *FAO Port State Measures Agreement*, Preamble.

[17] *FAO Port State Measures Agreement*, Art. 4(2).

7.3 Port State Measures to Combat IUU Fishing

Port State measures that may be applied to foreign fishing vessels include the requirement for fishing vessels to seek permission prior to their entry into ports; designation of ports where foreign flagged vessels may be admitted; the inspection of fishing vessels in ports; and enforcement actions such as the refusal to land or transship fish in ports if there is clear evidence that the fishing vessel has engaged in IUU fishing.[18] There is a further requirement for port State measures to be formulated in accordance with international law and implemented in a fair, transparent, and non-discriminatory manner.[19] In a case of *force majeure*, a port State has the obligation to give port access[20] to fishing vessels.[21]

It is generally accepted that once a vessel has voluntarily entered a port, it becomes subject to the laws, regulations and enforcement powers prescribed for permissible activities in the internal waters of the port State.[22] The full sovereignty of a State with respect to ports in its territory is provided under the LOSC. A port State has the right to take necessary steps to prevent any breach of the conditions associated with a port call.[23] A port State may also take action to investigate or institute proceedings with respect to any vessel discharge in violation of applicable international rules and standards when a vessel is voluntarily in its port or offshore terminal.[24]

The UN Fish Stocks Agreement contains a few but more direct provisions on port State control related to fishing vessels. Article 23(1) of the UN Fish Stocks Agreement provides for the right and duty of a port State to take measures to promote the effectiveness of subregional, regional and global conservation and management measures. A port State is given the right to inspect documents, fishing gear and catch on board fishing vessels when a fishing vessel is in its ports or offshore terminals.[25] The UN Fish Stocks Agreement also allows a port State to take enforcement actions such as the prohibition of landings and transshipments if it has been established that the catch has been taken in a manner which undermines the effectiveness of a conservation and management measure on the high

---

[18] *IPOA-IUU*, paras. 55–58; *FAO Port State Measures Agreement*, Parts II–IV.

[19] *IPOA-IUU*, para. 52.

[20] Port access means admission of foreign fishing vessels to ports or offshore terminals for the purpose of, inter alia, refuelling, re-supplying, transhipping, and landing. *See* IPOA-IUU, para. 53.

[21] *IPOA-IUU*, para. 54.

[22] George C. Kasoulides, *Port State Control and Jurisdiction: Evolution of the Port State Regime* (Netherlands: Martinus Nijhoff Publishers, 1993), at 23.

[23] *LOSC*, Art. 25.2.

[24] *LOSC*, Art. 218.

[25] Agreement for the Implementation of the Provisions of the United Nations Convention on the Law of the Sea of 10 December 1982 relating to the Conservation and Management of Straddling Fish Stocks and Highly Migratory Fish Stocks, concluded on 4 September 1995, in force 11 December 2001, 34 ILM 1542 (1995); 2167 UNTS 88. *Hereinafter referred to as* UN Fish Stocks Agreement. Art. 23(2).

seas.[26] Similar to the IPOA-IUU, the LOSC, UN Fish Stocks Agreement, and the FAO Code of Conduct for Responsible Fisheries[27] provide for the application only of non-discriminatory port State measures to vessels.[28]

The following sections examine specific port State measures embodied in international fisheries instruments and national regulations.

7.3.1 *Advanced Notice of Port Entry*

The IPOA-IUU and the FAO Port State Measures Agreement require fishing vessels and vessels involved in fishing related activities to provide a reasonable advance notice of their entry into port, a copy of their authorisations to fish, details of their fishing trip and quantities of fish on board.[29] The FAO Model Scheme and the FAO Port State Measures Agreement provide a list of specific information which States may require from foreign fishing vessels prior to their entry into ports. The list include details related to the identity of the vessel, purpose of port access, details on fishing authorisation, information about the trip, and information on species caught. Other information may also be required of support ships and carrier vessels to provide an initial assessment of whether or not they have been involved in IUU fishing. A number of port States require prior notification from fishing vessels in their national legislation. Examples include New Zealand,[30] Côte d'Ivoire,[31] Canada,[32] Gambia,[33] and Turkey.[34]

International fisheries instruments do not specify what constitutes a reasonable advance notice of entry, this being left to individual national determination. For example, Canada, Fiji, and India require at least 24 hours advanced notification of entry for foreign fishing vessels before arrival[35] while New Zealand and Gambia require a 72 hour notice.[36]

---

[26] *UN Fish Stocks Agreement*, Art. 23(3).

[27] FAO, *Code of Conduct for Responsible Fisheries,* Adopted at the 28th Session of the FAO Conference, Rome, Italy, 31 October 1995. *Hereinafter referred to as* FAO Code of Conduct.

[28] *LOSC*, Art. 25(3); *UN Fish Stocks Agreement*, Art. 23(1); *FAO Code of Conduct*, para. 8.3.1; *See also* LOSC, Arts. 119(3) and 227.

[29] *IPOA-IUU*, para. 55; *FAO Port State Measures Agreement*, Art. 8.

[30] New Zealand, *Fisheries Act 1996, (No. 88 of 1996)*, Sec. 113ZD(1).

[31] *Protocol setting out the fishing opportunities and the financial contribution provided for by the Agreement between the European Community and the Republic of Côte d'Ivoire on fishing off the coast of Côte d'Ivoire for the period from 1 July 2007 to 30 June 2013*, Annex Chapter VII, Number 10(1) and 10(2).

[32] *Canada Marine Act* (1998, c. 10), Sec. 56(2).

[33] Gambia, *Fisheries Act 2007*, Sec. 41.

[34] Turkey, *Notification 1/1 Regulating Commercial Fishing (2007–43)*, Art. 12(2)(e).

[35] *See* Canada, *Coastal Fisheries Protection Regulations*, C.R.C., c. 413, [SOR/94–362, s. 4(F)], Art. 12(1); Fiji, *Marine Spaces (Foreign Fishing Vessels) Regulations (Chapter 158-A, Section 20)*, Art. 21; Gabon, *Fisheries Regulations of 1995*, Sec. 6; India, Maritime Zones of India (Regulation of Fishing by Foreign Vessels) Rules, 1982, Reg. 5.

[36] New Zealand, Ministry of Fisheries, New Zealand Plan of Action to Prevent, Deter and Eliminate Illegal, Unregulated and Unreported Fishing, hereinafter referred to as New Zealand NPOA-IUU, May 2004, Sec. 5.1; Gambia, *Fisheries Act 2007*, Sec. 41(1).

7.3.2 *Designation of Ports*

Since foreign fishing vessels are normally only admitted to certain ports which can cater to their landing and transshipment needs, it is imperative that a State designates ports where such foreign fishing vessels may be admitted. This also enables port States to ensure proper monitoring and control of the activities of foreign fishing vessels calling at their ports. Included in the responsibility to designate ports where foreign fishing vessels may be permitted admission is the need to publicise such ports, and more importantly, ensure that such ports have the capacity to conduct inspections. Examples of States which provide for the designation of ports in their national legislation are Turkey,[37] Slovenia,[38] the Philippines,[39] Guinea,[40] Gabon,[41] and Côte d'Ivoire.[42] In Slovenia, the landing of fish in ports not designated by the Minister of Fisheries is considered an offence under national law.[43]

7.3.3 *Inspection of Fishing Vessels*

A port State is required to carry out inspections of foreign fishing vessels for the purpose of monitoring compliance with relevant conservation and management measures. There are different elements in the inspection of foreign fishing vessels. These elements include the procedure for inspection, what to inspect, the precautions that need to be taken when inspecting vessels, information that needs to be collected, reporting of information to relevant authorities, and safeguarding and confidentiality of information.

The FAO Model Scheme provides an inspection procedure for foreign fishing vessels. The procedure starts with vessel identification, inspection of authorisation to fish and other documentation, and examination of fishing gear and fish and fishery products.[44] Vessel identification involves the verification of the validity of the identity documents and confirmation, through appropriate contacts with flag States and RFMOs, that the flag, external identification number and international radio call sign are correct. It also includes an examination of whether or not the vessel has changed flag and its port of registration and ownership.[45] The FAO

---

[37] Turkey, *Notification 1/1 Regulating Commercial Fishing (2007–43)*, Art. 7(1)(d).

[38] Slovenia, *Marine Fisheries Act (ZMR-2)*, Art. 24(1).

[39] Philippines, Department of Agriculture (DA), Fisheries Administrative Order (FAO) 199, *Guidelines on Fish Transshipment*, 04 April 2000, Sec. VII.

[40] Guinea, *Arrêté n° 0602/95/MPA/CAB portant réglementation de l'exercice de la pêche industrielle*, Art. 16.

[41] Gabon, *Loi n° 015/2005 portant Code des pêches et de l'aquaculture en République Gabonaise*, Art. 32.

[42] Côte d'Ivoire, *Code des Douanes Loi n° 64–291 du 1er Août 1964 telle que modifiée par l'ordonnance n°88–225 du 2 Mars 1988*, Art. 56.

[43] Slovenia, *Marine Fisheries Act (ZMR-2)*, Art. 51(32).

[44] *FAO Model Scheme*, Appendix B.

[45] *FAO Model Scheme*, Appendix B, para. 1.

Model Scheme requires that the inspection procedures must be publicised, including a stipulation of the rights of the master of the vessel during the inspection, as well as his or her obligations in providing all the necessary assistance and information to the inspector.

Under the FAO Port State Measures Agreement, each party is required to inspect the number of vessels in its ports required to reach an annual level of inspections sufficient to achieve the objectives of the Agreement.[46] Parties of the agreement are also required to seek the minimum levels for inspection of vessels through RFMOs. In inspecting vessels, parties need to give priority to vessels that have been denied entry or use of the port, those vessels which are requested to be inspected by other States, and other vessels for which there are clear grounds for suspecting that they have engaged in IUU fishing.[47]

In general, an inspection is conducted on the vessel, documents, fishing gears and other equipment, and fish and fish products onboard the vessel. Such inspection is necessary to ensure the compatibility of the information obtained from the vessel's prior notice of entry and vessel identification process and information collected during inspection. Apart from the fishing authorisation, inspectors may also review the fishing logbook, stowage plans, drawings or descriptions of fish holds, VMS reports, safety certificates, and if applicable, catch certificates. The inspection of fishing gears and fish and fishery products onboard the fishing vessel ascertains that the gears have been used and fish harvested in accordance with the conditions of the fishing authorisation.

There are a number of precautions that need to be taken during inspection. First, inspections would need to be carried out by properly qualified persons who need to make sure that they present appropriate identity documents to the master of the vessel. Second, inspectors need to make all possible efforts to avoid unduly delaying a vessel, ensure that the vessel suffers minimum interference and inconvenience, and prevent degradation of the quality of the fish. Third, inspectors would need to be accompanied by an interpreter of the language of the inspected foreign vessel. Fourth, a port State would need to ensure that inspections are not conducted in a manner that would constitute harassment of any fishing vessel. Such precautions signify the need for port States to provide proper training to their inspectors in areas of information collection, catch inspection of fish holds and gears, fish identification and measurement, boarding of vessels, training in relevant languages, and the collection, evaluation, and preservation of evidence.

After port inspection of a foreign fishing vessel, the reporting of the results of the inspection follows. A port State needs to ensure that the results of the

---

[46] *FAO Port State Measures Agreement,* Art. 12(1).

[47] *FAO Port State Measures Agreement,* Art. 12(3).

inspection are presented to the master of the vessel and signed by both the inspector and the master. The master would also need to be provided the opportunity to add any comment to the report and contact the relevant authorities of the flag State if he or she is encountering serious difficulties in understanding the report.[48] The port State needs to report the results of its inspections to the flag State of the inspected vessel, other relevant States, and to relevant RFMOs.[49] The basic information that may be remitted to the flag States and other relevant authorities include the flag State of the vessel and identification details, name, nationality and qualification of the master and crew, fishing gear, catch onboard, and total landed and transshipped catch.[50] Such details, together with other information that may be required by RFMOs must be provided with due regard to the confidentiality of information.[51] This implies that port States need to establish domestic guidelines not only on the inspection of fishing vessels but also on the type of information and the manner by which such information may be released to third parties.

7.3.4 *Enforcement Actions*

If there are reasonable grounds to believe that the vessel has engaged in or supported IUU fishing, several measures may be undertaken by a port State. Enforcement actions that may be taken against a fishing vessel are based on the presumption that the vessel has undertaken IUU activities. Such presumption would need to be established under national laws, based on domestic regulations, measures adopted by relevant RFMOs, and applicable international conservation and management measures. The FAO Model Scheme has provided a list of fishing activities which may constitute as IUU fishing and can provide a basis for detecting a fisheries violation. The list of IUU fishing activities include fishing without a licence, failure to maintain accurate records, fishing in closed areas, falsifying or concealing vessel marking, and failure to comply with VMS requirements.[52] The FAO Port State Measures Agreement, on the other hand, does not provide a list of activities that may be considered IUU fishing but refers back to the scope and nature of IUU fishing provided under the IPOA-IUU.[53] Another basis for ascertaining that a vessel has engaged in or supported IUU fishing is if it is listed in an RFMO IUU list.[54] Apart from direct forms of IUU fishing, other actions may also constitute a fisheries violation, and may warrant enforcement

---

[48] *FAO Model Scheme*, para. 3.
[49] *FAO Model Scheme*, para. 6; *FAO Port State Measures Agreement*, Art. 18(1)(a).
[50] *IPOA-IUU*, para. 58.
[51] *IPOA-IUU*, para. 60; *FAO Model Scheme*, para. 7.
[52] See also discussions in Chapter 2 on the scope and nature of IUU fishing.
[53] *FAO Port State Measures Agreement*, Art. 1(e).
[54] This matter is further discussed in Chapter 9.

actions, such as the obstruction of authorised port inspectors in carrying out their duties. The failure to provide required information and falsification of documents may also be grounds for undertaking enforcement actions against fishing vessels.

Following an inspection, the measures that may be taken against presumed IUU vessels include the denial of entry into port and prohibition of landing and transshipment of fish. Other means of enforcement such as monetary penalties for related fisheries violations are discussed in Chapter 9.

7.3.4.1 *Denial of Landing and Transshipment of Fish in Ports*

The prohibition of landing and transshipment of fish is the most common port State measure undertaken against vessels which have undermined international and regional conservation and management measures. It may also be considered as one of the most effective measures in deterring IUU fishing due to the fact that a restriction of this nature directly impacts on the marketing or trade of fish. There are examples of States which prohibit the landing and transshipment of fish by vessels engaged in IUU fishing activities. Iceland bans the landing and transshipment of catch in its ports by foreign fishing vessels found to be in violation of agreements on utilisation and preservation of living marine resources to which Iceland is a party.[55] Canada also prohibits the loading, unloading, and transshipment of fish and fish products, or carrying out repairs in its ports if the foreign fishing vessel is not in compliance with relevant conservation and management measures.[56] Other States which provide similar port enforcement actions against illegal fishing vessels in their national laws are Gambia[57] and Ghana.[58]

7.3.4.2 *Denial of Entry into or Use of Port*

The denial of port access as an enforcement action against vessels which have failed to comply with fisheries laws and regulations is not expressly set out in international fisheries agreements such as the LOSC and the UN Fish Stocks Agreement. However, the lack of direct reference to such legal measures in international agreements does not detract from the rights of a State to deny fishing vessels entry into its ports.[59] States, especially those belonging to RFMOs have

---

[55] *See* Iceland, Act No. 228 April 1998 concerning fishing and processing by foreign vessels in Iceland's exclusive fishing zone, in Terje Lobach, 'Measures to be Adopted by the Port State in Combating IUU Fishing', in FAO, Report of and Papers Presented at the Expert Consultation on Illegal, Unreported and Unregulated Fishing, Sydney, Australia, 15–19 May 2000, *FAO Fisheries Report No. 666* (Rome: FAO, 2001), at 93.

[56] Canada, *Coastal Fisheries Protection Regulations*, C.R.C., c. 413, Sec. 5.1.12(b), 5.1(a)(vii), 5.1(a)(viii), and 5.1(a)(ix).

[57] Gambia, *Fisheries Act 2007*, Sec. 64(3).

[58] Ghana, *Fisheries Act 2002* (Act No. 625 of 2002), Sec. 133(1).

[59] Lobach, *supra* note 55, at 93.

denied port entry to vessels engaged in IUU fishing. In 2001, a New Zealand flagged vessel was denied entry into a Uruguayan port on the ground that it has been involved in IUU fishing.[60] Australia also refused the entry of two Belize flagged fishing vessels on the ground that these vessels failed to show that the catch of Patagonian toothfish landed previously in Mauritius had been caught in accordance with conservation measures of CCAMLR.[61]

There are other States which deny the entry of vessels engaged in IUU-related activities. Norway denies port access to foreign fishing vessels which have conducted unregulated fishing on the high seas.[62] The U.S. also denies port privileges to fishing vessels taking part in large scale driftnet fishing on the high seas.[63] Under New Zealand law, if a foreign fishing vessel which has undermined international conservation and management measure has already entered the internal waters or a port of New Zealand, it may be asked to leave those waters or that port, except in cases of *force majeure*.[64]

Port denial can be justified on the basis of the general principle that the fishing vessel is subject to the full sovereignty of a State while the vessel is in its ports. However, there is a need for a port State to ensure that the conditions for the denial of fishing vessels and other port State procedures are duly published.[65] In accordance with international law, a vessel cannot be denied port access in case of *force majeure* or distress.[66]

A principal concern in the implementation of port State control for fishing vessels is the formulation of fair, transparent, and non-discriminatory port State measures. There are no criteria to determine what constitutes a "fair, transparent, and non-discriminatory measure". The vagueness of this clause may provide the basis of divergent practices. The Organisation for Economic Cooperation and Development (OECD) High Seas Task Force argues that a port State measure should be able to avoid unjustifiable discrimination between foreign vessels.[67] In order to ensure that only non-discriminatory measures are implemented, the grounds upon which suspected IUU vessels may be detained, released and prosecuted, or whose entry may be denied or landing and transshipment prohibited, need to be made clear in the development of port State schemes.

---

[60] Rachel Baird, 'CCAMLR Initiatives to Counter Flag State Non-enforcement in Southern Ocean Fisheries', 36 *Victoria University of Wellington Law Review* (2005), at 745.

[61] *Ibid.*

[62] Lobach, *supra* note 55, at 3.

[63] United States of America, *Magnuson-Stevens Fishery Conservation and Management Act*, An Act to provide for the conservation and management of the fisheries and for other purposes, Public Law 94–265. As amended through 11 October 1996, Sec. 206(a). Foreign fishing vessels under U.S. law are not allowed to call on U.S. ports. *See* Sec. 251 of this Act.

[64] New Zealand, *Fisheries Act 1996*, (No. 88 of 1996), Sec. 113ZD (2),(3), and (5).

[65] *LOSC*, Art. 25(3).

[66] *IPOA-IUU*, para. 54.

[67] High Seas Task Force, *Promoting Responsible Ports*, *supra* note 2, at 3.

The question on the application of non-discriminatory measures also brings into question the treatment of foreign fishing vessels vis-à-vis national fishing vessels in ports. Since the requirements in international fisheries instruments on port State measures only apply to foreign fishing vessels, an issue may arise as to whether or not enforcement actions against foreign fishing vessels are similar with those applied to national fishing vessels and if disparity in the application of such measures may constitute a discrimination between those vessels. This subject is not clearly addressed in the current framework for port State measures for fisheries. However, in the interest of effectively combating IUU fishing in ports, it makes sense to increase efforts in applying stringent measures for all types of fishing vessels, whether foreign or national flagged, and in public and private ports, if practicable.

Compared to other port State measures such as advanced notice of port entry and inspection of fishing vessels, an enforcement action similar to the prohibition of fish landings may incite more objections from other States because it also takes the form of a market or trade restriction.[68] In the dispute between Chile and the EU on swordfish fisheries in the South Eastern Pacific Ocean, the EU claimed that Chile has prescribed and implemented a measure in a discriminatory manner when it prohibited the unloading of swordfish in its ports.[69] There has been no definitive ruling on the Chile-EU swordfish case yet; however, it can be argued that the burden to prove that the measure applied is non-discriminatory in nature will rest on Chile as a port State. Therefore it is crucial that port States not only prescribe clear national laws in controlling fishing vessels in ports but also determine the scope of application of such laws and regulations. There is also a need for port States to formulate procedures to establish clear evidence that a fishing vessel has engaged in IUU fishing before an enforcement action is undertaken.

7.3.5 *Cooperation with Flag States*

While the exercise of state jurisdiction over its territories is provided under international law, the extent of port State jurisdiction over the actions of a foreign vessel outside its territory is subject to some uncertainty. Hence, it has been essential to empower a port State to undertake actions in such circumstances through treaties as recognised under the IPOA-IUU and the FAO Model Scheme which provide for the requirement for port States to cooperate with flag States in undertaking some of its measures. Two areas where such cooperation is imperative are in undertaking enforcement actions and the exchange of information.

---

[68] Issues related to market restrictions are discussed in the next chapter.

[69] International Tribunal on the Law of the Sea (ITLOS), *Case Concerning the Conservation and Sustainable Exploitation of Swordfish Stocks in the South-eastern Pacific Ocean (Chile v. European Community)*, List of Cases No. 7, Constitution Chamber Order of 20 December 2000, para. 2.3(d).

Under the IPOA-IUU, the right to take enforcement action against an IUU vessel which has been granted port access is accompanied by the obligation to report the matter to the flag State of that vessel, especially if it has been established that the vessel has conducted IUU fishing in areas beyond the jurisdiction of the port State.[70] In such a case, the port State may take other actions with the consent of, or upon the request of the flag State.[71] The port State may also invite the flag State to participate in port inspections subject to appropriate arrangements.[72]

Cooperation with flag States is also provided in the FAO Compliance Agreement. This agreement states that when a fishing vessel is voluntarily in its port and there are reasonable grounds for believing that the fishing vessel has been used for an activity that undermines the effectiveness of international conservation and management measures, the port State is required to promptly notify the flag State.[73] The flag and port States could enter into arrangements that would allow the port State to conduct investigations as may be considered necessary to establish the violation by the fishing vessel.[74] Similarly, the FAO Code of Conduct encourages a port State to provide assistance when it has been requested to establish the non-compliance of a fishing vessel of a flag State.[75] The U.S. has adopted the relevant provisions of the FAO Compliance Agreement in its domestic legislation, which specifically provides that when a foreign fishing vessel which has undermined international conservation and management measures is voluntarily in its ports, it shall promptly notify the flag State, and if requested by the flag State, make arrangements to undertake investigatory measures necessary to establish the violation of the vessel.[76]

The exchange of port inspection information with the flag State of the suspected IUU vessel is necessary not only to inform the flag State of the activities of the vessel but also to enable it to take punitive or corrective actions against its vessel. A recent example of this is the de-registration of the vessel *Paloma V* from the Namibian registry on the basis of the information obtained from the port inspection conducted by New Zealand authorities which proved that the vessel has been involved in IUU fishing. In May 2008, *Paloma V*, a fishing vessel flying the flag of Namibia after having fished in the Southern Ocean, had entered New Zealand to access its port. An inspection was carried out on the vessel and

[70] *IPOA-IUU*, para. 56.

[71] *IPOA-IUU*, para. 59.

[72] *FAO Model Scheme*, para. 3.6.

[73] FAO, Agreement to Promote Compliance with International Conservation and Management Measures by fishing Vessels on the High Seas, Rome, Italy, concluded on 24 November 1993, entered into force 24 April 2003, 33 ILM 968. *Hereinafter referred to as* the FAO Compliance Agreement, Art. V(2).

[74] *FAO Compliance Agreement*, Art. V(2).

[75] *FAO Code of Conduct*, para. 8.3.2.

[76] 16 U.S.C.A. § 5504. *United States Code Annotated Currentness*, Title 16, Conservation. Chapter 75, High Seas Fishing Compliance, § 5504.

fisheries officers found evidence that the vessel has been engaged in activities that breached CCAMLR conservation and management measures and the New Zealand *Fisheries Act 1996*.[77] After New Zealand authorities informed the Namibian government of the conduct of its vessel, it undertook to de-register the fishing vessel.[78]

Another area where cooperation of port States is crucial is when an inspecting State deems it necessary to bring the vessel to the nearest port if, following a boarding and inspection, it has clear grounds for believing that the vessel has committed a serious fisheries violation and the flag State has failed to respond or take action against the vessel.[79] In such a case, the port State has the obligation to render its assistance to the inspecting State, including taking all necessary steps to ensure the well-being of the crew. These measures adopted under international fisheries agreements further illustrate the role of port States in assisting in the implementation of other enforcement measures, which may be taken by flag and coastal States.

7.3.6 *Other Port State Measures*

Based on State practice, there are a number of other measures currently applied to foreign fishing vessels calling into ports which are not directly specified under international fisheries instruments. These measures include the requirement to provide a declaration of compliance, authorisation for port entry, authorisation for landing and transshipment, prohibition of access to port facilities, and port departure requirements.

Apart from an advanced notice of entry, some States also require foreign fishing vessels to declare in writing that they have not engaged or supported IUU fishing. Such requirement is more common among members of RFMOs, but also among EU member States.[80] For such a requirement, port entry or access is

---

[77] Omunkete Fishing (Pty) Limited, the owner of *Paloma V*, applied for a judicial review on the conditions of approval imposed by the Ministry of Fisheries, the search of the vessel, and the subsequent actions taken and proposed action to be undertaken by the Ministry of Foreign Affairs and Trade to report to CCAMLR the alleged violation of the vessel on the basis of the information and documents obtained by the Ministry of Fisheries. The High Court of New Zealand ordered that the Ministry of Foreign Affairs and Trade do not take further steps in relation to obtained information except to notify CCAMLR that it has obtained relevant information. See High Court of New Zealand Wellington Registry, CIV 2008-485-1310 Under the Judicature Amendment Act 1972 in the matter of an application for judicial review between Omunkete Fishing (Pty) Limited and the Minister of Fisheries and the Minister of Foreign Affairs and Trade, Judgment, 01 July 2008.

[78] 'Govt strips Paloma V of Namibian flag', *The Namibian*, 24 July 2008, www.namibian.com.na. Accessed on 10 January 2009.

[79] *UN Fish Stocks Agreement*, Art. 21(8).

[80] See for example, EU, Council Regulation 601/2004 of 22 March 2004 laying down certain control measures applicable to fishing activities in the area covered by the Convention on the conservation of Antarctic marine living resources and repealing regulations (EEC) No. 3943/90, EC 66/98 and (EC) 1721/1999, Art. 27(2).

usually dependent on a declaration that a vessel has not engaged in IUU fishing. If it has been established by the port State that a vessel has been involved in IUU fishing contrary to its declaration, the port State may undertake enforcement actions against such vessels.

Some States also require vessels entering into their ports to obtain authorisation for port entry, such as the U.S.[81] and Canada.[82] Another measure is the requirement for a foreign fishing vessel to acquire a licence or authorisation to land or transship catch, as provided in the national regulations of Ghana,[83] New Zealand,[84] Guinea,[85] Benin,[86] and Côte d'Ivoire.[87] A number of States also provide for certain requirements prior to departure such as obtaining clearance for departure and submission of information such as its intended time and position of departure.[88]

In terms of undertaking actions against IUU vessels, some port States also prohibit access to port service facilities from such vessels, in addition to denying the landing and transshipment of fish caught through illegal means.[89] Denying access to supplies and services in ports is another effective measure that may be applied by a State to curtail IUU fishing. However, such a measure would need to be subject to certain exemptions such as in cases of *force majeure* or when the vessel is in distress.

There are a whole range of measures that may be applied by a port State to address IUU fishing within the rights accorded to it under international law. What is critical is that the requirements for port access and procedures for inspection and enforcement are incorporated in domestic legislation or bilateral and multilateral treaties. Port State measures may also be developed within a system of other measures, such as monitoring, control and surveillance. In the case of the Pacific Island States, port access and inspection are an integral part of the regional harmonised minimum terms and conditions for fishing access, the implementation of which has been considered as a contributing factor to the effectiveness of port related measures despite the limited capacity of most of these States.[90]

---

[81] U.S., 19 C.F.R. § 101.4, Code of Federal Regulations Currentness, Title 19, *Customs Duties*, Chapter I, Bureau of Customs and Border Protection, Department of Homeland Security; Department of the Treasury (Refs & Annos), Part 101, General Provisions (Refs & Annos), § 101.4.

[82] *Canada Marine Act (1998, c. 10)*, Sec. 58(3).

[83] Ghana, *Fisheries Act, 2002* (Act No. 625 of 2002), Sec. 133(1).

[84] New Zealand, *Fisheries (Commercial Fishing) Regulations 2001*, Reg. 19(2).

[85] Guinea, *Loi n° L/95/13/CTRN du 15 mai 1995 portant Code de la pêche maritime*, Art. 29.

[86] Benin, *Code des Douanes de la République du Bénin du 21 Novembre 1967*, Article 62(2).

[87] Côte d'Ivoire, *Code des Douanes Loi n° 64–291 du 1er Août 1964 telle que modifiée par l'ordonnance n°88-225 du 2 Mars 1988*, Art. 59(3°).

[88] New Zealand, *Fisheries (Commercial Fishing) Regulations 2001*, Reg. 17(10); *Canada Marine Act* (1998, c. 10), Sec. 58(3); and Philippines, DA-FAO 199, *Guidelines on Fish Transshipment*, 04 April 2000, Sec. VII.6.

[89] Gambia, *Fisheries Act 2007*, Sec. 64(3).

[90] FAO, Report of the FAO/FFA Regional Workshop to Promote the Full and Effective Implementation of Port State Measures to Combat Illegal, Unreported and Unregulated Fishing, Nadi, Fiji, 28 August-01 September 2006, *FAO Fisheries Report No. 810* (Rome, FAO, 2007), at para. 61.

Port State schemes, specifically those adopted by RFMOs, are also essential components of national port State regimes, and are discussed in Chapter 9. A successful implementation of a port State regime for fisheries also involves effective collaboration between relevant national authorities with fisheries, navy, customs, health and immigration functions, which can only be achieved through interagency cooperation and legal and administrative arrangements.

Chapter Eight

Trade and Market Measures

8.1 International Fish Trade and IUU Fishing Concerns

The FAO estimates that 45 per cent of the world fish catch is now traded internationally. In 2006, the total world exports of fish and fish products reached USD85.9 billion, which represents an increase of 55 per cent from 2000.[1] Similarly, the value of imports in the same period reached USD89.6 billion or an increase of 49 per cent.[2] The most important fish and fish products that are traded internationally are shrimp, groundfish, salmon, tuna, and fishmeal.[3] While developed States such as Norway, U.S., Denmark, Canada, Spain, Netherlands, Japan, United Kingdom, Italy, France, and Germany are the top exporters and importers of fish and fish products, the significant contribution of developing States to the international trade of fish is undeniable. The net exports of fish by developing States have shown a continuous increasing trend over the decades, which is estimated at 49 per cent in value and 59 per cent in quantity of the total fishery exports in 2006.[4] For many developing States, fish trade represents a significant source of food security, employment and income generation, and foreign currency earnings. This overall trend is primarily driven by China, which has now become the world's largest exporter of fish.[5] At the regional level, the EU is the world's biggest market for fish, reaching USD41.3 billion in 2006, and is continuously expanding its dependence on imports for fish supply.[6]

The vast expansion of trade is attributed to the use of refrigeration and improved transportation and communication. In general, the rising trade values and volumes for all fish commodities (except fish meal volumes) reflect the increasing globalisation of the fisheries value chain, in which production processing is being outsourced to Asia, as well as Central and Eastern Europe and North Africa.[7]

---

[1] Food and Agriculture Organization (FAO), *Fact Sheet: The International Fish Trade and World Fisheries*, June 2008, www.fao.org. Accessed on 20 June 2008, at 1.

[2] *Ibid.*

[3] *Ibid.*

[4] *Ibid.*

[5] *Ibid.*

[6] *Ibid.*

[7] FAO, Committee on Fisheries, Sub-Committee on Fish Trade, Eleventh Session, Bremen, Germany, 2–6 June 2008, 'Status and Important Recent Events Concerning International Trade in Fishery Products,' *COFI:FT/XI/2008/3* (Rome: FAO, 2008), at para. 12.

Outsourcing of fish processing takes place both at the regional and global levels, depending on the product form, labour costs and transportation time. Many species, such as salmon, tuna, catfish, and tilapia are increasingly traded in their processed form.[8] In addition, there has been a growth in the international or global distribution channels through large retailers.[9] It is this transformation and complexity of the international fish trade that makes it more challenging for States to address concerns on the sustainability of fisheries resources, which has centred mostly on the production aspect of fisheries. One needs to bear in mind that between the fishing vessel that catches the fish and the end consumer, a long chain of stakeholders are involved in the fish trade, including fish processors, brokers and agents, and wholesalers and retailers. Ideally, these stakeholders would need to be involved in the regulation of fishing activities in order to achieve the objective of sustainable use and management of fisheries resources.

One of the concerns about the international trade of fish and fish products is IUU fishing. There are no accurate data on how much of the fish and products traded internationally are derived from IUU fishing. Without proper controls in place, it is particularly difficult for States to determine whether or not internationally traded fish, or fish products which have undergone post-harvest processing or those which have been subject to re-exportation, have been caught through IUU means. Existing estimates and studies of the extent of illegal and unreported fishing only lead to basic conclusions that some fish species that enter international trade, such as tuna, toothfish, and swordfish, have been widely targeted by IUU fishers.

The development of the IPOA-IUU took place against the backdrop of a variety of multilateral efforts to restrict international trade in fish and fishery products harvested through IUU fishing.[10] Thus, to appreciate how trade measures may be applied to address the IUU fishing problem, an understanding of the relationship between international trade law and international fisheries law is necessary. This chapter examines the legal framework for the international trade of fish and fish products, the issues being discussed within FAO with respect to promoting responsible fish trade, and concerns related to the application of the Convention on International Trade in Endangered Species of Wild Flora and Fauna (CITES) on commercially exploited aquatic species. It also analyses the internationally agreed market related measures adopted under the IPOA-IUU, as well as some of the recent developments at the international level to strengthen market and trade measures to combat IUU fishing, such as catch certification and trade

[8] *Ibid.*

[9] *Ibid.*

[10] FAO Fisheries Department, Implementation of the International Plan of Action to Prevent, Deter and Eliminate Illegal, Unreported and Unregulated Fishing, *FAO Technical Guidelines for Responsible Fisheries No. 9* (Rome: FAO, 2002), at 47.

documentation, trade restrictive measures, traceability systems, use of harmonised commodity coding systems, eco-labelling, and other market related measures.

8.2. Legal Framework for International Trade and Fisheries

There are three interrelated issues being dealt with in the international trade of fish and fishery products: food safety and security, sustainability of fisheries resources, and globalisation of fisheries. Dealing with these issues requires an approach in the regulation of fish trade that takes into account all relevant fisheries, environment, and trade related instruments. There are two areas in the framework for international trade and fisheries which require examination. The first involves the international trade of fish and fishery products within the framework of the World Trade Organization (WTO) and the FAO. The second is in regard of the international trade of endangered species, which includes a small number of aquatic species, under the legal framework of CITES.

8.2.1 *International Trade in Fish and Fishery Products under the WTO and FAO*

The international trade in fish and fishery products is primarily governed by the international trade rules developed under the General Agreement on Tariffs and Trade (GATT) and related agreements of the WTO. The Preamble to the Marrakesh Agreement establishing the WTO, provides that

> ...trade and economic endeavour should be conducted with a view to raising standards of living, ensuring full employment and large and steadily growing volume of real income and effective demand, and expanding the production of and trade in goods and services, while allowing for the optimal use of the world's resources in accordance with the objective of sustainable development, seeking both to protect and preserve the environment and to enhance the means for doing so in a manner consistent with their respective needs and concerns at different levels of economic development.[11]

Similarly, the FAO, in pursuing its objective of effective conservation, management and development of marine living resources, promotes adherence to the principles, rights and obligations established in WTO Agreements. It is therefore the agreements, codes, and guidelines established under the purview of WTO and FAO which form part of the regime for the global trade in fishery products.

---

[11] This policy principle was further supported at the World Food Summit in 1996, a meeting that took place shortly after the formation of the WTO, which aims at halving the number of undernourished people in developing countries (estimated at 800 million) by 2015 using all manner of strategies including that of international trade. The World Food Summit specifically pledged to guarantee that policies on food and agriculture trade would contribute to the development of food security for all concerned through a system of fair and global market oriented trade.

The framework for the international trade in fish and fishery products is guided by the FAO Code of Conduct for Responsible Fisheries. One of the general principles under the FAO Code of Conduct is Article 6.14 which states that "(i)nternational trade in fish and fishery products should be conducted in accordance with the principles, rights and obligations established in World Trade Organization (WTO) Agreement and other relevant international agreements." This makes it clear that international fish trade would need to achieve the fundamental goals of GATT, which includes the substantial reduction of tariffs and other barriers to trade, as well as the elimination of discriminatory treatment in international trade.

GATT has adopted two principles to ensure that multilateral trade is conducted on a non-discriminatory basis. These principles are the most favoured nation and national treatment principles provided in Articles I, II, and III. In general, GATT requires its Members to treat other members the way they treat their 'most favoured' members or the States with which they have the most unrestricted trade. It also requires WTO members to treat imported products from other WTO members no less favourably than their domestic products. GATT has also adopted the general system of preferences as an exemption to the most favoured nation principle, allowing some level of preferential treatment for developing WTO Members. The practice of the general system of preferences vary in terms of product coverage, benefits to developing members, the level of tariff cuts, rules of origin, or application of non-tariff barriers.[12] Other provisions of GATT 1994 which are relevant to fisheries trade also point to the principle of non-discrimination, particularly Articles VIII, X, XI, and XIII which deal with rules on the importation of products. Even the general exceptions in Article XX of GATT 1994, which is of most relevance to fisheries conservation and management and trade, would need to be applied in a non-discriminatory manner.[13]

Article 6.14 of the FAO Code of Conduct also provides the requirement for States to ensure that their policies, programmes and practices related to trade in fish and fishery products do not result in obstacles to fisheries trade, environmental degradation or negative social, including nutritional impacts. However, with the application of a number of related WTO agreements on fisheries, concerns have been raised that the importation of fish products result in other forms of

[12] M. Rafiquil Islam, *International Trade Law of the WTO* (Oxford: Oxford University Press, 2006), at 16.

[13] Article XX(b) and (g) of GATT 1994 provides that "(S)ubject to the requirement that such measures are not applied in a manner which would constitute a means of arbitrary or unjustifiable discrimination between countries where the same conditions prevail, or a disguised restriction on international trade. Nothing in this Agreement shall be construed to prevent the adoption or enforcement by any contracting party of measures:...(b) necessary to protect human, animal or plant life or health;...(g) relating to the conservation of exhaustible natural resources if such measures are made effective in conjunction with restrictions on domestic production or consumption."

barriers to trade apart from import duties, such as control or sanitary measures, technical requirements, as well as trade distortions through subsidies and dumping.[14] As discussed in Chapter 3, the major WTO Agreements which are relevant to fisheries trade are the Agreement on the Application of Sanitary and Phytosanitary Measures (SPS Agreement), Agreement on Technical Barriers to Trade (TBT Agreement), Agreement on Implementation of Article VI of GATT 1994 (Anti-dumping), Agreement on Import Licensing Procedures, Agreement on Safeguards, and Agreement on Trade-related Aspects of Intellectual Property.

Article 11 of the FAO Code of Conduct elaborates on the principle under Article 6.14 by outlining two aspects of trade in fish that would need to be taken into account: post harvest practices and international trade. The specific provisions under Article 11 address four goals with respect to post harvest and international trade: (a) safety and quality of fish; (b) food security; c) sustainable development and utilisation of fisheries resources; and (c) international trade consistent with WTO Agreements. It can be noted that Article 11 of the FAO Code of Conduct does not specifically address concerns about IUU fishing, but nevertheless forms part of the general framework under which international trade in fish and fishery products must be conducted.

8.2.1.1 *Maintaining Food Safety and Quality*

One of the general principles of the FAO Code of Conduct is that the harvesting, handling, processing and distribution of fish must be carried out in a manner which will maintain the nutritional value, quality and safety of the products, reduce waste, and minimise negative impacts on the environment.[15] The FAO Code of Conduct provides a number of measures that promote the safety and quality of fish. It requires States to establish and maintain effective national safety and quality assurance systems to ensure the right of consumers to safe fish and fishery products.[16] Traditional quality control programmes were based on establishing effective hygiene control through inspection of facilities to ensure adherence to established and generally accepted Codes of Good Hygiene Practices (GHP) and Good Manufacturing Practices (GMP).[17] Examples of more recent safety and quality assurance methods and systems include Hazard Analysis Critical Control Point (HACCP), Quality control (QC), International Organization for Standardization (ISO) Quality Assurance (QA)/Quality Management (QM)

---

[14] FAO Websites, World Trade Organization Agreements, www.fao.org. Accessed on 11 July 2008.

[15] FAO, *Code of Conduct for Responsible Fisheries,* Adopted at the 28th Session of the FAO Conference, Rome, Italy, 31 October 1995, Art. 6.7.

[16] *FAO Code of Conduct*, Art. 11.1.1 and 11.1.2.

[17] H.H Huss, L. Ababouch, and L. Gram, Assessment and Management of Seafood Safety and Quality, *FAO Fisheries Technical Paper No 444* (Rome: FAO, 2003), at 7.

Standards, Quality Systems, and Total Quality Management (TQM).[18] These programmes are supposed to be implemented throughout the industry in compliance with the quality standards agreed within the context of the FAO/WHO Codex Alimentarius Commission and other relevant organisations and arrangements.[19] The Codex Alimentarius is a collection of standards, codes of practice, guidelines and other recommendations which deal with detailed requirements relating to a food or food group, operation and management of production processes, or the operation of government regulatory systems for food safety and consumer protection.[20]

A number of standards have been formulated under the Codex Alimentarius related to fisheries, such as standards for frozen, canned, smoked, and other forms of processed fish and other fish products.[21] Some of these standards are now included in the combined Code of Practice for Fish and Fishery Products, which has further been modified to incorporate guidance on the use of HACCP.[22] This Code covers technological guidelines and the essential requirements of hygiene in the production of fish, shellfish and their products to ensure that such products are safe for human consumption and meet the requirements of the appropriate Codex product standards.[23]

---

[18] *See Ibid.*, at 10–12.

[19] *FAO Code of Conduct*, Art. 11.1.3.

[20] *See* Secretariat of the Joint FAO/WHO Food Standards Programme, *Understanding the Codex Alimentarius* (Rome: WHO and FAO, 2006).

[21] *See* Recommended International Code of Practice – General Principles of Food Hygiene, CAC/RCP 1-1969, Codex Standard for Quick Frozen Finfish, Uneviscerated and Eviscerated, Codex Stan 36-1981, Rev. 1-1995; Codex Standard for Quick Frozen Fish Sticks (Fish Fingers), Fish Portions and Fish Fillets-Breaded or in Batter, Codex Stan 166-1989, Rev. 2-2004; Codex Standard for Canned Salmon, Codex Stan 3-1991, Rev.1-1995; Codex Standard for Quick Frozen Blocks of Fish Fillet, Minced Fish Flesh and Mixtures of Fillets and Minced Fish Flesh, Codex Stan 165-1989, Rev.1-1995; Codex Guidelines for the Sensory Evaluation of Fish and Shellfish in Laboratories, CAC-GL 31-1999; Model Certificate for Fish and Fishery Products, CAC/GL 48-2004; Recommended International Code of Practice for Smoked Fish, CAC/RCP 25-1979; Recommended International Code of Practice for Crabs, CAC/RCP 28-1983; Code of Practice for Fish and Fishery Products, CAC/RCP 52-2003; Codex Standard for Salted Fish and Dried Salted Fish of the Gadidae Family of Fishes, Codex Stan 167-1989, Rev. 2-2005; Standard for Crackers from Marine and Freshwater Fish, Crustacean and Molluscan shellfish, Codex Stan 222-2001; Codex Standard for Canned Tuna and Bonito, Codex Stan 70-1981, Rev. 1-1995; Codex Standard for Canned Crab Meat, Codex Stan 90-1981, Rev. 1-1995; Codex Standard for Quick Frozen Shrimps or Prawns, Codex Stan 92-1981, Rev. 1-1995; Codex Standard for Canned Sardines and Sardine-type Products, Codex Stan 94-1981; Codex Standard for Quick Frozen Lobsters, Codex Stan 95-1981, Rev 2-2004; Codex Standard for Canned Finfish, Codex Stan 119-1981, Rev. 1-1995; Codex Standard for Boiled Dried Salted Anchovies, Codex Stan 236-2003; and Recommended International Code of Practice for Lobsters, CAC/RCP 24-1979.

[22] The Hazard Analysis Critical Control Point (HACCP) is a science based system which is aimed to prevent food safety problems from occurring rather than reacting to non-compliance of the finished product. The HACCP system accomplishes this through the identification of specific hazards and the implementation of control measures. *See* Code of Practice for Fish and Fishery Products, Sec. 5.

[23] *Code of Practice for Fish and Fishery Products*, CAC/RCP 52-2003.

Other relevant organisations for the promotion of food safety include the World Organisation for Animal Health (OIE) through its Aquatic Animal Health Code and WTO through the SPS Agreement. The OIE has adopted the Aquatic Animal Health Code and the Manual of Diagnostic Tests for Aquatic Animals which aim to ensure the sanitary safety of international trade in aquatic animals (fish, molluscs, crustaceans and amphibians) and their products. This Code specifies health measures that may be used by veterinary authorities of importing and exporting countries to avoid the transfer of agents pathogenic for animals or humans, while avoiding unjustified sanitary barriers.[24] Similarly, the SPS Agreement provides for the establishment of national sanitary and phytosanitary regulations which are not more trade restrictive than required. The agreement specifically provides that in assessing the risk to animal or plant life or health and determining the measure to be applied for achieving the appropriate level of sanitary or phytosanitary protection from certain risks like pests and diseases, WTO Members must take into account available relevant economic factors such as the potential damage in terms of loss of production or sales in the event of the entry, establishment or spread of a pest or disease, the costs of control or eradication in the territory of the importing Member, and the relative cost effectiveness of alternative approaches to limiting the risks.[25]

One of the challenges in implementing the codes of practice and standards on food safety and quality is that most of them are non-mandatory in nature. They are not established to replace national regulations that address technological and hygienic problems which might be unique to a particular area or specific fishery. These standards and guidelines only aim at encouraging harmonisation of national sanitary measures, consistent with the FAO Code of Conduct, SPS Agreement and TBT Agreement.[26] The FAO Code of Conduct recognises that it is unlikely that identical sanitary measures and certification programmes in different States can be achieved given the diversity of industry, governments, and legal systems among States.[27] Therefore it recognises the principle of equivalence which "provides that as long as the programme and methods can be seen to be clearly defined, enforced, scientifically valid and verifiable as to their effectiveness then States should recognise each other's systems as being 'equivalent'.[28]

---

[24] *See* World Organisation for Animal Health Website, www.oie.int. Accessed on 13 July 2008.

[25] WTO, *Agreement on the Application of Sanitary and Phytosanitary Measures*, concluded on 15 December 1993, in force 01 January 1995, GATT Doc. MTN/FA Ii-A1A-4, 33 ILM XXX, Art. 2(2). *Hereinafter referred to as SPS Agreement*. Art. 5(3).

[26] *FAO Code of Conduct*, Art. 11.1.4; *SPS Agreement*, Art. 3; and WTO, *Agreement on Technical Barriers to Trade*, Geneva, concluded on 12 April 1979, in force 1 January 1980, 1186 UNTS 276. *Hereinafter referred to as* TBT Agreement. Art. 2.2.6.

[27] FAO, Responsible Fish Utilisation, *FAO Technical Guidelines for Responsible Fisheries No. 7* (Rome: FAO, 1998), at para. 57.

[28] *Ibid.*, para. 59.

Another challenge is with respect to the general nature of some of the standards and guidelines on food safety and quality. These standards often do not consider requirements specific to particular food products, leaving the interpretation of their application to the food inspectors,[29] thus creating diversity in practice. Some of these codes of practice would need to be modified to take into account local conditions and specific consumer requirements. However, any modification would need to be in keeping with the principle of equivalence on food safety and quality. States would also need to ensure that any formulation of independent national food laws and standards do not become unnecessary barriers to trade.

8.2.1.2 *Ensuring Food Security*

There are a variety of uses for fisheries resources; however, the most common is as a source of food. About 75 per cent of global fish production is used for direct human consumption.[30] Fish represents a valuable source of protein and nutrition in the diets of the nationals of many States. Apart from increasing the availability of local food, particularly in low income food-deficit countries, the trade in fish contributes to the income and foreign exchange of States.[31] The earnings from international trade in fish contribute to ensuring food security at the aggregate level.[32]

In recognition of the vital role of fish and its contribution to food security, the FAO Code of Conduct places emphasis on the role of post-harvest activities and international trade in ensuring food security. Under Article 11.1 of the FAO Code of Conduct, States are encouraged to use fish for human consumption and promote the consumption of fish whenever appropriate.[33] States are also required to cooperate in order to facilitate the production of value added products by developing countries,[34] through transfer of technology and joint ventures. It is

---

[29] *FAO Fisheries Technical Paper No 444*, *supra* note 17, at 7.

[30] *FAO State of World Fisheries and Aquaculture 2006* (Rome: FAO, 2007), at 34.

[31] In a study conducted by FAO and the Norwegian Agency for International Development (NORAD), the positive and negative impact of international fish trade on food security, including food availability and accessibility, employment, distribution of benefits and hard currency earnings, of 11 States were assessed. The countries are Brazil, Chile, Fiji, Ghana, Kenya, Namibia, Nicaragua, the Philippines, Senegal, Sri Lanka, and Thailand. This study indicated that in eight out of 11 countries, international trade has had a positive impact on food security. However, it was noted that international trade in food products was determined to have had a negative impact on the fish resources for all countries. *See* J Kurien, Responsible Fish Trade and Food Security, *FAO Fisheries Technical Paper No 456* (Rome: FAO, 2005).

[32] For example, earnings from fish exports can make important contribution to the increasing total food import bills of low-income food-deficit countries. *See ibid.*, at 36. *See also* FAO State of World Fisheries and Aquaculture 2006, *supra* note 30, at 114.

[33] Apart from using post-harvest activities to encourage the human consumption of fish, this provision also refers to the utilisation of fish for other purposes such as the production of animal feeds and fertilisers to ensure that fish is not lost from the food chain altogether. *See FAO Technical Guidelines for Responsible Fisheries No 7*, *supra* note 27, para. 119.

[34] *FAO Code of Conduct*, Art. 11.1.10.

therefore imperative to reduce the amount of fish that is lost between capture and consumption to help ensure fish food security, preserve the integrity of the fishing industry, and assist in the sustainable exploitation of fisheries resources. This also calls for the harmonisation of standards in order to prevent post-harvest loss of thousands of tonnes of fish and fishery products which are detained, rejected or destroyed each year at the borders of many importing regions in the world for failure to comply with national safety and quality requirements.[35]

8.2.1.3 *Sustainable Development and Utilisation of Fisheries Resources*

While ensuring food safety and food security through post-harvest activities and international trade, States are also required to promote the sustainable development of fisheries resources. The FAO Code of Conduct outlines some specific provisions related to post-harvest practices in this regard. In particular, the FAO Code of Conduct requires States to ensure that processing, transporting, and storage methods are environmentally sound.[36] States are also required to encourage those involved in fish processing, distribution and marketing to reduce post-harvest losses and wastes, improve the use of by-catch consistent with responsible fisheries management practices, and use resources, such as water and energy, in an environmentally sound manner.[37] However, laws, regulations, and policies addressing the environmental effects of post-harvest activities would need to be adopted without creating any market distortion.[38] The FAO Code of Conduct also calls for improving the identification of the origin of fish and fishery products traded.[39] This is discussed in greater detail in the succeeding sections.

According to FAO, most of its members have taken steps to reduce post-harvest losses in processing, distribution and marketing of fish. The main measures taken include improving handling and conservation methods, enacting food safety regulation, procedures and standards, HACCP implementation and training, and improved use of by-catch in processing, by-product utilisation and diversification of product use.[40] FAO members have also reported that they have taken steps to eliminate the processing of and trading in, illegally harvested fish through enhanced control and inspection measures and introduction of mandatory reporting schemes.[41]

---

[35] *FAO State of World Fisheries and Aquaculture 2006, supra* note 30, at 136.

[36] *FAO Code of Conduct*, Art. 11.1.7.

[37] *FAO Code of Conduct*, Art. 11.1.8.

[38] *FAO Code of Conduct*, Art. 11.1.12.

[39] *FAO Code of Conduct*, Art. 11.1.10.

[40] FAO, Committee on Fisheries, Sub-Committee on Fish Trade, Eleventh Session, Bremen, Germany, 2–6 June 2008, Monitoring Implementation of Article 11 of the Code of Conduct for Responsible Fisheries, *COFI:FT/XI/2008/9* (Rome: FAO, 2008), at 2.

[41] *Ibid.*, at 3.

With respect to fish trade, the FAO Code of Conduct only outlines general provisions to the effect that the international trade in fish and fishery products should not compromise the sustainable development of fisheries and responsible utilisation of living aquatic resources.[42] This is underpinned by Article XX of GATT 1994 which allows States, as an exception to the application of other GATT provisions, to adopt or enforce measures relating to the conservation of exhaustible natural resources, such as restrictions on domestic production or consumption as long as such measures are not applied in a discriminatory manner. States are also required to cooperate in promoting responsible international fish trade,[43] and not to undermine conservation measures for living aquatic resources in order to gain trade or investment benefits.[44] The requirement to cooperate with respect to the promotion of responsible international trade and adherence to multilateral conservation and management measures for fisheries emphasises the need to avoid the application of unilateral measures that would affect trade in an unjustifiable manner.

8.2.1.4 *Promoting International Trade Consistent with WTO Agreements*

Article 11.2 of the FAO Code of Conduct elaborates on the need to abide by the principles, rights, and obligations established under WTO agreements. The FAO Code of Conduct provides a number of measures consistent with the principles of fair and non-discriminatory trade practices and harmonisation of trade rules on fisheries. One of the measures in this regard is the requirement to liberalise trade in fish and fish products and eliminate barriers and distortions to trade such as duties, quotas and non-tariff barriers.[45] Tariff and non-tariff barriers applied to fish products are known to reduce the opportunities of States to maximise their benefit and comparative advantage from a liberalised trade which will increase the cost of fish and fish products for consumers. The FAO Code of Conduct also requires States not to directly or indirectly create unnecessary or hidden barriers to trade which limit consumers' freedom of choice of supplier or that restrict market access.[46] States are required not to condition access to markets and resources and purchase of specific technology or sale of other products, except those which may come under the framework of bilateral and multilateral fishing agreements.[47] As an example, distant water fishing nations which are unsuccessful in gaining access to fisheries in the EEZ of developing coastal States should not withhold market access from that coastal State. Furthermore, part of market

---

[42] *FAO Code of Conduct*, Art. 11.2.2.
[43] *FAO Code of Conduct*, Art. 11.2.11, 11.2.13, and 11.2.15.
[44] *FAO Code of Conduct*, Art. 11.2.12.
[45] *FAO Code of Conduct*, Art. 11.2.5.
[46] *FAO Code of Conduct*, Art. 11.2.6.
[47] *FAO Code of Conduct*, Art. 11.2.7 and 11.2.8.

liberalisation is the elimination of measures that distort trade, such as subsidies that contribute to overcapacity in the fishing industry, overfishing, and IUU fishing. Lastly, similar to regulations on post-harvest activities, the FAO Code of Conduct requires States to adopt measures on fish trade to comply with the SPS and the TBT Agreements.[48]

The FAO Code of Conduct also contains provisions with respect to the laws and regulations to be adopted by States relating to fish trade. Article 11.3 requires States to ensure that laws, regulations, administrative procedures and measures affecting international trade in fish and fishery products are transparent and based on scientific evidence.[49] Apart from harmonising standards applicable to international fish trade, States are required under the FAO Code of Conduct to simplify, as well as periodically review laws and regulations. The development of laws and regulations on international fish trade would need to be undertaken with appropriate consultation with industry and environmental and industry groups.[50] A State would also need to provide sufficient information and time to States and producers affected by any change to the legal requirements affecting trade.[51]

In order to support the implementation of fish trade related provisions of the FAO Code of Conduct, particularly Articles 11.2 and 11.3, Technical Guidelines for Responsible Fish Trade have been adopted by the FAO Sub-Committee on Fish Trade.[52] These guidelines have no formal legal status and are intended to provide general advice and assist in further dissemination and understanding of the relevant provisions of the FAO Code of Conduct. The technical guidelines explain the rationale for implementing the provisions of the FAO Code of Conduct and provide examples of measures and practices that may be adopted by States to promote responsible international fish trade. The measures provided in the technical guidelines also attempt to link more effectively the international trade in fish and fishery products with the sustainable conservation and management of fisheries resources. An example of such measures is the requirement for States to take into account the increasing demand for verification that fish products in international trade are originating from legal fishing operations and sustainable fisheries (and aquaculture).[53] Means of verification include catch documentation and trade documentation schemes developed by RFMOs and ecolabelling of products, which are discussed in succeeding sections.

In the drafting of the FAO Technical Guidelines on Responsible Fisheries, a number of issues have been raised with respect to responsible international fish

---

[48] *FAO Code of Conduct*, Art. 11.2.4.

[49] *See also* FAO Code of Conduct, Art. 11.2.3.

[50] *FAO Code of Conduct*, Art. 11.3.2.

[51] *FAO Code of Conduct*, Art. 11.3.4.

[52] FAO, Report of the Technical Consultation on Technical Guidelines for Responsible Fish Trade, Rome, 5–7 November 2007, *FAO Fisheries Report No 854* (Rome: FAO, 2007).

[53] *Ibid.*, para. 30.

trade. Some FAO Members have emphasised that the technical guidelines should further address issues such as the requirements of developing States to participate actively in international trade, reducing barriers to trade, facilitating marketing of fish and fishery products caught in a manner consistent with fisheries conservation and management measures and implications of WTO case law, particularly with respect to the interpretation of Article XX of GATT 1994.[54] As will be discussed in the subsequent sections, these issues are also relevant to addressing IUU fishing using trade related measures.

8.2.2 *The International Trade of Endangered Aquatic Species under CITES and FAO*

The issue of trade in threatened wildlife is addressed primarily by the Convention on International Trade in Endangered Species of Wild Flora and Fauna.[55] Although there are only a few aquatic species included in the list of threatened species, States have started to recognise the increasing role of CITES in relation to resources exploited in marine and large freshwater areas. Some of the commercially exploited aquatic species of fish, molluscs and echinoderms listed under CITES Appendices are the paddlefish and sturgeons (Acipenseriformes species), Caribbean queen conch *Strombus gigas*, whale shark Rhincodon typus, basking shark *Cetorhinus maximus*, Seahorses *Hippocampus spp*, European eel, saw fishes *Pristidae*, humphead (Napoleon) wrasse, Mediterranean date mussel, pipefishes and seahorses, and sea cucumber.[56] A number of non-fish species which are important either as targeted species in marine fisheries or taken as by-catch, such as a number of whale species and marine turtle species, are listed on Appendix 1.[57]

When the FAO Code of Conduct was adopted in 1995, the role of CITES in aquatic species was considered by making provision for cooperation in complying with relevant international agreements regulating trade in endangered species.[58] The FAO Technical Guidelines for Responsible Fish Trade also supplement the requirement under the FAO Code of Conduct by calling for the participation and cooperation of States in the development, implementation and enforcement

---

[54] FAO Committee on Fisheries, Report of the Tenth Session of the Sub-Committee on Fish Trade, Santiago de Compostela, Spain, 30 May–02 June 2006, *FAO Fisheries Report No 807* (Rome: FAO, 2006), at 9.

[55] See preliminary discussions presented in Chapter 3. *Convention on International Trade in Endangered Species of Wild Fauna and Flora*, Washington D.C., USA, concluded on 3 March 1973, in force 1 July 1975, 27 UST 1087; TIAS 8249; 993 UNTS 243. *Hereinafter referred to as* CITES. Amended at Bonn, on 22 June 1979 and Gaborone on 30 April 1983.

[56] FAO, *Convention on International Trade in Endangered Species of Wild Fauna and Flora (CITES) Implications to Fisheries*, www.fao.org. Accessed on 11 July 2008.

[57] FAO, *Convention on International Trade in Endangered Species of Wild Fauna and Flora (CITES)*, www.fao.org. Accessed on 11 July 2008.

[58] *FAO Code of Conduct*, Art. 11.2.9.

of measures to regulate trade in endangered species, such as those under CITES and other measures developed by relevant organisations such as RFMOs.[59] To emphasise the importance of cooperation between CITES and RFMOs, a number of resolutions have also been adopted. As an example, CITES recommended the adoption of CCAMLR *Dissostichus* Catch Documentation scheme by its Parties, and implementation of its requirements to verify the introduction into, export from, or transit through territories under their jurisdiction of Patagonian toothfish.[60] CITES further called on its Parties to cooperate with CCAMLR in combating IUU fishing in toothfish and gathering information in connection with IUU fishing. The International Commission for the Conservation of Atlantic Tunas (ICCAT) has also adopted a resolution calling on Parties to CITES to consult fully with ICCAT in reaching conclusions on proposals for listing any relevant marine species and for the revision of the criteria for listing species on the CITES Appendices.[61] These resolutions are rather general in nature and have yet to fully take into account ongoing discussions between the FAO and CITES on the trade in commercially exploited and endangered aquatic species.

Under CITES, trade in specimens of species under CITES Appendices, which include aquatic and associated non-fish species, require prior grant and presentation of appropriate export, import, and re-export permits and introduction from sea certificates.[62] The issuance of these permits is subject to a number of conditions, including the requirement that the specimen was not obtained in contravention of CITES or the laws of a State for the protection of fauna and flora. As discussed in Chapter 3, trade in CITES aquatic species contrary to these requirements are considered illegal fishing.

There are a number of legal and technical issues related to the regulation of CITES species based on these requirements for listing and granting of permits.

---

[59] FAO, Responsible Fish Trade, *Technical Guidelines for Responsible Fisheries No. 11* (Rome; FAO, 2009), at paras. 59 and 62. *See for example* ICCAT, Resolution by ICCAT on Cooperation with the Food and Agriculture Organization of the United Nations (FAO) with Regard to Study on the Status of Stocks and By-catches of Sharks Species, 21 December 1995; Resolution by ICCAT on Cooperation with the Convention on International Trade in Endangered Species of Wild Fauna and Flora (CITES), *93-08 MISC*; Resolution by ICCAT Concerning the Composition of the Delegations of ICCAT Contracting Parties to CITES 93-09 MISC. *See also* CITES Conf.12.4, Cooperation between CITES and the Commission for the Conservation of Antarctic Marine Living Resources Regarding Trade in Toothfish. See also discussions in Chapter 9.

[60] CITES Conf.12.4, Cooperation between CITES and the Commission for the Conservation of Antarctic Marine Living Resources Regarding Trade in Toothfish, *ibid.*

[61] ICCAT, Resolution by ICCAT on Cooperation with the Food and Agriculture Organization of the United Nations (FAO) with Regard to Study on the Status of Stocks and By-catches of Sharks Species, 21 December 1995; Resolution by ICCAT on Cooperation with the Convention on International Trade in Endangered Species of Wild Fauna and Flora (CITES), *93-08 MISC*; Resolution by ICCAT Concerning the Composition of the Delegations of ICCAT Contracting Parties to CITES 93-09 MISC.

[62] *CITES*, Art. III.

FAO and CITES have been working closely[63] to address some of these issues. How these issues are resolved would largely determine what rules would apply to fishing for CITES aquatic species and what may be called IUU fishing in this respect. One of the issues that require the cooperation of FAO and CITES is the need to refine the criteria and guidelines for listing species on CITES Appendices to reflect the specific characteristics of aquatic resources, as well as the evaluation of proposals to amend CITES Appendices containing these species. CITES has adopted Res.Conf.9.24 (Rev.CoP14) which provides the revised guidelines for the listing criteria for commercially exploited aquatic species. These criteria take into account factors such as the decline in population or habitat, vulnerability to intrinsic and extrinsic factors, and the need to regulate trade in order to avoid reaching a level in population at which survival might be threatened.[64] The listing of commercially exploited aquatic species under CITES also draws attention to the need to establish management strategies to control possible overfishing of some of these species brought about by illegal international trade, such as in the case of sea cucumbers and humphead wrasse.[65]

The legal interpretation of the term "introduction from the sea" under CITES has also been a focus of debate among States. Under Article 1 of CITES, "Introduction from the sea" is considered as trade, and has further been defined as "transportation into a State of specimens of any species which were taken in the marine environment not under the jurisdiction of any State". It was agreed under CITES Resolution Conf.14.6 that "the marine environment not under the jurisdiction of any State" under the definition of introduction from the sea means those marine areas beyond the areas subject to the sovereignty or sovereign rights of a State consistent with international law, as reflected in the LOSC. However, it

---

[63] *See* Appendix F on 'FAO Workplan on CITES and Commercially-exploited Aquatic Species', in Report of the Twenty-Fifth Session of the Committee on Fisheries, Rome, 24–28 February 2003, *FAO Fisheries Report No 702* (Rome: FAO, 2003). A Memorandum of Understanding between FAO and CITES was adopted by the 10th Session of the COFI Sub-committee on Fish Trade in June 2006 and signed by FAO and CITES during the 54th Meeting of the Standing Committee of CITES in October 2006. A Trust Fund Project on "CITES and commercially-exploited species, including the evaluation of listing proposals (GCP/INT/987/JPN) funded by Japan is also being implemented to contribute to FAO's efforts to ensure that CITES provisions are applied appropriately and responsibly to commercially exploited aquatic species.

[64] CITES, Conf. 9.24 (Rev.CoP14), *Criteria for amendment of Appendices I and II*, 2007. See also CITES, Fourteenth Meeting of the Conference of the Parties, The Hague, Netherlands, 3–15 June 2007, the Interpretation of Annex 2 a (Criteria for the Inclusion of Species in Appendix II in accordance with Article II, paragraph 2(a), of the Convention) and Annex 5 (Annex 5 definitions, explanations and guidelines) or Resolutions Conf.9.24 (Rev.CoP13) in relation to commercially exploited aquatic species, Submitted by the Food and Agriculture Organization, CoP14 Inf.64.

[65] FAO Committee on Fisheries, Sub-committee on Fish Trade, Eleventh Session, Bremen, Germany, 2–6 June 2008, CITES Issues with Respect to International Fish Trade, *COFI:FT/XI/2008/4* (Rome: FAO, 2008), at paras. 23–26. *See also* CITES, Technical Workshop on the conservation of sea cucumbers in the families *Holothuridae* and *Stichopodidae*, Kuala Lumpur, Malaysia, 1–3 March 2004.

remains uncertain as to how "transportation into a State" should be interpreted.[66] CITES also provides that the introduction from the sea of any specimen of a species included in Appendices I and II requires prior grant of a certificate from a management authority of the State of introduction. Such a certificate may only be issued upon meeting certain conditions such as the assurance that the introduction will not be detrimental to the survival of the species and that living specimens will be handled so as to minimise the risk of injury, damage to health or cruel treatment.[67] These provisions do not really clarify how the CITES process of making a non-detrimental finding and issuing an introduction from the sea certificate should be implemented.

8.3 Internationally Agreed Market Related Measures to Address IUU Fishing

It is within the legal framework for the international trade in fish and fishery products, primarily through the link between international trade and the sustainable development and utilisation of fisheries resources, that measures under the IPOA-IUU would need to be considered and implemented. The IPOA-IUU reiterates the principles of the FAO Code of Conduct and specifically provides that measures adopted under paragraphs 66 to 76 "...are to be implemented in a manner which recognises the right of States to trade in fish and fishery products harvested in a sustainable manner and should be interpreted and applied in accordance with the principles, rights and obligations established in the World Trade Organization, and implemented in a fair, transparent and non-discriminatory manner".[68] These measures are what the IPOA-IUU refers to as "Internationally Agreed Market-Related Measures". Although the IPOA-IUU does not define what a "market-related measure" is, the term is generally understood to encompass several types of controls on the importation and exportation of goods.

Similar to the FAO Code of Conduct, the IPOA-IUU has, on a number of provisions, stressed the need to apply internationally agreed market related measures in a fair, transparent, and non-discriminatory manner.[69] In addition, the IPOA-IUU

---

[66] *See* F. Meere, G. Geen, Q. Hanich, M. Lack, G. Sant, and M. Tsamenyi, 'CITES 'Introduction from the Sea': A Practical Way Forward', Proceedings of a Joint TRAFFIC/ANCORS Workshop, 21 *TRAFFIC Bulletin* (2008). In this workshop, it was argued that for the purpose of CITES, specimens can be considered as having been transported into a State when the species is either landed on the vessel as long as the vessel is flagged (and not Stateless) and it is therefore under the control of a State, or landed in a port.

[67] *CITES*, Articles III(5) and IV(6).

[68] FAO, *International Plan of Action to Prevent, Deter, and Eliminate Illegal, Unreported, and Unregulated Fishing*, adopted on 23 June 2001 at the 120th Session of the FAO Council. *Hereinafter referred to as* IPOA-IUU. Para. 65.

[69] *IPOA-IUU*, paras. 65, 66, 68, 69, 73, and 74.

requires a State to ensure that measures relating to international trade in fish and fishery products are transparent and where applicable, based on scientific evidence.[70] The IPOA-IUU further emphasises that "(t)rade related measures should only be used in exceptional circumstances, where other measures have proven unsuccessful to prevent, deter and eliminate IUU fishing, and only after prior consultation with interested States.[71] Hence, trade measures against IUU fishing need to be applied as a last resort.

This section examines examples of internationally agreed market or trade related measures[72] which the IPOA-IUU has adopted to address IUU fishing. These measures include the traceability of fish and fishery products, trade related RFMO measures, trade restrictive measures, and Harmonised Commodity Description and Coding System. This section also discusses other relevant market measures such as the eco-labelling and restricting business with IUU fishers.

8.3.1 *Traceability of Fish*

Traceability is defined by the ISO as the ability to trace the history, application or location of an entity by means of recorded identification.[73] As a tool in food inspection and certification system, traceability or product tracing means the ability to follow the movement of a food through specified stages of production, processing and distribution.[74] It is also a tool to protect consumers against deceptive marketing practices and ensure fair practices in food trade on the basis of accurate product description. In fisheries, this implies the development of systems giving information on the different stages of the chain of fish and fishery products from the boat, sea, or farm to the table or fork.

The Codex Alimentarius Commission adopted principles for the application of traceability or product tracing. It provides that traceability or product tracing should be practical, technically feasible and economically viable within a food inspection and certification system. Such system should also not be made more

---

[70] *IPOA-IUU*, para. 68.

[71] *IPOA-IUU*, para. 66.

[72] Trade related measures are border controls that allow a State to regulate, restrict or prohibit trade. Examples of trade measures include landing actions, certification, labelling, or size requirement, among others. *See* Linda A. Chaves, 'Illegal, Unreported and Unregulated Fishing: WTO-Consistent Trade Related Measures to Address IUU Fishing', in FAO, Report of and Papers Presented at the Expert Consultation on Illegal, Unreported and Unregulated Fishing, Sydney, Australia, 15–19 May 2000, *FAO Fisheries Report No. 666* (Rome: FAO, 2001).

[73] ISO 8402:1994, as cited in FAO, Committee on Fisheries, Subcommittee on Fish Trade, Eighth Session, Bremen, Germany, 12–16 February 2002, Item 8 of the Provisional Agenda, Traceability of Products from Fisheries and Aquaculture, *COFI:FT/VIII/2002/7*, (Rome: FAO, 2002), at para 1.

[74] Codex Alimentarius, *Principles for Traceability/Product Tracing as a Tool within a Food Inspection and Certification System*, CAC/GL 60-2006.

trade restrictive than necessary, and should be implemented on a case by case basis. The Codex Alimentarius Commission identified a number of factors that may be taken into account in formulating and applying traceability requirements. These factors include the objectives or outcomes of the food inspection and certification systems, assessed food safety risks, characteristics of the potential deceptive marketing practices being addressed, capabilities of developing States, and provision of assistance to the exporting State.[75] The basic characteristics of traceability responds to the first aim of the general framework for the international trade in fish and fishery products, which is to maintain food safety and quality.

Traceability would need to be distinguished from certification, in that traceability is a mechanical process documenting all the stages of production and distribution that food products go through, while certification is more of a statement ensuring that certain specifications have been undertaken in compliance with environmental, social, or food safety and quality standards.[76] Traceability systems also serve as a form of protection for companies, in the sense that if a problem arises at any stage of production or processing, it would be possible to recall only the batch affected and not the entire production.[77]

A traceability system may either be voluntary or mandatory, depending on its objectives and the associated regulations and standards that it hopes to implement. A traceability system may also be a public or private standard.[78] An example of a mandatory and public standard for the traceability of fish is that which is being implemented by the EU. The Council Regulation (EC) No 104/2000 on the common organisation of markets in fishery and aquaculture products provides a mandatory traceability requirement for fishery products made available in the market in large quantities.[79] Its main aim is to ensure food safety. Fishery products may not be offered for retail sale to the final consumer unless appropriate marking or labelling indicates the commercial designation of the species, the production method (for instance caught at sea or in inland waters or farmed), and the catch area.[80] The EU also implemented a programme called the "TraceFish" from 2000–2002, which provided the basis for the current implementation of similar

---

[75] *Ibid.*, paras. 14–19.

[76] FAO, *COFI:FT/VIII/2002/7, supra* note 73, para 2.

[77] Iciar Martinez, David James, and Henri Loréal, Application of Modern Analytical Techniques to Ensure Seafood Safety and Authenticity, *FAO Fisheries Technical Paper No 455* (Rome: FAO, 2005), at 28.

[78] FAO, Committee on Fisheries, Sub-committee on Fish Trade, Tenth Session, Traceability and Labelling in Fish Trade, Santiago de Compostela, Spain, 30 May-02 June 2006, *COFI:FT/X/2006/6*, (Rome: FAO, 2006), at 2.

[79] European Union (EU), Council Regulation (EC) No. 104/2000 of 17 December 1999 on the common organisation of the markets in fishery and aquaculture products, *OJ L 17, 21.1.2000*.

[80] EC No. 104/2000, Art. 4(1).

traceability programmes for seafood products.[81] TraceFish formulated standard guidelines for trading partners in the EU to be able to interchange traceability data in a consistent and fast manner.[82]

The U.S. implements mandatory food traceability requirements under its *Food and Drug Administration Act*, which requires producers, distributors, importers, transporters and packers of food in the U.S. to establish and maintain records to identify immediate sources and recipients of food.[83] While this legislation aims to protect consumers against food borne illness, the U.S. also extended the application of food traceability to protect against deliberate contamination of food.[84] Prior notification is required for imported seafood.[85] The notice should include information on the manufacturer and shipper, the grower, the country of origin, the country from which the article is shipped, and the anticipated port of entry. If such notice is not provided, the shipment is refused admission into the U.S.

Most States do not have equivalent mandatory traceability requirements for food products such as in the EU and U.S. Different States have different traceability programs and subscribe to various suppliers of food traceability systems. Some of the most common traceability providers specific to the seafood industry and fishing vessel operations are Wisefish, Trace 2000, and the Astra System.[86]

A number of issues arise with respect to the implementation of traceability programmes. There have been differing views among States with respect to the scope of application of food traceability, as it addresses both food safety and non-food safety issues. Some States argue that product tracing principles in food import and export inspection and certification systems should be formulated primarily as a risk assessment tool to ascertain food safety.[87] Other States maintain

---

[81] European Commission, Community Research, Quality of Life and Management of Living Resources, *Traceability of Fish Products*, www.tracefish.org, Accessed on 12 February 2006.

[82] European Commission, Community Research, Quality of Life and Management of Living Resources, *TraceCore-XML Standard Guidelines*, www.tracefish.org, Accessed on 12 February 2006.

[83] U.S., *Food and Drug Administration Act*, Title 3, Section 306, Maintenance and Inspection of Records for Food, 9 December 2004.

[84] U.S., *Public Health Security and Bioterrorism Preparedness Response Act of 2002*. Public Law 107–188. 2002.

[85] U.S., *Public Health Security and Bioterrorism Preparedness Response Act of 2002*, Section 307.

[86] See Arni Petersen and David Green, *Seafood Traceability: A Practical Guide for the U.S. Industry*, Report Prepared for the North Carolina Sea Grant, no date. www.ncseagrant.org. Accessed on 23 August 2008.

[87] Joint FAO/WHO Food Standards Programme, Codex Alimentarius Commission, Twenty-eighth Session, Rome, Italy, 04–09 July 2005, Report of the Thirteenth Session of the Codex Committee on Food Import and Export Inspection and Certification Systems, Melbourne, Australia, 06–10 December 2004, *ALINORM 05/28/30*, para 86.

that traceability systems should also be used to provide more information for consumers and could be used to achieve specific conservation objectives.[88]

Traceability systems also have their utility in addressing IUU fishing. Both the FAO Code of Conduct and the IPOA-IUU call for the improvement of the traceability of fish and fish products. Article 11.1.11 of the FAO Code of Conduct provides the requirement for States to ensure that international and domestic trade in fish and fishery products are consistent with sound conservation and management practices through improving the identification of the origin of fish and fishery products traded. This is reiterated in paragraph 71 of the IPOA-IUU which provides that States should take steps to improve the transparency of their markets to allow the traceability of fish or fish products.[89] However, neither provision stipulates the actions that may be taken to enforce traceability requirements. It can only be assumed that for the purpose of addressing IUU fishing, the traceability of fish and fishery products means tracing back to the area of harvest and imposing supplementary requirements that will ensure that the fish or a fishery product has not been sourced through IUU means. What a State would do if a fish or fishery product has been traced back to an IUU activity will depend on its domestic policy and law.

Another related limitation is the application of diverging traceability systems. Similar to SPS measures, States have different levels of implementation of requirements for the traceability of fish. This makes it difficult to harmonise measures across different fish trading States, with the potential to create barriers to trade.[90] There are also financial implications of implementing traceability requirements for producers and producing countries, particularly for small scale fisheries and small to medium scale fisheries related business entities.

8.3.2 *Harmonized Commodity Description and Coding System*

The IPOA-IUU encourages States to work towards using the Harmonized Commodity Description and Coding System for fish and fishery products to promote the implementation of the IPOA-IUU.[91] However, the IPOA-IUU does not elaborate further on how such system can address IUU fishing.

---

[88] FAO, Committee on Fisheries, Report of the Eighth Session of the Sub-committee on Fish Trade, Bremen, Germany, 12–16 February 2002, *FAO Fisheries Report No. 673* (Rome: FAO, 2002), at para. 30.

[89] *IPOA-IUU*, para. 71.

[90] Joint FAO/WHO Food Standard Programme Codex Alimentarious Commission, Thirty-first Session, Geneva, Switzerland, 30 June-05 July 2008, Report of the Sixteenth Session of the Codex Committee on Food Import and Export Inspection and Certification Systems, Surfers Paradise, Queensland, Australia, 26–30 November 2007, *ALINORM 08/31/30*, para. 86.

[91] *IPOA-IUU*, para. 75.

The Harmonized Commodity Description and Coding System is a multipurpose nomenclature for international products developed by the World Customs Organization (WCO) and based on an international convention ratified by more than 130 States and regional economic integration organisations.[92] It provides a uniform classification for about 200,000 commodities or 5,000 commodity groups identified by a six digit code and arranged according to a legal and logical structure.[93] Fish and fishery products are covered under Chapter 3 of the Harmonized System Nomenclature with numbers from HS0301 to HS0307.[94]

The system is used by States for a number of purposes, including as a basis for customs tariff preferences and control, trade policies, monitoring of controlled goods, rules of origin, transport statistics, price monitoring, and economic research and analysis.[95] The harmonised codes provide customs officials with information to facilitate inspection of products and to regulate trade in controlled goods with environmental impact, such as hazardous wastes, endangered species, and ozone depleting substances.[96] The commodity codes have also been used by the WCO and national customs administrations as a tool for enforcement and compliance to help address transnational crime, including environment related violations of CITES and other multilateral environmental agreements.

The relevance of the Harmonized Commodity Description and Coding System to addressing IUU fishing has yet to be considered. However, the purpose and use of the Harmonized Commodity Description and Coding System provides a rationalisation of its use to identify fish caught through IUU means. Through the harmonised coding systems for fish and fishery products, States will be able to

---

[92] *International Convention on the Harmonized Commodity Description and Coding System*, Brussels, Belgium, concluded 14 June 1983, in force 1 January 1988, Protocol to the Amendment, Brussels, Belgium, concluded on 24 June 1986, in force 1 January 1988. ATS 1988 No. 30.

[93] World Customs Organization, *Overview: What is the Harmonized System?* Accessed on www.wco.org. Accessed on 16 January 2009.

[94] HS0301– Live fish; HS0302 – Fish, fresh or chilled, excluding fish fillets and other fish meat of heading 0304; HS0303 – Fish, frozen, including fish fillets and other fish meat of heading 0304; HS0304 – fish fillets and other fish meat (whether or not minced), fresh, chilled or frozen; HS0305 – Fish, dried, salted or in brine; smoked fish, whether or not cooked before or during the smoking process; flours, meals and pellets of fish, fit for human consumption; HS0306 – Crustaceans, whether in shell or not, live, fresh, chilled, frozen, dried, salted or in brine; crustaceans, in shell, cooked by steaming or by boiling in water, whether or not chilled, frozen, dried salted or in brine; flours, meals and pellets of crustaceans, fit for human consumption; HS0307 – Molluscs, whether in shell or not, live, fresh, chilled, frozen, dried, salted or in brine; aquatic invertebrates other than crustaceans and molluscs, live, fresh, chilled, frozen, dried, salted or in brine; flours, meals and pellets of aquatic invertebrates other than crustaceans, fit for human consumption. *See* WCO, *HS Nomenclature 2007 Edition*, Section 1 Chapter 3 Fish and crustaceans, molluscs and other aquatic invertebrates. www.wco.org. Accessed on 16 January 2009.

[95] World Customs Organization, *Overview: What is the Harmonised System?* www.wco.org. Accessed on 16 January 2009. See also WCO, *Position of Contracting Parties to the Harmonized System Convention and Non-Contracting Party Administrations*, 15 January 2009. www.wco.org. Accessed on 20 January 2009.

[96] WCO, *HS Multi-Purposes Tool*, www.wco.org. Accessed on 16 January 2009.

monitor the international trade in fishery goods crossing their borders, enabling them to impose additional measures on goods bearing such codes with respect to monitoring the legality of catches. In order to identify IUU caught fish, Harmonized Commodity Description and Coding System may be implemented together with catch certification requirements. The EU for example, has adopted a regulation to ensure that fishery products categorised under Chapter 3 and Tariff headings 1604 and 1605 of its Combined Nomenclature entering the EU market are not derived from IUU fishing.[97] Such fishery products must be accompanied by validated catch certificates when they enter the EU market.[98] Fisheries departments and customs administrations may also cooperate to exchange information on the trade in fishery products which fall under HS03 in order to identify which products are being traded illegally.

8.3.3 *Trade Related Measures Adopted by RFMOs*

The IPOA-IUU gives particular importance to multilateral trade related measures adopted by RFMOs that may be necessary to combat IUU fishing. In fact, trade related measures adopted by RFMOs may be considered as being at the heart of internationally agreed market related measures, particularly given the numerous references given to such measures in the IPOA-IUU.[99] The RFMOs are deemed to have served, and will continue to serve as the primary international bodies for developing and adopting market related measures to combat IUU fishing.[100]

There are two major multilateral trade related measures discussed in the IPOA-IUU. One measure is the prevention of trade in fish by vessels identified by RFMOs as having engaged in IUU fishing.[101] The second measure is the adoption of multilateral catch documentation and catch certification schemes.[102] These two measures are examined in Chapter 9.

8.3.4 *Trade Restrictive Measures*

The IPOA-IUU refers to trade related measures that take the form of a restriction, but only insofar as they have been agreed upon multilaterally. In addition to preventing the importation of fish from IUU vessels, the IPOA-IUU also provides

---

[97] EU, Council Regulation (EC) No 1005/2008 of 29 September 2008 establishing a Community system to prevent, deter and eliminate illegal, unreported and unregulated fishing, amending Regulations (EEC) No 2847/93, (EC) No 1936/2001 and (EC) No 601/2004 and repealing Regulations (EC) No 1093/94 and (EC) No 1447/1999, *OJ L 286/1 29.10.2008*, Art. 2(8).

[98] *See* EC No 1005/2008, Chapter III. A more thorough discussion of the ECU system to combat IUU fishing is presented in Chapter 10.

[99] *See for example* IPOA-IUU, paras. 66, 68, 69, 73, and 74.

[100] *FAO Technical Guidelines for Responsible Fisheries No 9*, *supra* note 10, at 48.

[101] *IPOA-IUU*, para. 66.

[102] *IPOA-IUU*, para 69.

for the adoption of "appropriate multi-laterally agreed measures such as import and export controls or prohibitions."[103] As will be discussed in the succeeding chapter, a number of RFMOs impose restrictions on the trade of fish derived from IUU fishing activities.

The imposition of trade restrictions on fishery products has resulted in disputes among States. The United States, for example, banned imports of yellowfin tuna from Mexico for failure to protect Eastern Pacific Tropical dolphins in accordance with the *Marine Mammal Protection Act*.[104] The GATT Panel concluded that prohibiting the imports of tuna products for the purpose of enforcing a domestic law is contrary to GATT rules even if the law aims to protect animal health or exhaustible natural resources.[105] It held that the U.S. could not embargo imports of tuna products from Mexico or intermediary States[106] simply because the production of tuna in that State did not satisfy U.S. regulations.[107]

One WTO case on unilateral trade prohibition is the U.S. embargo on shrimp and shrimp products from India, Malaysia, Pakistan, and Thailand for failing to mandate the use of turtle excluder devices in shrimp trawling or comparable measures that would prevent the adverse impact of commercial fishing operations on sea turtles.[108] The four States brought a joint complaint against the ban imposed by the U.S. in 1997. The WTO Panel ruled that the measure has been applied by the U.S. in a manner that constitutes arbitrary and unjustifiable discrimination between members of the WTO contrary to the requirements of the chapeau of Article XX of GATT 1994.[109] The WTO Panel further emphasised that the chapeau requirement also provides that measures adopted to serve environmental objectives cannot be a disguised restriction on international trade.[110]

---

[103] *IPOA-IUU*, para 69.

[104] World Trade Organization Website, *Environment: Disputes 4, Mexico, etc. versus U.S.: 'Tuna-Dolphin'*, www.wto.org. Accessed on 03 May 2006. The U.S. *Marine Mammal Protection Act 1972, As Amended* provides for the protection from the incidental taking of marine mammals by American fishing vessels and purse seine vessels of foreign States catching yellowfin tuna in the eastern tropical Pacific ocean (16 U.S.C. 1371).

[105] WTO, *Mexico, etc. versus U.S.: 'Tuna-Dolphin'*, *ibid.*, at 2.

[106] The embargo also applied to "intermediary States" or States handling tuna *en route* from Mexico to the U.S. [16 U.S.C. 1362(3)(c)(5)]. Intermediary States identified were Costa Rica, Italy, Japan, Spain, France, Netherlands Antilles, UK, Canada, Colombia, Republic of Korea and members of the Association of Southeast Asian Nations.

[107] WTO, *Mexico, etc. versus U.S.: 'Tuna-Dolphin'*, *supra* note 104, at 2. The Report of the Panel was circulated in 1999, but was not adopted.

[108] WTO, United States-Import Prohibition of Certain Shrimp and Shrimp Products, Report of the Panel, *WT/DS58/R*, 15 May 1998, para. 2.7. www.wto.org. Accessed on 15 January 2009. The relevant law in question is Section 609 of Public Law 101–102, codified at 16 United States Code (U.S.C.) § 1537.

[109] *Ibid.*, para. 7.44. Article XX(b) and (g) of GATT 1994 allow WTO Members to justify GATT-inconsistent measures if they are necessary to protect human, animal or plant life or health or if the measures relate to the conservation of exhaustible natural resources.

[110] *Ibid.*, para. 7.44. This decision is reiterated in the decision of the Appellate Body on the appeal made by the U.S. from certain issues of law and legal interpretations in the Panel Report. *See* WTO,

Restrictions impacting on trade which are taken unilaterally by States may easily be subject to disputes. Hence, it is necessary for a State to adopt trade restrictive measures consistent with its rights and obligations under relevant WTO Agreements.[111] The success in the application of import restrictions as a trade measure also lies in the efficiency of domestic import licensing regulations.[112] If domestic import regulations are formulated in a transparent and equitable manner and are not disguised as trade barriers, they are more likely to be accepted by States.

8.3.5 *Other Market Measures*

There are a host of other market measures which may be adopted to address IUU fishing. These measures include eco-labelling of fish products, support to sustainable fisheries by private companies, and increasing the awareness of those involved in the fishing industry and businesses to address IUU fishing.

8.3.5.1 *Eco-labelling of Fish Products*

A market related measure which is used progressively to address IUU fishing, although not directly provided under the IPOA-IUU, is eco-labelling of fish and fishery products. At the United Nations Conference for Environment and Development (UNCED), governments agreed to encourage the expansion of environmental labelling and other environmentally related product information programmes designed to assist consumers to make informed choices.[113] Eco-labels are seals of approval given to products that are deemed to have fewer impacts on the environment than functionally or competitively similar products.[114] If a particular fishery product bears a distinctive logo or eco-label, it is presumed to have been harvested in compliance with sustainable fisheries management measures. Based on the eco-labels awarded, and according to their environmental awareness and preference, consumers can then make the choice to purchase the product.

United States – Import Prohibition of Certain Shrimp and Shrimp Products, AB-1998-4, Report of the Appellate Body, *WT/DS58/AB/R*, 12 October 1998, para. 186.

[111] *IPOA-IUU*, para. 66.

[112] Ruangrai Tokrisna, 'WTO-Consistent Trade-Related Measures to Address IUU Fishing-Developing Countries Issues', in FAO, Report of and Papers Presented at the Expert Consultation on Illegal, Unreported and Unregulated Fishing, Sydney, Australia, 15–19 May 2000, *FAO Fisheries Report No. 666* (Rome: FAO, 2001), at 8.

[113] Agenda 21, Chapter 4, 'Changing Consumption Patterns', *hereinafter referred to as* Agenda 21, Rio de Janeiro, Brazil, 03–14 June 1992, para. 4.21.

[114] *See* James Salzman, *Environmental Labelling in OECD Countries* (OECD: Paris and Karen West, 1991), cited in Cathy Wessells, Kevern Cochrane, Carolyn Derree, Paul Wallis, and Rolf Willmann, Product Certification and Eco-labelling for Fisheries Sustainability, *FAO Fisheries Technical Paper No 422* (FAO: Rome, 2001), at 10.

There are three categories of eco-labelling. There are party labelling schemes which are considered "self declaration" and are established by individual companies based on their own product standards.[115] An example of this would be "dolphin safe" tuna labels of seafood companies.[116] The second category of party labelling schemes are established by industry associations for the products of their members, while third party labelling schemes are usually formulated by private initiators independent from the producers, distributors, and sellers of labelled products.[117] Labels range from "not overfished, to no marine mammal by-catch and not overfished, to no by-catch of any sort and not overfished, to ecosystem friendly where the entire ecosystem with its complicated system food chain is not harmed."[118] Eco-labelling schemes, however, do not directly address IUU fishing but the sustainability of fisheries resources in general.

8.3.5.1.1 *Marine Stewardship Council Eco-labelling Standard*

One of the most popular initiatives in eco-labelling is the formulation of the Marine Stewardship Council (MSC)[119] Principles and Criteria for Sustainable Fisheries. These principles consider the status of the target fish stocks, impact of the fishery on the ecosystem, and performance of the fishery management system.[120] Certain operational criteria under the MSC principles ensure that fishing activities are in compliance with all legal and administrative requirements of a State and that fish has not been caught through IUU fishing activities, such as the use of destructive fishing methods.[121] Fisheries which conform to these principles and criteria are certified. However, the MSC eco-labelling programme is voluntary and has a limited scope. Only about 7 per cent of the world's wild-capture fisheries are engaged in the programme, either as certified fisheries or in full assessment against the MSC standard for a sustainable fishery.[122] As at September 2008, there were more than 800 MSC-labelled seafood products sold

---

[115] *Ibid.*, at 11.

[116] E.g. Greenseas Dolphinsafe and John West Dolphin Friendly labels.

[117] Carolyn Deere, *Eco-labelling and Sustainable Fisheries* (FAO: Rome, 1999), at 6.

[118] Cathy Wessells, 'Barriers to International Trade in Fisheries,' *Discussion Paper Prepared for the First FAO E-mail Conference on Fish Trade and Food Security*, October-November 1998, as cited in Carolyn Deere, *Eco-labelling and Sustainable Fisheries*, *ibid.*, at 10.

[119] Marine Stewardship Council (MSC) is an independent, global, non-profit organisation which seeks to harness consumer purchasing power to generate change and promote environmentally responsible stewardship for the world's most important renewable food source. The MSC was first established by Unilever and World Wide Fund for Nature in 1997. See www.msc.org. Accessed on 17 March 2008.

[120] Marine Stewardship Council (MSC), *International Eco-labelling in Fisheries*, (London: MSC, 2004), at 4.

[121] *Ibid.*, at 17.

[122] MSC Website. *Certified Fisheries*. www.msc.org. Accessed on 03 September 2008.

in 36 countries worldwide.[123] This involves 36 fisheries accredited to the MSC standards and 75 others currently undergoing assessment.[124]

The MSC is also in the process of developing new technical guidelines to help fisheries with insufficient data to meet the MSC standards. The new guidelines introduce a risk assessment that will be initiated on small scale and data deficient fisheries to assess their performance and provide an alternative route to certification against MSC standards.[125] There are currently seven fisheries in Africa, South America, and Asia which are participating in the trials to test the new guidelines for the assessment of small scale and data deficient fisheries.[126] The application of these technical guidelines provides an opportunity for small scale fishers to participate in the eco-labelling programme and potentially be more competitive in the market.

There are various advantages in implementing eco-labelling schemes for fish and fishery products. Eco-labelling programmes can provide information about the environmental impact of products, provide consumers with the opportunity to express their environmental or ecological concerns through their purchasing behaviour, enhance incentives for producers to supply products that meet eco-labelling requirements, and encourage retailers and consumers to buy only fisheries products that come from sustainably managed resources.[127] For the purpose of combating IUU fishing, eco-labelling schemes may be used to distinguish between fish which have been caught contrary to fisheries conservation and management measures of a State or an RFMO and those which have been caught in a sustainable manner.

There are, however, challenges in the implementation of the MSC eco-labelling scheme. Questions have been raised about the application of the MSC eco-labelling criteria in certain fisheries. For example, there were criticisms that eco-labelling processes of the western rock lobster, New Zealand hoki and South Georgia toothfish were inaccurate and misleading, have failed to address the problem of IUU fishing, and have not complied with the MSC Principles and Criteria for Sustainable Fisheries.[128] Another challenge is the accountability in and transparency of the MSC certification process as well as the refinement and consistent interpretation and implementation of the MSC Principles and Criteria.[129]

---

[123] *Ibid.*

[124] *Ibid.*

[125] MSC Website, *Developing world fisheries embark on journey to MSC eco-label.* www.msc.org. Accessed on 17 March 2008.

[126] *Ibid.*

[127] Deere, *supra* note 117, at 7.

[128] Tavis Potts and Marcus Haward, 'International Trade, Eco-labelling, and Sustainable Fisheries—Recent Issues, Concepts, and Practices', *Environment, Development and Sustainability* (2006), at 12.

[129] *Ibid.*, at 13.

To address some of these concerns, it has been suggested that the implementation of the MSC eco-labelling process could be improved through the identification of critical indicators for failing a certification process such as the failure to follow scientific advice in management and the levels of IUU fishing in the fishery and by-catch levels.[130]

8.3.5.1.2 *FAO Eco-labelling Guidelines*

The utility of eco-labelling as a tool for promoting the sustainability of fisheries resources has also been recognised more formally in international fora. To provide equal opportunities for all States to participate in eco-labelling schemes, in 2005, the FAO introduced a voluntary instrument called Guidelines for the Eco-labelling of Fish and Fishery Products from Marine Capture Fisheries, to guide States in designing and promoting eco-labelling schemes.[131] These guidelines provide the minimum requirements and criteria for assessing whether an eco-label may be awarded to a fishery. The criteria for eco-labelling include the adequacy of data to evaluate the current state and trend of fish stocks, provision of scientific advice, use of data and information to identify the adverse impacts of the fishery on the ecosystem, and an effective management system and legal and administrative framework.[132] Some of these criteria are very similar to those formulated under the MSC eco-labelling programme, although the standards included in the guidelines are more comprehensive in terms of accreditation process and certification requirements. These criteria however, largely address the impacts of fishing activities on the ecosystem and can only deal indirectly with IUU fishing.

The FAO Guidelines for Eco-labelling of Fish and Fishery Products are based on principles consistent with those under the IPOA-IUU, such as the application of eco-labelling schemes in a transparent[133] and non-discriminatory manner, in a way that does not create unnecessary obstacles to trade,[134] and are based on best scientific evidence available.[135] The relationship between the two instruments, however, has not been elaborated. As the FAO Eco-labelling Guidelines and existing third party eco-labelling schemes deal with the certification of sustainable

---

[130] *Ibid.*

[131] FAO, *Guidelines for the Ecolabelling of Fish and Fishery Products from Marine Capture Fisheries*, (Rome: FAO, 2005). The Expert Consultation convened to discuss these guidelines agreed that minimum substantive requirements should be developed separately for inland and marine capture fisheries. *See* FAO, Report of the Expert Consultation on the FAO Guidelines for Eco-labelling, Rome, Italy, 03–04 March 2008, *FAO Fisheries Report No 864* (Rome: FAO, 2008), at para. 7(a).

[132] There has been a view that eco-labelling schemes should contain additional criteria with respect to fisheries where decent employment conditions are practiced. *See* M. Ben-Yami Column, How to Attain Decent Conditions of Work? World Fishing, April 2008. www.benyami.org.

[133] *FAO Guidelines for Ecolabelling of Fish and Fish Products, supra* note, 131, para. 2.4.

[134] *Ibid.*, para. 2.5.

[135] *Ibid.*, para. 2.10.

fisheries, it can only be assumed that eco-labelling partly addresses the problem of unregulated fishing.

8.3.5.2 *Market Measures by Private Companies*

The seafood industry has also adopted measures to promote sustainable fishing, particularly by espousing eco-labelling schemes. As an example, Unilever has committed to buy its fish from sustainable sources and supports the MSC standard for fish certification. In 1996, Unilever wrote to all of its suppliers asking them to confirm that their fish were legally caught in specified FAO statistical areas and has stopped sourcing products from suppliers who could not offer that confirmation.[136]

Unilever has also established its own assessment tool known as the "traffic light system", where each fishery is assessed according to five indicators: fisheries research, quota system, regulatory tools, control systems, and long term management plan.[137] In this assessment tool, the effect of fishing on marine ecosystems is also taken into account. The assessment results are graded based on three colours—red, green, and yellow. A fishery that gets all green colours is deemed sustainable and Unilever would recommend its certification under MSC standards.[138] A fishery that shows a mix of green and yellow is deemed managed and progressing, and a fishery that gets one or more red is considered poorly managed.[139] The fishery is deemed unmanaged if its assessment scores red in all five indicators.[140] Unilever does not obtain fish from unmanaged fisheries and supports those which are making progress towards sustainability.[141]

A number of companies are now supporting sustainable fisheries. Retailers like Wal-Mart and ASDA pledge to switch to 100 per cent MSC-certified fish within 3 to 5 years.[142] J Sainsbury plc of the UK, the Royal Ahold (Netherlands), and the Whole Foods Market of the U.S. have all committed to sourcing fish from sustainable sources and closely work with the MSC.[143] The widening campaign against IUU fishing is progressively establishing a trend towards buying legally caught fish, which could result in loss of market for fishing companies which cannot comply with international and regional certification processes.

---

136 Unilever, *Fishing for the Future II: Unilever's Fish Sustainability Initiative (FSI)*, FF/RLD/SP/5000/1003, at 6. www.unilever.com. Accessed on 17 March 2008.

137 *Ibid.*

138 *Ibid.*

139 *Ibid.*

140 *Ibid.*

141 *Ibid.*

142 MSC Website, www.msc.org. Accessed on 17 March 2008.

143 Cathy Roheim and Jon G Sutinen, 'Trade and Marketplace Measures to Promote sustainable Fishing Practices', *Natural Resources, International Trade and Sustainable Development*, Issue Paper No 3, May 2006, at 30–31.

8.3.5.3 *Measures to Deter Business with IUU Fishers*

The IPOA-IUU provides that States should ensure that all those involved in fisheries related businesses, (*i.e.* fishers, importers, transhippers, buyers, consumers, equipment suppliers, bankers, insurers, other service suppliers) and the public, are aware of the detrimental effects of doing business with IUU vessels and take measures to deter such business.[144] According to the IPOA-IUU, these measures may include legislation that makes it a violation to conduct such business or to trade in fish or fish products derived from IUU fishing.[145] These measures have been adopted mostly by regional fisheries organisations in their resolutions. RFMOs such as the Northeast Atlantic Fisheries Commission (NEAFC), Inter-American Tropical Tuna Commission (IATTC), Indian Ocean Tuna Commission (IOTC), CCAMLR, and ICCAT, encourage importers, transporters and other sectors concerned, to refrain from transaction in and transshipment of tuna and tuna-like species caught by vessels included in the IUU list.[146] The Lake Victoria Fisheries Organisation (LVFO) also calls on its Partner States to enact legislation that makes it a violation to conduct business or to trade in fish or fish products derived from IUU fishing in Lake Victoria.[147]

Deterring business with IUU vessels provides another layer of measure that would ideally assist States in preventing IUU fishing. However, adopting legal measures to accommodate such concerns may not be as straightforward as the IPOA-IUU suggests. While regulating entities directly related to fisheries, such as those involved in post-harvest fisheries production may be easy to achieve, there are entities within the fisheries business chain which do not deal exclusively with fishing vessels or companies and are not directly covered under international or national fisheries law, such as banks, equipment suppliers, insurers and service suppliers. Imposing restrictions on how these entities conduct their businesses is beyond the scope of fisheries management and policy and may require adjustments in different areas of law such as commercial law, banking and finance, and corporation law.

---

[144] *IPOA-IUU*, para. 73 and 74.

[145] *IPOA-IUU*, para. 73 and 74.

[146] ICCAT, *Recommendation by ICCAT to Establish a List of Vessels Presumed to Have Carried Out Illegal, Unreported and Unregulated Fishing Activities in the ICCAT Convention Area*, 02-23 GEN, 04 June 2003, para. 9(f); NEAFC, *Scheme of Control and Enforcement 2009*, Art. 45(2)(f); CCAMLR, *Conservation Measure 10-07 (2006),* Scheme to Promote Compliance by Non-Contracting Party Vessels with CCAMLR Conservation Measures, Art. 11(h); CCAMLR, *Conservation Measure 10-06 (2008),* Scheme to Promote Compliance by Contracting Party Vessels with CCAMLR Conservation Measures, Art. 18(ix); IATTC, *Resolution C-05-07,* Resolution to Establish a List of Vessels Presumed to have Carried out Illegal, Unreported and Unregulated Fishing Activities in the Eastern Pacific Ocean, 24 June 2005, Art. 9(f); IOTC, *Resolution 09/03* On Establishing a List of Vessels Presumed to have Carried out Illegal, Unregulated and Unreported Fishing in the IOTC Area, para. 12.

[147] Lake Victoria Fisheries Organization, *Regional Plan of Action to Prevent, Deter, and Eliminate Illegal, Unreported and Unregulated (IUU) Fishing on Lake Victoria and Its Basin*, Bagamoyo, Tanzania, 27 May 2004, Art. 3.4(iv).

Chapter Nine

RFMO Measures to Address IUU Fishing

9.1 The Duty to Cooperate and Regional Fisheries Organisations

Cooperation among States is one of the fundamental legal principles underpinning the long term conservation and sustainable use of marine fisheries resources in modern times. The LOSC mandates cooperation, either directly or through appropriate subregional or regional organisations in several instances: to ensure the conservation and development of the same stocks or stocks of associated species which occur within the EEZs of two or more coastal States;[1] to agree on measures necessary for the conservation of stocks or stocks of associated species which occur within EEZs and in areas beyond and adjacent to EEZs;[2] and to ensure the conservation and promotion of optimum utilisation of highly migratory species within and beyond EEZs.[3] Cooperation is also required with respect to anadromous stocks[4] and catadromous species.[5] Additionally, Article 123 of the LOSC requires cooperation in the management of the living resources in semi-enclosed seas. The duty to cooperate is also reiterated in Articles 117 and 118 of the LOSC with respect to the conservation and management of living resources in areas of the high seas. Article 117 imposes a duty on all States "to take, or to cooperate with other States in taking, such measures for their respective nationals as may be necessary for the conservation of the living resources of the high seas". Under Article 118, States whose nationals harvest identical living resources on the high seas are obliged to cooperate to take measures to conserve and manage these resources and to establish subregional or regional fisheries organisations to this end.

The duty to cooperate through global, regional, subregional or bilateral arrangements has been reinforced in a number of post-LOSC fisheries instruments, including Chapter 17 of Agenda 21, FAO Code of Conduct for Responsible Fisheries, FAO Compliance Agreement, and the UN Fish Stocks Agreement. States can give effect to their duty to cooperate either by establishing and

---

[1] United Nations Convention on the Law of the Sea, Montego Bay, Jamaica, concluded on 10 December 1982, in force 16 November 1994, 1833 UNTS 3; 21 ILM 1261 (1982), *hereinafter referred to as* LOSC. LOSC, Art. 63(1).

[2] *LOSC*, Art. 63(2).

[3] *LOSC*, Art. 64.

[4] *LOSC*, Art. 66.

[5] *LOSC*, Art. 67.

becoming members of subregional or regional fisheries management organisations (RFMOs) or by agreeing to apply the conservation and management measures of such organisations.[6,7] More fundamentally, non-members of or non-participants in RFMOs are not discharged from the obligation to cooperate in the conservation and management of straddling fish stocks and highly migratory fish stocks.[8]

There are now over forty regional fisheries organisations[9] operating worldwide—ten have been established under FAO and the rest created under international agreements between three or more contracting Parties.[10] Regional fisheries organisations have different mandates, functions, structures, and competences[11] and can be classified into three categories based on their main functions,

---

[6] *Agreement for the Implementation of the Provision of the United Nations Convention on the Law of the Sea of 10 December 1982 Relating to the Conservation and Management of Straddling Fish Stocks and Highly Migratory Fish Stocks*, New York, concluded on 4 August 1995, in force 11 December 2001, 34 ILM 1542 (1995); 2167 UNTS 88. *Hereinafter referred to as UN Fish Stocks Agreement.* Arts. 8(3) and 8(5); Food and Agriculture Organization (FAO), *Agreement to Promote Compliance with International Conservation and Management Measures by Fishing Vessels on the High Seas*, Rome, Italy, concluded on 24 November 1993, entered into force 24 April 2003, 33 ILM 968. *Hereinafter referred to as FAO Compliance Agreement.* Art. V(3); FAO, *Code of Conduct for Responsible Fisheries,* Adopted at the 28th Session of the FAO Conference, Rome, Italy, 31 October 1995. *Hereinafter referred to as* FAO Code of Conduct. Art. 7.1.3; United Nations Conference on Environment and Development (UNCED), Agenda 21, Chapter 17, *Protection of the Oceans, All Kinds of Seas, Including Enclosed and Semi-enclosed Seas, and Coastal Areas and the Protection, Rational Use and Development of their Living Resources*, Rio de Janeiro, Brazil, 03–14 June 1992. *Hereinafter referred to as* Agenda 21. Para. 17.49.

[7] In current practice, those which agree to apply conservation and management measures adopted by RFMOs generally include cooperating non-members, participants, and fishing entities.

[8] *UN Fish Stocks Agreement*, Art. 17; *FAO Compliance Agreement*, Preamble; *FAO Code of Conduct*, Article 7.1.5; *Agenda 21*, par. 17.60. This also pertains to States with real interest in the fisheries concerned. *See* UN Fish Stocks Agreement, Art. 8(3). For further discussion, see Erik Jaap Molenaar, 'The Concept of 'Real Interest' and Other Aspects of Co-operation through Regional Fisheries Management Mechanisms', 15 *The International Journal of Marine and Coastal Law* (2000), at 475–530. *See also* Erik Jaap Molenaar, 'Participation, Allocation and Unregulated Fishing: The Practice of Regional Fisheries Management Organisations', 18 *The International Journal of Marine and Coastal Law* (2003), at 457–480.

[9] The FAO uses the term "Regional Fishery Body" which is defined as referring to "a mechanism through which three or more States or international organisations that are parties to an international fishery agreement or arrangement collaboratively engage each other in multilateral management of fishery affairs related to transboundary, straddling, highly or high seas migratory stocks, through the collection and provision of scientific information and data, serving as technical and policy forum, or taking decisions pertaining to the development and conservation, management and responsible utilisation of the resources. *See* FAO Website, *Regional Fishery Bodies and Arrangements*, www.fao.org. Accessed on 20 January 2009.

[10] *See* FAO Website, Regional Fisheries Bodies Home Page. www.fao.org. Accessed on 15 January 2009. For information on these regional fisheries bodies, *see* Michel J. Savini, Summary Information on the Role of International Fishery Bodies with Regard to the Conservation and Management of Living Resources of the High Seas, *FAO Fisheries Circular No. 835, Rev. 1* (Rome, FAO, 1991); S.H. Marashi, The Role of FAO Regional Fishery Bodies in the Conservation and Management of Fisheries, *FAO Fisheries Circular No. 916* (Rome, FAO, 1996).

[11] *See* Appendix E, Major Issues Affecting the Performance of Regional Fishery Bodies, in FAO, Report of the Meeting of FAO and Non-FAO Regional Fishery Bodies or Arrangements, Rome, Italy, 11–12 February 1999, *FAO Fisheries Report No 597* (Rome: FAO, 1999).

namely, scientific research organisations,[12] advisory and regional coordination organisations,[13] and management organisations.[14,15] About half of the total number of regional fisheries organisations have been established since the negotiation of the LOSC.[16] Regional fisheries organisations created in more recent times address gaps in the international fishery regulatory framework, particularly on the high seas.

9.2 The Role of Regional Fisheries Organisations in Addressing IUU Fishing

In addition to establishing regional fisheries organisations as "vehicles for good fishery governance",[17] the role of such organisations, particularly those with

---

[12] Examples of scientific research organisations are Advisory Committee on Fishery Research (ACFR), Coordinating Working Party on Fisheries Statistics (CWP), International Council for the Exploration of the Sea (ICES), Network of Aquaculture Centres in Asia-Pacific (NACA), North Pacific Marine Science Organization (PICES), and the Secretariat of the Pacific Community (SPC).

[13] Examples of regional fisheries organisations with advisory and coordination functions are Asia Pacific Fisheries Commission (APFIC), Bay of Bengal Programme Inter-governmental Organisation (BOBP-IGO), Regional Fisheries Advisory Committee for the Southwest Atlantic (CARPAS), Fishery Committee for the Eastern Central Atlantic (CECAF), Committee for Inland Fisheries for Africa (CIFAA), Ministerial Conference on Fisheries Cooperation among African States Bordering the Atlantic Ocean (COMHAFAT), Commission for Inland Fisheries of Latin America (COPESCAL), Joint Technical Commission for the Argentina/Uruguay Maritime Front (CTMFM), Regional Fisheries Committee for the Gulf of Guinea (COREP), Permanent Commission for the South Pacific (CPPS), European Inland Fisheries Advisory Commission (EIFAC), Fishery Committee of the West Central Gulf of Guinea (FCWC), Forum Fisheries Agency (FFA), Mekong River Commission (MRC), North Atlantic Marine Mammal Commission (NAMMCO), Latin American Organization for Fisheries Development (OLDEPESCA), Southeast Asian Fisheries Development Centre (SEAFDEC), Sub-regional Fisheries Commission (SRFC), Southwest Indian Ocean Fisheries Commission (SWIOFC), and Western Central Atlantic Fishery Commission (WECAFC).

[14] Examples of regional fisheries management organisations are Commission for the Conservation of Antarctic Marine Living Resources (CCAMLR), the Convention on the Conservation and Management of the Pollock Resources in the Central Bering Sea (CCBSP), Commission for the Conservation of Southern Bluefin Tuna (CCSBT), General Fisheries Commission for the Mediterranean (GFCM), Inter-American Tropical Tuna Commission (IATTC), International Commission for the Conservation of Atlantic Tunas (ICCAT), Indian Ocean Tuna Commission (IOTC), International Pacific Halibut Commission (IPHC), International Whaling Commission (IWC), Lake Victoria Fisheries Organization (LVFO), Northwest Atlantic Fisheries Organization (NAFO), North Atlantic Salmon Conservation Organization (NASCO), Northeast Atlantic Fisheries Commission (NEAFC), North Pacific Anadromous Fish Commission (NPAFC), Pacific Salmon Commission (PSC), Regional Commission for Fisheries (RECOFI), Southeast Atlantic Fisheries Organisation (SEAFO), South Indian Ocean Fisheries Agreement (SIOFA), Western and Central Pacific Fisheries Commission (WCPFC), and yet to be established South Pacific Regional Fisheries Organisation (SPRFMO).

[15] Are K. Sydnes, 'Regional Fisheries Organizations: How and Why Organizational Diversity Matters', 32 *Ocean Development and International Law* (2001), at 354.

[16] *See* Judith Swan, 'Regional Fishery Bodies and Governance: Issues, Action and Future Directions', *FAO Fisheries Circular No. 959* (Rome, FAO, 2002).

[17] FAO, Report of the Meeting of FAO and Non-FAO Regional Fishery Bodies or Arrangements, Rome, Italy, 11–12 February 1999, *FAO Fisheries Report No 597* (Rome: FAO, 1999).

management functions as vehicles for the development and implementation of sustainable fisheries management principles to combat IUU fishing has been recognised since the inception of the IPOA-IUU. As discussed in Chapter 2, CCALMR was the driving force behind the development of the IUU fishing concept. Other regional fisheries organisations were also involved in the first global review of IUU fishing held in 2000.[18] Continuous cooperative efforts are being undertaken by regional fisheries organisations to address common IUU fishing concerns.

The IPOA-IUU recognises the critical role of RFMOs in combating IUU fishing. Since RFMOs have the mandate to adopt resolutions or conservation and management measures which are binding on their members, they are in a more prominent position to develop and implement mechanisms to combat IUU fishing than those organisations exclusively with advisory or scientific functions. Recognising this critical role, the IPOA-IUU devotes an entire section to the implementation of the IPOA-IUU through RFMOs or by "States acting through relevant RFMOs".[19]

9.3 RFMO Measures to Combat IUU Fishing

As RFMOs continue to strengthen their capacity to address IUU fishing, it is increasingly acknowledged that dealing effectively with the IUU fishing problem is not only about establishing a list of IUU vessels and applying measures against such vessels, but also implementing a range of complementary measures with respect to authorising vessels to fish, regulating fishing activities at sea and in port, and ensuring compliance with conservation and management measures.

9.3.1 *Authorisations to Fish and to Transship Fish*[20]

As noted in Chapter 6, in the EEZ, the responsibility to issue authorisation to fish lies with the coastal State in whose EEZ the fishing operation is conducted. On the high seas, particularly in areas regulated by RFMOs, the flag State member of the RFMO has the primary duty to ensure that only vessels it has authorised are able to fish in the area of competence of the RFMO. The flag State also has the

[18] *See* Kevin Bray, 'A Global Review of Illegal, Unreported, and Unregulated Fishing', in FAO, Report of and Papers Presented at the Expert Consultation on Illegal, Unreported and Unregulated Fishing, Sydney, Australia, 15–19 May 2000, *FAO Fisheries Report No. 666* (Rome: FAO, 2001).

[19] *See* IPOA-IUU, paras. 80, 81, and 83. Based on these provisions, "States, acting through relevant regional fisheries management organisations" may be interpreted as referring to both members, cooperating non-members, and other participants of RFMOs which are applying measures adopted by such RFMOs to address IUU fishing.

[20] Transshipment is the process of unloading or transfer of fish from a fishing vessel to another fishing vessel either at sea or in port.

primary duty to ensure that vessels flying its flag do not undertake activities which undermine the effectiveness of conservation and management measures adopted by the RFMO in question. The scope and content of an authorisation to fish in RFMO areas are derived from the domestic laws of the flag State of the vessel and the conservation and management measures adopted by the RFMO.

The responsibility to issue authorisation to fish rests with the flag State and not the RFMO. A number of RFMOs, including NAFO, NEAFC, WCPFC, and SEAFO require their members to authorise the use of vessels flying their flags only where they are able to discharge effectively their duties in respect of such vessels.[21] Some RFMOs require their flag State members to take effective control of their vessels through fishing authorisation.[22] It is a requirement under the WCPF Convention that a vessel may only conduct fishing within areas under the jurisdiction of other members only when it holds a licence, permit or authorisation required by the coastal State.[23] CCAMLR members are required to include certain conditions in the fishing licences they issue. These include the specific areas, species and time periods for which fishing is authorised and the timely notification by the vessel to its flag State of exit from and entry into any port in the Convention area; movement of the vessel between areas, sub-areas and divisions; reporting of catch data in accordance with CCAMLR requirements; and operation of a VMS.[24]

Other RFMOs have more general authorisation to fish requirements. For example, SEAFO, NEAFC, IOTC, NAFO, and ICCAT recognise the importance of authorisation to fish in ensuring that only licensed fishing vessels undertake fishing activities in their management areas.[25] Licences are required to be carried on

[21] Convention on the Conservation of Highly Migratory Fish Stocks in the Western and Central Pacific Ocean, Honolulu, Hawaii, concluded 5 September 2000, entry into force 19 June 2004, [2004] ATS 15. *Hereinafter referred to as* WCPFC Convention. Art. 24(2); Convention on the Conservation and Management of Fishery Resources in the Southeast Atlantic Ocean, Windhoek, Namibia concluded on 20 April 2001, in force 13 April 2003; 41 ILM 257. *Hereinafter referred to as* SEAFO Convention. Art. 14(2); *NAFO Conservation and Enforcement Measures 2009*, NAFO/FC Doc. 09/1, Art. 18(1); Northeast Atlantic Fisheries Commission (NEAFC), *Scheme of Control and Enforcement*, February 2009, Art. 4(1).

[22] *SEAFO Convention*, Art. 14(3)(b).

[23] *WCPF Convention*, Art. 24(3)(a).

[24] CCAMLR, *Conservation Measure 10-02 (2008)*, Licensing and Inspection Obligation of Contracting Parties with Regard to their Flag Vessels Operation in the Convention Area, paras. 1 and 2.

[25] *SEAFO Convention*, Art. 14(3)(b) and Annex, Section One; NAFO, *Conservation and Enforcement Measures*, Art. 13; NEAFC, *Scheme of Control and Enforcement*, Art. 4; ICCAT, *Resolution by ICCAT Concerning the Measures to Prevent the Laundering of Catches by Illegal, Unreported and Unregulated (IUU) Large-scale Tuna Longline Fishing Vessels*, 02-25 GEN, 04 June 2003, para. 1; ICCAT, *General Outline of Integrated Monitoring Measures Adopted by ICCAT*, 02-31 GEN, para. 1(i)(b); ICCAT, *Recommendation by ICCAT Concerning the Duties of Contracting Parties and Cooperating Non-parties, Entities, or Fishing Entities in Relation to Their Vessels Fishing in the ICCAT Convention Area*, 03-12 GEN, 19 June 2004, para. 1.

board vessels and must contain at the very least, basic information such as the vessel name, port in which and the number under which it is registered, international call sign, name and address of owners and charterers, overall length, engine power, commercial fishing gears, vessel activity, number of crew, and gross tonnage.[26]

In addition to regulating activities of vessels authorised to fish in their areas of competence, RFMOs also regulate the transshipment operations of fishing vessels. All transshipments of fish in RFMO areas are required to take place in designated ports in order to prevent the laundering of fish,[27] except in certain circumstances.[28] ICCAT for example exempts small-scale albacore longline vessels from the prohibition on transshipment at sea[29] while IATTC provides a general exemption for vessels which transship fresh fish at sea.[30] IOTC members are given the responsibility to determine whether or not to permit their vessels to transship at sea.[31] In the case of the CCSBT, both members and cooperating non-members are required to ensure that transshipment activities are conducted in accordance with the transshipment procedures established by the Commission.[32] In WCPFC, flag States of vessels seeking an exemption from transshipment at sea in accordance with established criteria need to submit an application for exemption to the Commission.[33]

The distinction in RFMO practice between transshipment operations which may be allowed at sea and those which may only take place in port, would seem

[26] IOTC, *Resolution 01/02 Relating to Control of Fishing Activities*, para. 2; IPHC, *Licensing Procedures for Area 2A*, no date; ICCAT, *Recommendation by ICCAT Concerning the Establishment of an ICCAT Record Over 24 Meters Authorised to Operate in the Convention Area*, 02-22 GEN, 04 June 2003, para 5(d).

[27] SEAFO for example has adopted an interim prohibition on all transshipments at sea in the SEAFO Convention area. *See* SEAFO, *Conservation Measure 03/06 on an Interim Prohibition of Transshipments at Sea in the SEAFO Convention Area and to Regulate Transshipments in Port*, Approved 04/10/2006.

[28] IOTC, *Resolution 08/02 on Establishing a Programme for Transshipment by Large-scale Fishing Vessels*, Sec. 1.1; *Recommendation by ICCAT to Establish Multi-Annual Recovery Plan for Bluefin Tuna in the Eastern Atlantic and Mediterranean,* 06-05 BFT, 13 June 2007, para. 35; ICCAT, *Recommendation by ICCAT Establishing a Programme for Transshipment*, 06-11 GEN, 13 June 2007, paras. 1, 3-5, and Annexes 1 and 2; *WCPF Convention*, Art. 29(5); WCPFC, Conservation and Management Measure on the Regulation of Transshipment, *Conservation and Management Measure 2009-06*, paras. 25 and 35.

[29] ICCAT, *Recommendation by ICCAT Establishing a Programme for Transshipment*, 06-01 GEN, 13 June 2007, Sec. 2.3.

[30] IATTC, *Resolution C-08-02,* Resolution on Establishing a Program for Transshipments by Large-Scale Fishing Vessels, Sec. 1.3.

[31] IOTC, *Resolution 08/02 on Establishing a Programme for Transshipment by Large-scale Fishing Vessels*, Sec. 2.4.

[32] *See* CCSBT, *Resolution on Establishing a Program for Transshipment by Large-scale Fishing Vessels*, Adopted at the Fifteenth Annual Meeting, 14–17 October 2002.

[33] WCPFC, *Conservation and Management Measure 2009-06*, paras. 26 and 35.

to be missing the crucial point that from the perspective of combating IUU fishing, the critical issue in transshipment regulation is not so much where the transshipment takes place, but whether there are adequate and verifiable procedures to ensure that such transshipment activities, regardless of where they take place, are properly monitored to minimise the occurrence of IUU fishing.

The practice of RFMOs on transshipment monitoring covers three specific requirements. The first is that all carrier vessels transhipping at sea must carry observers in accordance with any observer programme established by the RFMO.[34] The second requirement relates to the inspection of transshipment declaration and verification of other relevant documentation, such as applicable catch documents or statistical documents relating to the transshipment declaration of the species covered under the documentation scheme.[35] The third is the requirement on members of RFMOs to report annually to their respective commissions the quantities of species transshipped and the list of vessels which conducted the transshipment.[36] In the case of CCAMLR for example, each contracting flag State member is required to notify the Secretariat at least 72 hours in advance if any of its vessels intend to transship within the Convention Area. The flag State may also permit the vessel to provide such notifications directly to the Secretariat.[37]

The transshipment requirements imposed on fishing vessels by RFMOs include notification of transshipment and relevant information prior to transshipment, as well as the completion and submission of transshipment declaration. Fishing vessels are required to notify their flag State authorities at least 24 hours prior to the commencement of transshipment, including provision of information on the vessel and intended transshipment.[38] Transshipment declarations must be completed and submitted to the flag State not later than 15 days after the transshipment has

[34] IOTC, *Resolution 08/02 on Transshipment by Large-scale Fishing Vessels*, Sec. 4.16; ICCAT, *Recommendation by ICCAT Establishing a Programme for Transshipment*, paras. 15 and 16 and Annex 2; IATTC, *Resolution C-08-02 Establishing a Program for Transshipments by Large-Scale Fishing Vessels*, para. 16-17; WCPFC, *Conservation and Management Measure 2009-06*, para.13.

[35] IOTC, *Resolution 08/02 on Transshipment by Large-scale Fishing Vessels*, Sec. 5.18; ICCAT, *Recommendation by ICCAT Establishing a Programme for Transshipment*, para. 17 and Annex 1; IATTC, *Resolution C-08-02 Establishing a Program for Transshipments by Large-Scale Fishing Vessels*, para. 16-17.

[36] IOTC, *Resolution 08/02 on Transshipment by Large-scale Fishing Vessels*, Sec. 5.19; ; ICCAT, *Recommendation by ICCAT Establishing a Programme for Transshipment*, para. 18; WCPFC, *Conservation and Management Measure 2009-06*, Annex II.

[37] CCAMLR, *Conservation Measure 10-09 (2009)*, Notification System for Transshipments within the Convention Area, para. 2.

[38] IOTC, *Resolution 08/02 on Transshipment by Large-scale Fishing Vessels*, Sec. 4.12; IATTC, *Resolution C-08-02 Establishing a Program for Transshipments by Large-Scale Fishing Vessels*, para. 13. For CCAMLR, the period for advance notification of transshipment is at least 72 hours. *See* CCAMLR, *Conservation Measure 10-09 (2009)*.

been completed.[39] The master of the receiving carrier vessel is required to complete and submit a transshipment declaration to the secretariat of the commission of the relevant RFMO and the flag State member of the RFMO within 24 hours of the completion of the transshipment.[40] If the vessel wishes to land its fish after conducting transshipment at sea, it is required to transmit a transshipment declaration to the port State within 48 hours before landing the fish.[41] In NEAFC and NAFO, landing may only commence after authorisation has been given by the port State. Such authorisation is only provided after confirmation from the flag State that the fishing vessel possesses sufficient quota for the species declared; that the quantity of fish on board has been duly reported and taken into account for calculation of any catch or effort limitation that may be applicable; that the fishing vessel had authorisation to fish in the areas where fish has been caught; and that the presence of the vessel in the area of catch declared has been verified by VMS data.[42]

In relation to transshipment of fish in port, fishing vessels are required to submit information on the vessel and the transshipment activity, as well as a transshipment declaration.[43] The role of port States is to verify the accuracy of the information received from the fishing vessels and cooperate with the flag States to ensure that landings are consistent with the reported catches.[44] In NEAFC, members are required to ensure that reports for transshipments are communicated by fishing vessels to Fisheries Monitoring Centres by electronic means.[45] The masters of the fishing vessels are required to cooperate and assist in the inspection, including providing access to any areas, decks, rooms, catch, nets or other gear or equipment, and provide any relevant information which the port State may request.[46]

---

[39] IOTC, *Resolution 08/02 on Establishing a Programme for Transshipment by Large-scale Fishing Vessels*, Sec. 4.13; *Recommendation by ICCAT to Establish Multi-Annual Recovery Plan for Bluefin Tuna in the Eastern Atlantic and Mediterranean*, 06-05 BFT, 13 June 2007, para. 35 and Annex 3; *Recommendation by ICCAT Establishing a Programme for Transshipment*, 06-01 GEN, 13 June 2007, Sec. 4.12; IATTC, *Resolution C-08-02 Establishing a Program for Transshipments by Large-Scale Fishing Vessels*, para. 13.

[40] IOTC, *Resolution 08/02 on Transshipment by Large-scale Fishing Vessels*, Sec. 4.14; *Recommendation by ICCAT Establishing a Programme for Transshipment*, 06-01 GEN, 13 June 2007, Sec. 4.13.

[41] IOTC, *Resolution 08/02 on Transshipment by Large-scale Fishing Vessels*, Sec. 4.15; ICCAT, *Recommendation by ICCAT to Establish Multi-Annual Recovery Plan for Bluefin Tuna in the Eastern Atlantic and Mediterranean*, 06-05 BFT, 13 June 2007, para. 35; IATTC, *Resolution C-08-02 Establishing a Program for Transshipments by Large-Scale Fishing Vessels*, para. 15.

[42] *NEAFC Scheme of Control and Enforcement*, 2009, Art. 23; NAFO, *Conservation and Enforcement Measures*, Art. 47.

[43] IOTC, *Resolution 08/02 on Transshipment by Large-scale Fishing Vessels*, Annex I - Conditions Relating to In-Port Transshipment by LSTVs, Secs. 1-3.

[44] IOTC, *Resolution 08/02 on Transshipment by Large-scale Fishing Vessels*, Annex I, Sec. 5.

[45] NEAFC, *Scheme of Control and Enforcement*, Art. 13.

[46] See for example NAFO, *Conservation and Enforcement Measures*, Art. 48.

9.3.2 *Record of Fishing Vessels and Vessels Authorised to Tranship Fish*

It is now common practice to establish and maintain records covering fishing vessels authorised by flag or coastal States to fish for species regulated by RFMOs. With the exception of SEAFO and the WCPFC,[47] other RFMOs do not directly require their members (and cooperating non-members) to establish national records of fishing vessels. However, most RFMOs require their members to provide information on vessels authorised to fish within their areas of competence similar to those required under Article VI of the FAO Compliance Agreement.[48] Members of RFMOs are also required to provide information on any additions or deletions to the records and the reason for deletion.[49] From the information provided by members and cooperating non-members, a regional record of fishing vessels is established by RFMOs. In general, any fishing vessel not entered into the RFMO record of fishing vessels is deemed not to be authorised to fish for, retain on board, transship or land species regulated by the RFMO.[50]

The IATTC, CCSBT, ICCAT, IOTC and WCPFC require their members and cooperating non-members to ensure that large scale tuna fishing vessels included on the regional record have no history of IUU fishing activities or that, if those vessels have such history, the new owners have provided sufficient evidence demonstrating that the previous owners and operators have no legal, beneficial, or financial interest in, or control over those vessels.[51]

---

[47] *See* SEAFO Convention, Art. 14(3)(c); WCPFC Convention, Art. 24(4); WCPFC, Record of Fishing Vessels and Authorisation to Fish, *Conservation and Management Measure 2004-01*, Part B.

[48] ICCAT, *Recommendation of ICCAT Concerning the Establishment of an ICCAT Record of Vessels*; IATTC, *Resolution C-03-07*, Resolution on the Establishment of a List of Longline Fishing Vessels Over 24 Meters (LSTLFVs) Authorised to Operate in the Eastern Pacific Ocean, 27 June 2003; CCSBT, *Resolution on Illegal, Unregulated and Unreported fishing (IUU) and Establishment of a CCSBT Record of Vessels over 24 Meters Authorised to Fish for Southern Bluefin Tuna*, 10 October 2003; IOTC, *Resolution 07/02* Concerning the Establishment of an IOTC Record of Vessels Authorised to Operate in the IOTC Area, 2007; *WCPF Convention*, Annex IV.

[49] *WCPFC Convention*, Art. 24(6). Members of the WCPFC are required to promptly inform the Commission of any deletions of the record by reason of voluntary relinquishment or non-renewal of the fishing authorisation by the fishing vessel owner or operator; the withdrawal of the fishing authorisation; the fact that the fishing vessel concerned is no longer entitled to fly its flag; the scrapping, decommissioning or loss of the fishing vessel concerned; and any other reason.

[50] IOTC, R*esolution 07/02 on the IOTC Record of Vessels*, para. 2; WCPFC, *Conservation and Management Measure 2004-01*, Part B.

[51] ICCAT, *Recommendation of ICCAT Concerning the Establishment of an ICCAT Record of Vessels*, para. 5(d); IATTC, *Resolution on the Establishment of a List of Longline Fishing Vessels*, para. 4; CCSBT, *Resolution on IUU Fishing and the Establishment of a CCSBT Record of Vessels*, para. 6; IOTC, *Resolution 07/02* on the IOTC Record of Vessels, para. 4; WCPFC, *Conservation and Management Measure 2004-01*, para. 1(f) and 1(h).

ICCAT requires its members to take measures to maintain a register of all high seas fishing vessels of more than 24 metres in length authorised to fly their flag and fish in the ICCAT Convention area.[52] ICCAT also encourages non-members to provide the names of vessels flying their flag which conduct fishing operations in the ICCAT management area. Fishing vessels not entered into the ICCAT register are deemed not to be authorised to fish in the regulatory area, and are consequently considered IUU vessels.[53]

NAFO mandates the establishment and maintenance of a register of all fishing vessels of more than 50 gross tonnes authorised to fish in the NAFO regulatory area.[54] The information contained in the NAFO Register of Fishing Vessels, however, is limited to the present identification and characteristics of fishing vessels. As a result, the history of such fishing vessels may not be easily determined from the NAFO Record.[55]

Aside from maintaining a record of vessels authorised to fish, RFMOs also maintain records of vessels authorised to transship at sea. Similar to conditions placed on vessels not included in the record of fishing vessels, RFMO conservation and management measures provide that carrier vessels not entered on the record are deemed not to be authorised to receive fish in transshipment operations at sea.[56] ICCAT members are required to ensure that fishing vessels and mother vessels flying their flag only transfer or receive at-sea transshipment of ICCAT species from members or cooperating non-members.[57] The record of carrier vessels authorised to conduct transshipment at sea includes information on the vessel and the time period authorised for transhipping. These vessels are also required to comply with transshipment procedures and requirements relating to VMS and observers.

9.3.3 *IUU Vessel Listing*

The creation of a list of vessels engaged in IUU fishing is one of the most effective measures adopted by RFMOs to combat IUU fishing.[58] A number of RFMOs,

---

[52] ICCAT, *Resolution by ICCAT Regarding the Agreement to Promote Compliance with International Conservation and Management Measures by Fishing Vessels on the High Seas*, 94-08 GEN, 23 January 1995.

[53] *Ibid.*

[54] NAFO, *Conservation and Enforcement Measures*, Art. 15.

[55] Information required for the NAFO Register of Fishing Vessels includes vessel name; radio call sign; flag State; internal reference number or unique Contracting Party vessel number, external registration number or side number of the vessel; port name; vessel owner; vessel charterer; vessel type; vessel gear; vessel capacity; vessel power; and fishing license details. *See* NAFO, Conservation and Enforcement Measures, Annex IV.

[56] IOTC, *Resolution 08/02* on Establishing a Programme for Transshipment by Large-scale Fishing Vessels, para. 5.

[57] ICCAT, *Recommendation by ICCAT on Transshipments and Vessel Sightings*, 97-11 GEN, 13 June 1998, para. 1; IATTC, *Resolution C-08-02,* Resolution on Establishing a Program for Transshipments by Large-scale Fishing Vessels, para. 6.

[58] *IPOA-IUU*, para. 81.4.

including ICCAT, IATTC, IOTC, WCPFC, SEAFO, NEAFC, NAFO, and CCAMLR have established IUU Vessel Lists which are published on their respective websites. ICCAT, IOTC, NEAFC, and NAFO only require the listing of non-member vessels whereas CCAMLR, IATTC, WCPFC and SEAFO provide for the listing of IUU vessels flying the flags of members, cooperating non-members and non-members. Non-tuna RFMOs such as NAFO, NEAFC, and SEAFO have provisions for the mutual recognition of the IUU lists of other RFMOs. Tuna RFMOs such as ICCAT, IATTC, IOTC, and WCPFC are also working towards the reciprocal recognition of IUU vessel lists and have provided links to IUU vessel lists of all other tuna RFMOs on their respective websites.

The RFMO measures establishing IUU vessel lists cover the identification of IUU vessels and procedures for the listing and deletion of the vessel on the list. Although there are minor variations in the procedures adopted by RFMOs, in general, placing a vessel on the IUU vessel list starts with the sighting of a vessel presumed to have undermined the effectiveness of conservation and management measures instituted by the relevant RFMO. The RFMO member whose inspecting authority has sighted an alleged IUU vessel provides the Secretariat of the RFMO with information on the vessel. The Secretariat of the RFMO then provides all the members and cooperating non-members as well as the flag State of the vessel, where the flag State is not a member of the RFMO, the provisional list of presumed IUU vessels, together with all supporting evidence. The members and cooperating non-members of the RFMO then transmit their comments to the Secretariat on the provisional IUU vessels list. The non-member flag State of the vessel is given the opportunity to provide evidence that the vessel in question neither fished in the area nor fished in contravention of the RFMO conservation measures. The members and cooperating non-members of the RFMO closely monitor the activities of the vessels on the provisional IUU list, including changes of name, flag, and registered owner. The RFMO examines the provisional IUU vessels list, including all submitted evidence and further information by its members and cooperating non-members and the flag State of the vessel, and makes a decision to either approve the list or delete any vessel from the list. After that decision is made, the Secretariat of the RFMO ensures that the IUU vessels list is publicised. Members and cooperating non-members of RFMOs are then required to take all necessary measures under their applicable legislation to ensure that their vessels do not conduct any fishing related business with such vessels, and that proper actions are taken against those vessels.

RFMO IUU listing schemes are relatively new and the complexities of the legal and policy issues involved in their implementation are only starting to emerge. Some of the key issues which require careful consideration include the listing of non-member vessels, the types of activities classified as IUU fishing, the procedures for including and deleting vessels on the IUU list, the relationship between IUU vessel listing and the record of fishing vessels, and reciprocal listing of IUU vessels among RFMOs.

9.3.3.1 *Listing of Non-Member Vessels*

CCAMLR has adopted a separate IUU listing measure for vessels flying the flags of its members, or what is called the CP-IUU Vessel List. A provisional CP-IUU Vessel list is created based on information provided by other contracting Parties, trade statistics, and information gathered by port States.[59] This procedure is similar to that established for creating a non-contracting Party vessels list or NCP-IUU Vessel List under CCAMLR Conservation Measure 10–07 (2009).

The WCPFC also adopted a conservation and management measure on IUU vessel listing in 2006, and amended in 2007, which is applicable to vessels flying the flags of members, cooperating non-members, and non-members[60] and has since included member vessels on its IUU Vessel List.[61] The IATTC and SEAFO have also adopted measures similar to those of the WCPFC which apply to all members and non-member vessels.[62]

However, it has been a common practice among some RFMOs to include on their respective IUU lists only vessels flying the flag of non-members.[63] In most RFMOs, vessels of non-members sighted fishing in their respective convention areas are automatically presumed to be undermining the effectiveness of their conservation and management measures;[64] thereby shifting the burden on to the non-member vessel to prove that its fishing activities are not in contravention of applicable conservation and management measures adopted by the RFMO.

---

[59] CCAMLR, *Conservation Measure 10-06 (2008)*, Scheme to Promote Compliance by Contracting Party Vessels with CCAMLR Conservation Measures, paras. 2 and 3.

[60] WCPFC, Conservation and Management Measure to Establish a List of Vessels Presumed to have carried out Illegal, Unreported and Unregulated Fishing Activities in the WCPO, *Conservation and Management Measure 2007-03*, 07 December 2007.

[61] WCPFC, Technical and Compliance Committee, Fourth Regular Session, 2–7 October 2008, Pohnpei, Federated States of Micronesia, Draft IUU Vessel List and Current WCPFC IUU Vessel List, Prepared by the Secretariat, *WCPFC-TCC4-2008/11*, 15 September 2008.

[62] IATTC, *Resolution C-05-07*, Resolution to Establish a List of Vessels Presumed to Have Carried Out Illegal, Unreported and Unregulated Fishing Activities in the Eastern Pacific Ocean, para. 1; SEAFO, *Conservation Measure 08/06* Establishing a List of Vessels Presumed to Have Carried Out Illegal, Unreported and Unregulated (IUU) Fishing Activities, para. 1.

[63] As at 13 January 2009, there are three vessels with unknown flags listed as IUU vessels in the IOTC area, www.iotc.org; ICCAT has listed 22 vessels deemed to be conducting IUU fishing in its area of competence, most of which have unknown registry. www.iccat.es. There are 22 listed IUU vessels fishing in the IATTC area, only one of which is flagged under a Member, www.iattc.org; In NAFO and NEAFC Convention areas, 22 vessels of non-members are listed on the IUU vessel list. Only two vessels fly the flag of a member of NEAFC. www.nafo.int and www.neafc.org; There are 16 vessels included in the CCAMLR NCP-IUU list. www.ccamlr.org.

[64] ICCAT, *Recommendation by ICCAT Amending the Recommendation by ICCAT to Establish a List of Vessels Presumed to have Carried out IUU Fishing Activities in the ICCAT Convention area*, 06-12, 2007, para. 1; NEAFC, *Scheme of Control and Enforcement*, Art. 44(1); IOTC, *Resolution 09/03* on Establishing a List of Vessels Presumed to Have Carried Out IUU Fishing in the IOTC Area, para. 1; CCAMLR, *Conservation Measure 10-07 (2009)*, Scheme to Promote Compliance by Non-Contracting Party Vessels with CCAMLR Conservation Measures, para. 3; GFCM, *Recommendation GFCM/2006/4*, Establishment of a List of Vessels Presumed to Have Carried Out Illegal, Unreported and Unregulated Fishing Activities in the GFCM Area.

The listing only of non-member vessels on RFMO IUU lists focuses largely on the unregulated aspects of IUU fishing and does not essentially address the issues of illegal and unreported fishing particularly by member vessels. A major concern with the listing of IUU vessels of non-members of RFMOs is the lack of equivalent procedures for the sighting and listing of IUU vessels flying the flags of RFMO members and cooperating non-members. Some RFMOs do not make provision for the IUU listing of their own member vessels, a practice which runs the risk of being considered discriminatory under the IPOA-IUU.

The adoption of separate measures applicable to vessels of members and non-members of RFMOs implies that different actions may be taken against such vessels even if they have committed the same violation. Further, the sanctions imposed on RFMO member vessels may not necessarily be the same as those imposed on the IUU vessels of non-RFMO member vessels. The focus on IUU fishing activities undertaken by non-member vessels also prevents RFMOs from determining the precise extent of IUU fishing in their areas of competence which come from member vessels.

9.3.3.2 *Vessels Owned or Under the Control of Owners of IUU-listed Vessels*

As discussed in Chapter 2, there are different types of activities which constitute IUU fishing in RFMO areas. In general, these activities are vessel specific and include activities of identified vessels in violation of RFMO measures, including fishing activities conducted by vessels without nationality. However, the IATTC, WCPFC, and SEAFO have gone further by also including on their respective IUU Vessel List other vessels under the control of the owner of an IUU vessel.[65]

The IUU listing of the entire fleet of an owner or controller whose one vessel has been placed on an IUU vessel list may be deemed as necessary to compel such owners or operators who derive financial benefits from IUU fishing activities to assume corporate responsibility for their entire fleet of vessels. It can also be argued that this is a desirable progressive implementation of the IPOA-IUU for all States to take effective actions against nationals and beneficial owners.[66]

---

[65] WCPFC, *Conservation and Management Measure 2007-03*, para. 3(j); IATTC, *Resolution C-05-07*, para. 1(i); SEAFO, *Conservation Measure 08/06*, para. 3(j). The same provision can be found in Resolution A-04-07 of the Agreement on the International Dolphin Conservation Program (AIDCP), 20 October 2004. The IATTC provides the Secretariat for the AIDCP.

[66] The relevant paragraphs of the IPOA-IUU are paragraphs 18 and 19.

Paragraph 18 provides: "In the light of relevant provisions of the 1982 UN Convention, and without prejudice to the primary responsibility of the flag State on the high seas, each State should, to the greatest extent possible, take measures or cooperate to ensure that nationals subject to their jurisdiction do not support or engage in IUU fishing. All States should cooperate to identify those nationals who are the operators or beneficial owners of vessels involved in IUU fishing".

The counter argument is that placing an entire fleet on the IUU Vessel list for the violation of one vessel in the fleet may carry significant financial implications for a company. It is highly probable that an IUU vessel or vessel associated with an IUU vessel may not be able to fish in areas under the competence of other RFMOs because of the international stigma associated with IUU fishing. This approach may result in the inclusion of fishing vessels on an RFMO IUU vessels list that may not necessarily be fishing in its convention area.[67] Furthermore, such measure may over burden the RFMO in its implementation. It does not take into account the complexities of legal ownership and control of vessels, including beneficial ownership structures that may not be easily traceable. It may therefore encourage a quick transfer of ownership of vessels to avoid being placed on an IUU list, making it more difficult for the RFMO to ascertain the beneficial ownership of vessels.

The procedure for including vessels associated with IUU vessels by reason of ownership and control has not been clearly established in IATTC and SEAFO, and this issue has been particularly contentious in the WCPFC. Some members of the Commission proposed the deletion of the requirement[68] while other WCPFC Members support its implementation. As a compromise, the Fifth Regular Session of the WCPFC in 2008 approved a recommendation from the Technical and Compliance Committee for the criteria not be utilised in developing the Draft IUU Vessel List in 2009 and that additional procedures that will give effect to the provision would need to be developed for discussion at the 2009 meeting of the Technical and Compliance Committee for approval by the Commission.[69] The same decision was reached at the Sixth Regular Session of the WCPFC for application in 2010. While recognising the need to ensure that beneficial owners do not profit from the IUU fishing activities of their vessels, it is also important to ensure that the global fight against IUU fishing is undertaken in a fair and transparent manner and that innocent operators are not unduly punished. It is important therefore, that RFMOs develop the necessary procedures for and appropriate criteria in identifying vessels and owners involved in IUU fishing.

9.3.3.3 *Procedure for the Listing and Deletion of Vessels on the IUU List*

After vessels have been sighted and presumed to have conducted activities which undermine the conservation and management measures of RFMOs, they are

---

Paragraph 19 provides: "States should discourage their nationals from flagging fishing vessels under the jurisdiction of a State that does not meet its flag State responsibilities".

[67] WCPFC Technical and Compliance Committee, *Conservation and Management Measure 2007-03: Outstanding Issues from WCPFC4*, para. 8.

[68] WCPFC, *Summary Report of the Fourth Regular Session*, Tumon, Guam, 3–7 December 2007, paras. 310 and 311.

[69] WCPFC Technical Compliance Committee, *Summary Report of the Fourth Regular Session*, para. 115; Paragraph 108, 5th Regular Session of the WCPFC, Busan, Korea, 8 – 12 December 2008.

placed on a provisional IUU list. In order to be removed from the provisional or draft IUU list, flag States would need to demonstrate that their vessels sighted in the management areas of the RFMOs did not take part in any IUU fishing activities or that effective actions have been or are being taken in response to the IUU fishing activities in question.[70] Flag States have to prove that the vessel has changed ownership and that the previous owner no longer has any legal, financial or real interests in the vessel.[71] In other circumstances, flag States further have to demonstrate that the vessel was only fishing for unregulated resources and has fulfilled all relevant obligations.[72] For NEAFC and NAFO, vessels are removed from the IUU list if they have been sunk, scrapped, or permanently reassigned for purposes other than for fishing activities.[73] However, the information or evidence that the flag State would need to submit to meet these criteria is not specified.

The procedure for the inclusion of vessels on the IUU list and their deletion from the list is currently problematic. While steps have been taken by RFMOs to improve the efficiency of the process involved in the listing of IUU vessels, some of the elements of the process are still unclear. The areas of uncertainty include the criteria against which the effectiveness of measures undertaken by the flag State in response to IUU fishing activities may be assessed. The burden to prove that the vessel has not conducted or engaged in IUU fishing, or that the infringement of that vessel has been rectified lie on the vessel itself and its flag State. However, the RFMO members would still need to be satisfied that the evidence presented by the flag State is sufficient before a decision is made on the deletion of the vessel from the list. Where the IUU listing resulted from a violation against the laws of a coastal State member of the RFMO, it is unclear whether the severity of the penalty and actions taken by the flag State are to be determined by the flag State or the coastal State. The criteria for making such determinations are not clearly provided in existing IUU listing schemes.

In all RFMOs, decision making to place vessels on the IUU list is by consensus. Where the flag State of the nominated IUU vessel is not a member of the RFMO in question, consensus can easily be reached. However, where both the

[70] ICCAT, *Recommendation by ICCAT to Establish a List of Vessels Presumed to have Carried out IUU Fishing Activities,* para. 1; NEAFC, *Scheme of Control and Enforcement*, Art. 44(4); IOTC, *Resolution 09/03 on Establishing a List of Vessels Presumed to Have Carried Out IUU Fishing*, para. 1; CCAMLR, *Conservation Measure 10-07 (2006)*, para. 3; IATTC, *Resolution 05-07*, para. 6; SEAFO, *Conservation Measure 08/06*, para. 13.

[71] ICCAT, *Recommendation by ICCAT to Establish an IUU List*, para. 6; NEAFC, *Scheme of Control and Enforcement,* Art. 44(4); IOTC, *Resolution 09/03*, para. 9; CCAMLR, *Conservation Measure 10-07 (2006),* para. 10.

[72] ICCAT, *Recommendation by ICCAT to Establish an IUU List*, para. 6; NEAFC, *Scheme of Control and Enforcement,* Art. 44(4)(e); IOTC, *Resolution 09/03*, para. 9; CCAMLR, *Conservation Measure 10-07 (2006)*, para. 10.

[73] NEAFC, *Scheme of Control and Enforcement,* Art. 44(4)(f).

flag State and the State submitting the nomination are both members of the RFMO, experience in WCPFC has shown that consensus is often difficult to obtain. A possible solution to this problem is to exclude both the flag State and the State submitting the nomination from the final decision to include the vessel in or delete it from the IUU list.

There is also the issue of potential legal liability of RFMOs for listing vessels on their respective IUU lists. RFMOs possess international legal personality and can incur legal liability for wrongfully placing vessel on their IUU lists, which can result in financial loss to flag States and vessel owners. It is therefore imperative for RFMOs to formulate proper procedures and apply due process for the listing of IUU vessels.

9.3.3.4 *Record of Fishing Vessels vis a vis IUU Vessel List*

In some RFMOs, the relationship between the record of fishing vessels and the IUU vessel list would need to be clearly established. For example, under the WCPFC Conservation and Management Measure 2004-01, any vessel not included in the WCPFC Record of Fishing Vessels are deemed not authorised to fish for, retain on board, transship or land highly migratory fish stocks in the Convention Area beyond the national jurisdiction of the flag State.[74] However, neither Conservation and Management Measure 2007-03 nor Conservation and Management Measure 2004-01 provide for the deletion of IUU listed vessels from the WCPFC Record of Fishing Vessels. Consequently, the existing WCPFC measures create two separate lists of vessels which may not necessarily complement each other. This gap has been discussed in the meetings of the WCPFC and addressed in 2009 by amending WCPFC Conservation and Management Measure 2004-01 which require the deletion of IUU listed vessels from both member and WCPFC records of fishing vessels.[75]

9.3.3.5 *Reciprocal Recognition of IUU Vessel Lists*

IUU vessel listing schemes by non-tuna RFMOs, specifically NAFO, NEAFC, and SEAFO provide for the recognition of IUU lists of other RFMOs. These RFMOs exchange IUU lists and consult their members on whether or not to include such vessels on their respective IUU lists.[76] In NAFO, a non-member vessel that has been placed on the NEAFC IUU list is presumed to be undermining

[74] WCPFC, Record of Fishing Vessels and Authorisation to Fish, *Conservation and Management Measure 2004-01*, 10 December 2004, para 12.

[75] WCPFC, WCPFC Record of Fishing Vessels and Authorisation to Fish, *Conservation and Management Measure 2009-01*, para. 23.

[76] NEAFC, *Scheme of Control and Enforcement*, Art. 44(6); SEAFO, *Conservation 08/06 on IUU Vessel Listing*, para. 19.

the effectiveness of NAFO conservation and management measures.[77] Unless a NAFO member objects to the deletion from the NAFO IUU list of a NEAFC listed IUU vessel, that non-member vessel is added to the NAFO IUU list.[78] Where there is an objection to a NEAFC IUU-listed vessel being placed on or deleted from the NAFO IUU List, NAFO places that vessel on its Provisional IUU List.[79] The NAFO approach is in recognition of its shared boundary with NEAFC and the existence of stocks that straddle the two boundaries.

Tuna RFMOs are also moving toward mutual recognition of IUU vessels. Although tuna RFMOs have not adopted specific provisions recognising IUU vessel lists of other tuna RFMOs, their respective websites provide links to other RFMO IUU vessel lists to signify their cooperation in the global fight against IUU fishing.

While the reciprocal listing of IUU vessels is a positive and desirable development in the global fight against IUU fishing, current RFMO practice raises some legal and policy considerations which require careful consideration. Mutual recognition of IUU lists by RFMOs may result in the IUU listing of vessels which do not necessarily fish in the areas of competence of a particular RFMO and do not fly the flag of a member or cooperating non-member of the RFMOs undertaking the listing. The jurisdictional nexus between RFMOs and fishing vessels that do not necessarily engage in fishing activities in their areas of competence would need to be carefully considered and clearly established. It is unclear what actions can be taken against vessels included in reciprocal RFMO IUU vessel lists which do not fish in the area of competence of a particular RFMO, or whose flag State is not a member or cooperating non-member of the RFMO undertaking the IUU listing. RFMOs which recognise the IUU vessel lists of other RFMOs would therefore need to develop policy measures and procedures to address these issues.

9.3.4 *Other Schemes for Non-Member Vessels*

Aside from the listing of IUU vessels, RFMOs have implemented other schemes to deter IUU fishing activities of non-member vessels and to encourage such vessels to comply with regional conservation and management measures. These include undertaking individual or collective measures against IUU vessels of non-members and the development of cooperating non-member schemes.

In some RFMOs, if a vessel of a non-member is sighted in their area of competence, information on the vessel is distributed to members, and where appropriate it is inspected at sea and in port. The flag State of the vessel is also notified upon sighting. If it has been established that the vessel has engaged in IUU fishing,

---

[77] NAFO, *Conservation and Enforcement Measures*, Art. 52(2).
[78] NAFO, *Conservation and Enforcement Measures*, Art. 57(8).
[79] NAFO, *Conservation and Enforcement Measures*, Art. 57(8).

members and cooperating non-members are encouraged to take all necessary measures under applicable national legislation to respond to the IUU activity. The range of actions that may taken against IUU vessels include preventing such vessels, including mother ships and support vessels, from participating in any transshipment with IUU vessels, preventing the landing and transshipment of IUU vessels entering ports, and prohibiting the chartering of vessels included on the IUU list. Members and cooperating non-members of RFMOs are also required not to grant their flag to vessels included on the IUU list, except when sufficient evidence has been presented that the previous owner or operator has no further legal, beneficial or financial interest in, or control of the vessel, or that the granting of the vessel of its flag will not result in IUU fishing.[80]

Another effort aimed at combating IUU fishing is the cooperating non-member measures adopted by RFMOs.[81] These measures encourage non-members whose vessels fish in areas of the high seas managed by an RFMO to cooperate fully in the implementation of conservation and management measures by attaining the status of a cooperating non-member. A non-member may apply for the status of a cooperating non-member and needs to provide full information required by the RFMO in order for the status to be considered. The information required includes data on historical fisheries in the RFMO area, including nominal catches, details on fishing vessels involved, fishing effort and fishing areas, details on current fishing presence, information on any research programme conducted and results of such research, as well as all data required to be submitted by members of the RFMO.[82] The applicant is also required to confirm its commitment to apply the conservation and management measures adopted by the RFMO, as well as inform the organisation of the measures it has undertaken to ensure the compliance of its vessels with those measures. The data submitted to the RFMO by the non-member applicant, together with considerations relating to fishing capacity, are generally the basis for the granting of the status of cooperating non-member.

However, the cooperating non-member schemes of RFMOs do not fully resolve the problem of unregulated fishing on the high seas by vessels of non-members.[83]

---

[80] NAFO, *Conservation and Enforcement Measures*, Art. 55; ICCAT, *Recommendation to Establish an IUU Vessels List*, para. 9; IOTC, *Resolution 09/03*, para. 13; IATTC, *Resolution C-05-07*, para. 9; AIDCP, *Resolution A-04-07 on Establishing an IUU Vessels List*, para. 9; CCAMLR, *Conservation Measure 10-07 (2009)*, para. 22.

[81] IOTC, *Resolution 03/02* on Criteria for Attaining the Status of Co-operating Non-Contracting Party; WCPFC, Cooperating Non-members, *Conservation and Management Measure 2009-11*; IATTC, *Resolution C-07-02* Criteria for Attaining the Status of Cooperating Nonparty or Fishing Entity in IATTC.

[82] IOTC, *Resolution 03/02*, para. 3; NEAFC, *Scheme for Control and Enforcement*, Art. 34; IATTC, *Resolution C-07-02*, para. 3; WCPFC, Cooperating Non-members, *Conservation and Management Measure 2008-02*.

[83] See Chapter 2 of this book.

Because of different legal interpretations on the obligations of non-parties to regional fisheries management agreements,[84] RFMOs cannot guarantee that vessels of non-members whose applications for the status of cooperating non-member have been rejected will not continue to exercise their right to fish on the high seas. The cooperating non-member scheme simply supports the measure related to the presumption of IUU fishing in that any vessel of a non-member sighted as conducting fishing activities in an RFMO area is deemed to be an IUU fishing vessel. If a non-member flag State does not want its vessels to be placed on the IUU list, it has the option to obtain cooperating non-member status. However, the action that may be taken by the non-member if its application is not granted is not clear under this scheme.

The measure on cooperating non-member status does not provide any alternative for the fishing vessel except to be included in the IUU list if the application of its flag State is rejected by the RFMO. This may be interpreted as contrary to Article 119(3) of the LOSC on the conservation of living resources on the high seas which provides that "(S)tates concerned shall ensure that conservation measures and their implementation do not discriminate in form or in fact against the fishermen of any State".

Existing cooperating non-member schemes of RFMOs suggest that the criteria for obtaining such status have broadened. In addition to the commitment to apply regional conservation and management measures, more information is now demanded from non-members seeking the status of cooperating non-members.[85] It is plausible that more rigorous criteria for awarding the status of cooperating non-member, if adopted with inadequate guidelines or little transparency, may impede on the participation of non-members in RFMO efforts to effectively manage resources within their areas of competence. These challenges may contribute significantly to the increasing number of IUU vessels flying the flags of non-members in RFMO areas.

9.3.5 *Regulations on Chartering of Fishing Vessels*

In order to combat IUU fishing, some RFMOs regulate the operations of charter vessels, including the catch limit, reporting of catch, implementation of observer schemes, VMS, transshipment regulations, and port State measures. Other conditions may also apply in the chartering of fishing vessels in RFMO areas. For example, ICCAT provides the requirement that only its members may be

[84] Tore Henriksen, 'Revisiting the Freedom of Fishing and Legal Obligations on States Not Party to Regional Fisheries Management Organisations', 40 *Ocean Development and International Law* (2009), at 80–96.

[85] NEAFC, *Scheme of Control and Enforcement*, Art. 34; *See also* NEAFC, *Report of the 27th Annual Meeting of the North-east Atlantic Fisheries Commission*, 10–14 November 2008, NEAFC Headquarters, London, Volume I-Report, page 21.

chartering nations.[86] Chartered fishing vessels would need to be registered to responsible members and cooperating non-members, or to non-members which have agreed to apply the conservation and management measures of the organisation and enforce them on their vessels.[87] Such vessels would need to have a fishing licence from the chartering State and should not be on the IUU list.[88]

One of the issues with respect to the implementation of chartering arrangements is the question of jurisdiction between the chartering State and the flag State over chartered vessels. The ICCAT chartering scheme highlights the duty of flag State members to effectively exercise their duty to control their fishing vessels at the same time providing the responsibility of both the chartering member State and the flag member State in ensuring the compliance of chartered vessels with conservation and management measures of relevant RFMOs.[89] This provision does not provide clear guidelines as to how the chartering State and the flag State may exercise jurisdiction over chartered vessels in cases of IUU fishing. In the WCPFC, it was clearly stated that for the purpose of applying conservation and management measures on bigeye and yellowfin tuna, vessels operated under charter, lease or other similar mechanisms by developing island States and participating territories, which are integral part of their domestic fleet, shall be considered to be vessels of the host island State or territory.[90] This implies that the duty to exercise effective control over the activities of the vessel lies on the chartering State. Other areas of regional charter schemes where clarification may be required in order to ensure effective implementation includes the area in which charter vessels can operate and whether the scheme should be restricted to charter arrangements in the EEZ of WCPFC members.[91]

9.3.6 *Vessel Monitoring System*

RFMOs have recognised the importance of vessel monitoring systems (VMS) in addressing IUU fishing and have established measures on its operation in their areas of competence. These regulations stipulate the responsibilities of RFMO members and cooperating non-members to implement VMS. RFMOs have various VMS standards; however, they share some common characteristics such as technical specification for VMS equipment, protocols for the transmission of

---

[86] ICCAT, *Recommendation by ICCAT on Vessel Chartering*, 02-21 GEN, 13 June 2003, para. 2.

[87] ICCAT, *Recommendation by ICCAT on Vessel Chartering*, para. 3.

[88] ICCAT, *Recommendation by ICCAT on Vessel Chartering*, para. 9.

[89] ICCAT, *Recommendation by ICCAT on Vessel Chartering*, para. 4.

[90] WCPFC, Conservation and Management Measure for Bigeye and Yellowfin Tuna in the Western and Central Pacific Ocean, *Conservation and Management Measure 2008-01,* para. 2. A broader framework for chartering arrangements has yet to be developed by the WCPFC. *See* WCPFC, Charter Notification Scheme, *Conservation and Management Measures 2009-08*.

[91] WCPFC, Technical and Compliance Committee, Third Regular Session, 27 September-02 October 2007, Pohnpei, Federated State of Micronesia, WCPFC Fishing Vessel Charter Arrangements Scheme, WCPFC-TCC3-2007-23, 11 September 2007, para. 5.

VMS data related to fishing activities, including transshipment operations to a national VMS centre, database management by the national VMS centre, communication of messages from the national VMS centre to the secretariat of the relevant RFMO, and security and confidentiality of VMS data.

Vessels authorised to fish in RFMO areas are required to comply with VMS requirements, particularly if such vessels fish in areas beyond national jurisdiction. Tuna RFMOs such as ICCAT, IATTC, and IOTC require members, cooperating non-members, and fishing entities to install satellite based VMS onboard large scale longline vessels fishing for tuna in their areas of competence outside national jurisdiction.[92] Such vessels are required to transmit information such as vessel identification, position, and date and time of the position to land based fisheries monitoring centres.[93] Other RFMOs managing non-tuna species such as NAFO, CCAMLR, and NEAFC have very similar requirements on VMS but differ slightly on the provisions for satellite data reporting intervals, for instance two hours for NAFO, and four hours for CCAMLR.[94] NEAFC and NAFO also have a well established vessel position and catch reporting formats and standards known as the North Atlantic Format, which could be adopted as a model for developing an international standard for VMS position and catch reporting.[95] NEAFC and NAFO catch reports include catch on entry, catch on exit data, and transshipment reports.[96]

Concerns about the confidentiality of VMS information are being addressed in RFMOs. For example, CCAMLR provides specific measures to ensure the security and confidentiality of VMS data. These measures restrict the release and use of VMS reports and messages only for the purposes of active surveillance

---

[92] ICCAT, *Resolution by ICCAT Concerning a Management Standard for the Large-scale Tuna Longline Fishery*, 01-20 GEN, 22 February 2002, Attachment I, para. 1(iii); *ICCAT General Outline of Integrated Monitoring Measures,* para. 1(i)(vii); ICCAT, *Recommendation by ICCAT Concerning Minimum Standards for the Establishment of a Vessel Monitoring System in the ICCAT Convention Area*, 03-14 GEN, 19 June 2004, para 1; IOTC, *Resolution 06/03* on Establishing a Vessel Monitoring System Programme, para. 1; IATTC, *Resolution C-04-06*, Resolution on the Establishment of a Vessel Monitoring System, para. 1; *WCPFC Convention*, Art. 24(8) and 24(9); WCPFC, Commission Vessel Monitoring System, *Conservation and Management Measure 2007-02*.

[93] ICCAT, *Recommendation by ICCAT Concerning VMS*, para. 1(b); IOTC, *Resolution 06/03,* para. 2.

The IATTC also proposes the transmission of weekly data collected by observers related to at-sea reporting. *See* IATTC, *Resolution C-04-06*, Resolution on the Establishment of a Vessel Monitoring System.

[94] *NAFO Conservation and Enforcement Measures*, Art. 26; CCAMLR, *Conservation Measure 10-04 (2007)*, Automated Satellite-linked Vessel Monitoring Systems; NEAFC, *Scheme of Control and Enforcement*, Art. 11.

[95] FAO, Report of the Expert Consultation on Data Formats and Procedures for Monitoring, Control and Surveillance, Bergen, Norway, 25–27 October 2004, *FAO Fisheries Report 761* (Rome: FAO, 2005), at 11.

[96] NAFO, *Conservation and Enforcement Measures*, Art. 27; NEAFC, *Scheme of Control and Enforcement*, Arts. 12 and 13.

presence, and/or inspections by a contracting Party in a specified CCAMLR sub-area or division or for verifying the content of a *Dissotichus* Catch Document.[97] CCAMLR also implements appropriate measures to protect reports and messages from accidental or unlawful destruction or a loss, alteration, unauthorised disclosure or access, and all inappropriate forms of processing data.[98] Such measures address security issues such as system access, authenticity and data access, communication security, data security, and other security procedures.[99] NAFO provides for the security and confidentiality of all electronic reports, including those of VMS data.[100]

9.3.7 *Observer Programme*

RFMOs have established observer programmes not only to collect scientific data, but also to monitor the implementation of conservation and management measures. In general, observers may record and report fisheries data, collect verified catch data, observe and estimate catches with a view to identifying catch composition, discards, by-catches and undersized fish, record and describe gears and mesh sizes, and verify positions of vessels and entries made to logbooks. RFMO observer programmes specify a range of obligations of members and cooperating non-members in their implementation. These include the role of the organisation in training observers, receiving communications on observer reports, rights and obligations of observers in undertaking their functions, and rights and obligations of the master, operators, and crew in accepting observers onboard vessels.

RFMO observer programmes are also increasingly being utilised to monitor the implementation of conservation and management measures such as in SEAFO, ICCAT, and WCPFC.[101] Some RFMOs such as the CCSBT have adopted observer programmes which are more geared towards the collection of science related information such as details of the observed vessel, summary of the observed trip, and comprehensive catch, effort and environmental information.[102]

---

97 CCAMLR, *Conservation Measure 10-04 (2007)*, para. 18.

98 CCAMLR, *Conservation Measure 10-04*, Annex 10-04/B, Provisions on Secure and Confidential Treatment of Electronic Reports and Messages Transmitted Pursuant to Conservation Measure 10-04.

99 CCAMLR, *Conservation Measure 10-04 (2007)*, Annex 10-04/B, Provisions on Secure and Confidential Treatment of Electronic Reports and Messages.

100 *See* NAFO, *Conservation and Enforcement Measures*, Annex XIX, Rules on Confidentiality, Provisions on Secure and Confidential Treatment of Electronic Reports and Messages Transmitted Pursuant to Articles 26–27 of the Conservation and Enforcement Measures.

101 *SEAFO Convention*, Art. 14(3)(g); ICCAT, *Recommendation by ICCAT on the Establishment of a Closed Area/Season for the Use of Fish Aggregation Devices (FADs)*, 99-01 TRO, 15 June 2000, para. 6; WCPFC, Conservation and Management Measure for the Regional Observer Programme, *Conservation and Management Measure 2007-01*, para. 4.

102 CCSBT, *Scientific Observer Program Standards*, www.ccsbt.org. Accessed on 14 March 2004.

Other information collected through the CCSBT observer programme are catch information for each period of observation, biological measurements taken of individual Southern bluefin tuna, and information related to the Southern bluefin tuna tag recovery programme.[103]

NAFO implements a very comprehensive observer programme. It requires all fishing vessels to carry at least one observer at all times while fishing in the NAFO regulatory area.[104] Observers are mandated to monitor the compliance of fishing vessels with relevant conservation and enforcement measures by recording and reporting on the fishing activities of the vessels and verifying their position when they are engaged in fishing.[105] Observers also estimate catches with a view of identifying catch composition and monitoring discards, by-catches and the taking of undersized fish.[106] Observers record the type, mesh size, and gear attachments employed by the master, verify entries made to the logbooks, and collect catch and effort data for each haul.[107] Observers also monitor the functioning of the satellite tracking system and report any infringement to NAFO within 24 hours.[108] Similar functions are given to observers under the CCMALR observer programme. In addition, CCAMLR scientific observers also have power to collect and report factual data on sightings of fishing vessels in the Convention area.[109] In 2006, NAFO mandated the use of electronic observer reports to replace the manual process of submission and entering observer data in the database.[110] The use of electronic observer reports reduces the time that observers need to spend in the CCAMLR regulatory area.

One of the main issues associated with the implementation of observer programmes is their limited coverage. ICCAT, for example, directs its members to implement national observer programmes for around 5 to 10 per cent of the vessels operating in the North Atlantic, depending on the type of fishery.[111] Chartered vessels are also required to place observers on at least 10 per cent of the chartered vessels or during 10 per cent of the fishing time of the vessels.[112] At least 10 per cent of transshipment at sea in the ICCAT management area is also required to

---

[103] CCSBT, *Scientific Observer Program Standards*.

[104] NAFO, *Conservation and Enforcement Measures*, Art. 23(1).

[105] NAFO, *Conservation and Enforcement Measures*, Art. 23(4) and (5).

[106] NAFO, *Conservation and Enforcement Measures*, Art. 23(4) and (5).

[107] NAFO, *Conservation and Enforcement Measures*, Art. 23(4) and (5).

[108] NAFO, *Conservation and Enforcement Measures*, Art. 23(4) and (5).

[109] CCAMLR, *Scheme of International Scientific Observation*, Adopted at CCAMLR XI and amended at CCAMLR XVI.

[110] NAFO, *Conservation and Enforcement Measures*, Art. 61.

[111] ICCAT, *Recommendation by ICCAT Relating to the Rebuilding Program for North Atlantic Swordfish*, 02-02 SWO, 04 June 2003; ICCAT, *Recommendation by ICCAT to Establish a Plan to Rebuild Blue Marlin and White Marlin Populations*, 00-13 BIL, 26 June 2001.

[112] ICCAT, *Recommendation by ICCAT on Vessel Chartering*, 02-21 GEN, 04 June 2003, para 8.

be supervised by an observer on board.[113] The CCSBT requires an observer coverage of 10 per cent for catch and effort monitoring for each fishery.[114] A wider coverage for an observer programme is usually constrained by the lack of financial resources on the part of States and regional organisations. A smaller coverage of observer programmes leads to a limited amount of verified fisheries information which consequently limits the likelihood of detecting unreported fishing.

9.3.8 *Boarding and Inspection*

Boarding and inspection schemes entitle members of RFMOs to carry out boarding and inspection on the high seas of fishing vessels engaged in or reported to have engaged in a fishery regulated by the organisation. Each RFMO member has the obligation to ensure that vessels flying its flag accept boarding and inspection by authorised inspectors according to the procedures established by that organisation. Boarding and inspection schemes are generally based on the sightings of fishing vessels, or when there is reason to believe that the vessel is or has been operating in violation of the conservation and management measures of a particular RFMO and that such boarding and inspection is necessary to obtain or verify evidence of such a violation.[115] RFMOs also provide the rights and obligations of inspecting States, inspectors, and the master of fishing vessels in boarding and inspection schemes. These schemes establish inspection procedures on communication with fishing vessels prior to boarding and inspection, actual boarding of the vessel, carrying out the inspection, and report on the result of the inspection. Boarding and inspection procedures also provide for the precautions in conducting the inspection, use of force, and dispute settlement.

RFMOs such as NAFO, NEAFC, CCAMLR, and WCPFC have established boarding and inspection procedures that provide authority to an inspector to examine all relevant areas, gears, equipment, and documents to verify the compliance of vessels with regional fisheries management and conservation measures.[116] The procedure also includes measures that an inspector may take to establish the infringement committed by a fishing vessel.[117] These measures are

---

[113] ICCAT, *Recommendation by ICCAT on Vessel Chartering*, para 12.

[114] CCSBT, *Scientific Observer Program Standards*.

[115] NAFO, *Conservation and Enforcement Measures*, Art. 28; *NEAFC Scheme of Control and Enforcement*, Chapter IV; *NEAFC Scheme of Control and Enforcement*, Art. 17-25; WCPFC, Boarding and Inspection Procedures, *Conservation and Management Measure 2006-08*, para. 10, 42 and 43.

[116] NAFO, *Conservation and Enforcement Measures*, Art. 28; NEAFC, *Scheme of Control and Enforcement*, Art. 18; CCAMLR, *System of Inspection*, Adopted at CCAMLR VI, XII, XIII, XVI, and XVIII. www.ccamlr.org. Accessed on 13 December 2004; WCPFC, *Conservation and Management Measure 2006-08*, para. 22.

[117] NAFO, *Conservation and Enforcement Measures*, Arts. 31-32; NEAFC, *Scheme of Control and Enforcement*, Arts. 19 and 20.

consistent with the provisions of international fisheries agreements, and are applied to all vessels conducting fishing activities in the areas of competence of NAFO, NEAFC, and CCAMLR. In the case of ICCAT, the boarding and inspection scheme also applies to vessels without nationality.[118]

One of the key elements in the effectiveness of regional boarding and inspection schemes is the action taken by the flag State of the fishing vessel when the authorities of the inspecting vessel observe an activity that constitutes a violation of conservation and management measures of the relevant RFMO. The flag State may either assume the obligation to investigate and take enforcement action against the fishing vessel in question, or authorise the inspecting vessel to complete the investigation of the possible violation.[119] This includes taking the vessel into port for further inspection.

Without follow-up measures after conducting boarding and inspection, the activities of IUU fishing vessels cannot be deterred effectively. For example, through random at-sea inspections, NAFO establishes infringements of its regulations and collects evidence for prosecution within the legal system of each NAFO flag State. However, there has been a concern in NAFO on the increasing number of infringement cases with no follow up actions or lack of reporting by flag States on the status of actions taken against vessels involved in serious infringements contrary to NAFO Conservation and Enforcement Measures.[120] This problem is also affected by delayed submission of at-sea inspection and observer reports. Delayed at-sea reports in NAFO have been as high as 38 per cent while late observer reports have been as high as 80 per cent in 2007.[121]

The member of an RFMO is normally liable for the damage or loss attributable to their action in implementing the boarding and inspections procedures when such action is unlawful or exceeds what is reasonably required in the light of collected information.[122] Amongst the various RFMOs, the WCPFC provides specific provisions, albeit non-binding in nature, for the settlement of disagreements.[123] It provides that in the event of a disagreement concerning the interpretation, application or implementation of boarding and inspection procedures, the parties

---

[118] ICCAT, *Recommendation by ICCAT on Transshipments and Vessel Sightings*, 97-11 GEN, 13 June 1998, para. 2.

[119] WCPFC, *Conservation and Management Measure 2006-08*, para. 33.

[120] NAFO, "Annex 19, Annual Compliance Review 2008," in *Report of the Fisheries Commission*, 22–26 September 2008, NAFO/FC Doc. 08/20, at 8.

[121] NAFO, *Annex 19, Annual Compliance Review 2008*, page 8.

[122] WCPFC, *Conservation and Management Measure 2006-08*, para. 45.

[123] Paragraph 52 of the WCPFC Conservation and Management Measure 2006-08 provides that the provisions for the settlement of disagreements shall not prejudice the rights of any Member to use the dispute settlement procedure provided in the WCPF Convention. Article 31 of the WCPF Convention provides that "(t)he provisions relating to the settlement of disputes set out in Part VIII of the Agreement apply, *mutatis mutandis*, to any dispute between members of the Commission, whether or not they are also Parties to the (UN Fish Stocks) Agreement."

concerned would need to enter into consultation in order to resolve the disagreement.[124] If the disagreement remains unresolved following consultations, it is referred to the WCPFC Technical and Compliance Committee where a panel will be established to consider the matter.[125] The panel will prepare a report upon which the decision of the Commission is based.

9.3.9 *Trade Documentation and Catch Certification*

Two of the most commonly used schemes for documenting fish and fish products are trade documentation and catch certification. Trade documentation refers to "schemes established by RFMOs that require documentation to accompany particular fish and fish products through international trade identifying the origin of fish for the purpose of ascertaining levels of unreported fishing."[126] Catch certification is issued by relevant national authorities at the point of harvesting and covers all fish to be landed or transhipped, while a trade document is issued only with respect to products that enter international trade.[127] RFMOs such as ICCAT, IATTC, IOTC, and CCSBT have adopted trade documentation programmes, while CCAMLR has adopted a Catch Documentation Scheme for toothfish which is an amalgam of catch certification and trade documentation schemes. The Agreement on the International Dolphin Conservation Program (AIDCP) has also adopted a related certification called a dolphin safe tuna certification.

The Bluefin Tuna Statistical Document Program of ICCAT applies to all bluefin tuna imported into the territory of a Contracting Party. The Bluefin Tuna Statistical Document must contain information on imported, exported, or re-exported fish and fish products such as the name of the country issuing the document; description of vessel; name of the exporter and the importer; description of fish for re-export; area of harvest of the fish in the shipment; gear utilised to catch the fish; type of product and total weight; and point of export.[128] The document is

---

[124] WCPFC, *Conservation and Management Measure 2006-08*, para. 48.

[125] WCPFC, *Conservation and Management Measure 2006-08*, para. 49.

[126] FAO, Report of the Expert Consultation of Regional Fisheries Management Bodies on Harmonization of Catch Certification, La Jolla, United States of America, 09–11 January 2002, *FAO Fisheries Report No. 697* (Rome: FAO, 2002), at 1.

[127] *FAO Fisheries Report No. 697, ibid.*, at 1.

For example, IATTC states that bigeye tuna caught by purse seiners and baitboats and destined principally for canneries are not subject to the statistical document requirement. *See* IATTC, *Resolution C-03-01*, Resolution on IATTC Bigeye Tuna Statistical Document Program, 24 June 2003, para. 1.

[128] ICCAT, *Recommendation by ICCAT Concerning the ICCAT Bluefin Tuna Statistical Document Program*, 92-41 SDP, 25 July 1993, Appendix, para. 1; ICCAT, *Resolution by ICCAT Concerning the Effective Implementation of the ICCAT Bluefin Tuna Statistical Document Program*, 94-05 SDP, 23 January 1995, Addendum; ICCAT, *Recommendation by ICCAT Concerning the Implementation of the ICCAT Bluefin Tuna Statistical Document Program on Re-export*, 97-04 SDP, 12 December 1997, Attachment.

validated by a government official of the flag State of the vessel that harvested the tuna.[129] ICCAT also implements equivalent Statistical Document Programs for swordfish, bigeye tuna, and other species.[130] Other RFMOs such as the CCSBT, IATTC, and IOTC have comparable statistical document programmes and utilise very similar trade documents forms.[131]

For ICCAT, validation of a trade document is waived when the flag State of the vessel implements logbook and information retrieval systems that are accepted by the Commission.[132] For the logbook and information retrieval systems to be accepted, the flag State needs to present all government regulations requiring routine collection and provision of accurate information related to bluefin tuna harvests; apply sufficient penalties to deter non-compliance; and provide an outline of how the logbook system and other means would be used to provide evidence of the origin of exported fish at the final point of destination.[133] The linkage between trade documentation and the logbook system in identifying the origin of fish illustrates how such measures need to be coordinated in order to validate the accuracy of reports on fishing activities and address unreported fishing.

A specific scheme that addresses illegal by-catch of tuna is the certification of AIDCP Dolphin Safe Tuna and Tuna Products related to the IATTC.[134] An AIDCP Dolphin Safe Tuna Certificate is a document issued by the department of a national government which is responsible for implementing the procedures for the certification of AIDCP Dolphin Safe tuna.[135] An AIDCP Dolphin Safe Tuna Label which may be used on the packaging of the tuna certified under the programme,[136] is a graphic representation which distinguishes dolphin safe tuna and tuna products. This certification is implemented together with the System for

---

[129] ICCAT, Bluefin Tuna Statistical Document Program, Appendix, para. 1; ICCAT, *Resolution by ICCAT Concerning Validation by a Government Official of the Bluefin Tuna Statistical Document*, 93-02 SDP, 30 November 1993; ICCAT, *Recommendation by ICCAT on the Validation of Bluefin Statistical Documents between ICCAT Contracting Parties Which are Members of the European Community*, 96-10 SDP, 04 August 1997.

[130] ICCAT, *Recommendation by ICCAT on Establishing Statistical Document Programs for Swordfish, Bigeye Tuna, and Other Species Managed by ICCAT*, 00-22 SDP, 26 June 2001; ICCAT, *Recommendation by ICCAT Concerning the ICCAT Bigeye Tuna Statistical Document Program*, 01-21 SDP, 21 September 2002; ICCAT, *Recommendation by ICCAT Establishing a Swordfish Statistical Document Program*, 01-22 SDP, 21 September 2002.

[131] *See* CCSBT, *Southern Bluefin Tuna Statistical Document Program*, Updated October 2003, www.ccsbt.org. Accessed on 15 June 2005; IATTC, *Resolution C-03-01*, Resolution on IATTC Bigeye Tuna Statistical Document Program, 24 June 2003; IOTC, *Resolution 01/06*, Recommendation by IOTC Concerning the IOTC Bigeye Tuna Statistical Document Programme.

[132] ICCAT, *Recommendation Concerning Bluefin Tuna Statistical Document Program*, para. 1.

[133] ICCAT, *Resolution by ICCAT on Interpretation and Application of the ICCAT Bluefin Tuna Statistical Document Program*, 94-04 SDP, 23 January 1995, Addendum.

[134] IATTC, Agreement on the International Dolphin Conservation Program, *Procedures for AIDCP Dolphin Safe Tuna Certification*, amended, 20 October 2005.

[135] AIDCP, *Procedures for AIDCP Dolphin Safe Tuna Certification*, para 1.

[136] AIDCP, *Procedures for AIDCP Dolphin Safe Tuna Certification*, para 1.

Tracking and Verification of Tuna.[137] The purpose of the system for tracking and verifying tuna is to enable dolphin safe tuna to be distinguished from non-dolphin safe tuna from the time of capture, during unloading, storage, transfer, and processing, or when it is ready for retail sale.[138] Tuna which is positively identified by the IATTC as having been caught in contravention of IATTC tuna conservation and management measures is not eligible for AIDCP Dolphin-Safe Certificate.[139] This scheme not only addresses illegal fishing for tuna in the IATTC area, but also by-catch issues associated with IUU fishing.

Unlike the statistical document programmes, the Catch Documentation Scheme for *Dissostichus* spp. (CDS) adopted by CCAMLR aims not only at identifying the origin of toothfish imported into or exported from the territories of CCAMLR Contracting Parties, but also to determine whether the toothfish was harvested in a manner consistent with CCAMLR conservation measures.[140] Each landing of toothfish at the port of a CCAMLR member needs to be accompanied by a *Dissostichus* catch document (DCD) which contains information on the issuing authority; description of the vessel; fishing licence; weight, area, and date the catch was taken; date and port at which the catch was landed; and information on the recipients of the catch and amount of each species and product type received.[141] In addition, the DCD also requires information on landing and transshipment.[142]

NAFO has requirements for product labelling similar to catch certification. Article 23 of the NAFO Conservation and Enforcement Measure provides that when processed, all fish harvested in NAFO regulatory area are to be labelled in such a way that each species and product category is identifiable using an agreed code and marked as having been caught in the area.[143] Similarly, NEAFC requires all frozen fish caught in its management area to be clearly labelled or stamped with the species, production date, the sub-area and division where the catch was taken, and the name of the vessel which caught the fish.[144]

There are two major gaps in the implementation of trade documentation and catch certification schemes. First, there is no complete coverage of fisheries trade utilising these schemes. In the case of CCAMLR's CDS for example, among the 56 States trading for toothfish, only 35 States are believed to be complying with

---

[137] AIDCP, *Procedures for AIDCP Dolphin Safe Tuna Certification,* para. 4(a).

[138] AIDCP, *System for Tracking and Verifying Tuna*, para. 2.

[139] AIDCP, *Procedures for AIDCP Dolphin Safe Tuna Certification*, para 3.

[140] CCAMLR, *Conservation Measure 10-05 (2009)*, Catch Documentation Scheme for *Dissostichus* spp, Art. 1.

[141] CCAMLR, *Conservation Measure 10-0 (2004)*, Art. 6.

[142] CCAMLR, *Conservation Measure 10-0 (2004)*, Annex.

[143] In the case of shrimps and Greenland halibut harvested in particular area, the product would need to be marked in accordance with the stock area. *See* NAFO Conservation and Enforcement Measure, Art. 23.

[144] NEAFC, *Scheme of Control and Enforcement*, Art. 8.

CDS requirements.[145] This gap may create an opportunity for the trade in IUU caught toothfish. Second, statistical document programmes do not require statements that the catch was made in compliance with regional fisheries conservation and management measures and do not directly prohibit the importation of illegally harvested tuna. There is therefore a need to fill these gaps in the implementation of trade documentation and catch certification schemes in order to effectively address IUU fishing.

9.3.10 *Trade Restrictive Measures*

A number of RFMOs impose restrictions on the trade of fish derived from IUU fishing activities. ICCAT, NEAFC, CCAMLR, IATTC, and IOTC require their members and cooperating non-members to prohibit the imports of tuna and tuna-like species from vessels included in the IUU list.[146] ICCAT, in particular, has adopted resolutions prohibiting the importation of Atlantic bigeye tuna and all forms of its products from Belize, Cambodia, Honduras, St. Vincent and the Grenadines, Bolivia, Sierra Leone, Equatorial Guinea, and Georgia for failing to comply with the 1998 Resolution Concerning the Unreported and Unregulated Catches of Tuna by Large scale Longline Vessels in the Convention Area.[147] The import prohibitions are only lifted upon submission of satisfactory documentary evidence that the fishing practices of vessels flying the flags of members and cooperating non-members are being conducted in conformity with ICCAT conservation and management measures.[148] These trade restrictive measures are

---

[145] Kevin W. Riddle, 'Illegal, Unreported, and Unregulated Fishing', 37 *Ocean Development and International Law* (2006), at 282.

[146] ICCAT, *Recommendation by ICCAT to Establish a List of Vessels Presumed to have Carried Out IUU Fishing Activities in the ICCAT Convention Area*, 02-23 GEN, 04 June 2003, para. 9(e); NEAFC, *Scheme of Control and Enforcement*, Art. 45(2)(e); CCAMLR, *Conservation Measure 10-07 (2006)*, Art. 11(f); CCAMLR, *Conservation Measures 10-06 (2008)*, Art. 18(vii); IATTC, *Resolution C-05-07*, Art. 9(e); IOTC, *Resolution 09/03*, para. 12(e).

[147] ICCAT, *Recommendation by ICCAT Regarding Belize, Cambodia, Honduras, and St. Vincent and the Grenadines Pursuant to the 1998 Resolution Concerning the Unreported and Unregulated Catches of Tuna by Large-scale Longline Vessels in the Convention Area*, 00-15 SANC, 15 October 2001, para. 2; ICCAT, *Recommendation by ICCAT Regarding Bolivia Pursuant to the 1998 Resolution Concerning the Unreported and Unregulated Catches of Tuna by Large-scale Longline Vessels in the Convention Area*, 02-17 SANC, 04 June 2003, para. 1; ICCAT, *Recommendation by ICCAT for Trade Restrictive Measures on Sierra Leone*, 02-19 SANC, 04 June 2003; ICCAT, *Recommendation by ICCAT Concerning the Continuance of Trade Measures Against Equatorial Guinea*, 03-17 SANC, 19 June 2004; ICCAT, *Recommendation by ICCAT for Bigeye Tuna Trade Restrictive Measures on Georgia*, 03-18 SANC, 19 June 2004.

[148] ICCAT, *Recommendation Against Belize, Cambodia, Honduras, and St. Vincent and the Grenadines*, para. 5; *See also* ICCAT, *Recommendation by ICCAT Concerning the Importation of Bigeye Tuna and Bigeye Tuna Products from St. Vincent and the Grenadines*, 01-14 SANC, 21 September 2002; ICCAT, *Recommendation by ICCAT Concerning the Importation of Atlantic Bluefin Tuna, Atlantic Swordfish, and Atlantic Bigeye Tuna and their Products from Belize*, 02-16 SANC, 04 June 2003; ICCAT, *Recommendation by ICCAT Concerning the Importation of Bigeye*

consistent with paragraph 66 of the IPOA-IUU which calls on States to prevent fish caught by vessels identified by RFMOs to have been involved in IUU fishing from being traded or imported into their territories.

9.3.11 *Port State Measures*

RFMOs have adopted port State measures to combat IUU fishing. ICCAT, NAFO, NEAFC, CCAMLR, and IOTC have established port inspection programmes[149] while the WCPFC and SEAFO have yet to establish port State schemes for fishing vessels but embrace the provisions of the UN Fish Stocks Agreement on port State measures.[150] Most port State schemes adopted by RFMOs enumerate the obligations of port States and fishing vessels. For example, NEAFC has adopted various provisions for inspecting vessels of members, cooperating non-members, fishing entities, and non-members[151] while NAFO has highlighted the duties of flag States in implementing port State measures.[152] RFMO port State schemes have significantly increased the recognition of the role of port State measures in combating IUU fishing. Some of the elements of these port State measures are the designation of ports, advanced notice of entry, prior authorisation to land or tranship fish, inspection of fishing vessels, and port State enforcement.

9.3.11.1 *Designation of Ports*

One of the RFMO requirements for port State measures is the designation of ports to which fishing vessels may be permitted access for the purpose of landing and transshipment.[153] This includes requirements for the designation of competent authorities for the purpose of receiving prior notification.[154] RFMOs also

---

Tuna and Its Products from Honduras, 02-18 SANC, 04 June 2003; ICCAT, *Recommendation by ICCAT Concerning the Trade Sanction Against St. Vincent and the Grenadines*, 02-20 SANC, 04 June 2003.

[149] ICCAT, *Recommendation by ICCAT for a Revised ICCAT Port Inspection Scheme*, 97-10 GEN, 13 June 1998; *NAFO Conservation and Enforcement Measures*, Chapter V; NEAFC, *Scheme of Control and Enforcement*, Chapter V; IOTC, *Resolution 05/03* Relating to the Establishment of an IOTC Programme of Inspection in Port, 2005; CCAMLR, *Conservation Measure 10-03 (2008)*, Port Inspections of Vessels Carrying Toothfish; GFCM, *Recommendation GFCM/2008/1* on a Regional Scheme on Port State Measures to Combat Illegal, Unreported and Unregulated Fishing in the GFCM Area.

[150] *WCPF Convention*, Art. 27; SEAFO Convention, Art. 15. The WCPFC and SEAFO are still in the process of developing measures that may be adopted by port States to conserve and manage resources in their areas of competence.

[151] NEAFC, *Scheme of Control and Enforcement,* Art. 5.

[152] NAFO, *Conservation and Enforcement Measures*, Art. 47.

[153] NAFO, *Conservation and Enforcement Measures*, Art. 46(1); SEAFO, *Conservation Measure 09/07* to Amend and Consolidate Conservation Measure 02/05 Relating to Interim Port State Measures, para. 2.

[154] NAFO, *Conservation and Enforcement Measures*, Art. 46(3). NEAFC, *Scheme of Control and Enforcement*, Art. 21.

specifically provide for the establishment of a register of all ports where vessels are allowed port access and the requirement for the submission of relevant information, including associated conditions of entry and the period of notice required.[155]

9.3.11.2 *Advanced Notice of Port Entry*

RFMOs require their port member States to establish minimum period for prior notification, normally within three working days before the estimated time of arrival.[156] The period for prior notification is also adjusted depending on the distance between the fishing ground and the ports of the States where fish may be landed.

Among the RFMOs, only CCAMLR requires vessels of its members to provide notice of entry into port and at the same time convey a written declaration that they have not engaged in or supported IUU fishing.[157] This obligation is consistent with the requirements of the IPOA-IUU for vessels to provide a reasonable advance notice of their entry into port, a copy of their authorisations to fish, details of their fishing trip, and quantities of fish on board.[158]

Similar to the declaration that a vessel has not engaged in IUU fishing is the requirement for flag States to confirm the legality of the catch being landed in a port. Under Article 47 of the NAFO Conservation and Enforcement Measures, the flag State can confirm the legality of the catch by its fishing vessel if it has completed forms issued by the port State stating that the fishing vessel had sufficient quota for the species declared, that the declared quantity of fish had been duly reported by species taking into account any catch or effort limitations, that the vessel had an authorisation to fish in the areas declared, and that the presence of the vessel in the area where the catch was taken had been verified by VMS data.

9.3.11.3 *Prior Authorisation to Land or Tranship Fish*

In addition to the requirement for the advanced notice of entry, RFMO measures require prior authorisation as a precondition for landing and transshipment of fish in port. For NAFO and NEAFC, such authorisation may only be given after the port State authority has been given confirmation by the flag State that the catch to be landed or transhipped has been taken in compliance with the measures of the

---

[155] SEAFO, *Conservation Measure 09/07*, para. 4.

[156] NAFO, *Conservation and Enforcement Measures*, Art. 46(1) and (2); NEAFC, *Scheme of Control and Enforcement*, Art. 22; SEAFO, *Conservation Measure 09/07*, para. 5.

[157] CCAMLR, *Conservation Measure 10-03 (2008)*, Art. 2.

[158] *IPOA-IUU*, para. 55.

relevant RFMO.[159] Pending the receipt of such confirmation, NEAFC allows its member port States to authorise all or part of the landing provided that the fish consignment is kept in storage under control of the competent authorities. If the confirmation has not been received within 14 days of the landing, the port State may confiscate and dispose of the fish in accordance with national regulations.[160]

9.3.11.4 *Inspection of Fishing Vessels*

Port inspections are essential to verify compliance with agreed conservation and management measures. It is also during port inspections that possible infringements of national and regional fisheries regulations by fishing vessels are identified.[161] Several RFMOs make provision for the mandatory inspection of fishing vessels flying the flags of members and cooperating non-members which voluntarily enter designated ports.[162] ICCAT, NAFO, NEAFC, CCAMLR, IATTC, and IOTC require that fishing vessels flying the flags of non-members are not allowed to land or transship their fish until inspection has taken place.[163] Port inspectors are permitted to examine documents, logbooks, fishing gears, and catch on board vessels from non-members[164] and may also conduct thorough inspections of transhipped fish at the time of landing, as well as ensure the validity of certificate of transshipment.[165] Port inspection forms are used to collect detailed information on vessel identification, discharge of catches, and gear inspections.[166] Inspection

---

[159] NAFO, *Conservation and Enforcement Measure*, Art. 46(6); NEAFC, *Scheme of Control and Enforcement,* Art. 23(2).

[160] NAFO, *Conservation and Enforcement Measure*, Art. 46(7); NEAFC, *Scheme of Control and Enforcement,* Art. 24(3).

[161] Some RFMOs have established a list of serious infringements on the basis of which port States can undertake enforcement actions. See for example NEAFC, *Scheme of Control and Enforcement*, Art. 30; NAFO, *Conservation and Enforcement Measure*, Art. 37.

[162] ICCAT, *Port Inspection Scheme*, para. 2; CCAMLR, *Conservation Measure 10-03 (2008)*, Art. 1; IOTC, *Resolution 05/03*, para. 3; IPHC, *Pacific Halibut Fishery Regulations 2005*, Art. 20.

[163] ICCAT, *Recommendation by ICCAT Concerning the Ban on Landings and Transshipments of Vessels from Non-Contracting Parties Identified as Having Committed a Serious Infringement*, 98-11 GEN, 21 June 1999, para. 2; NAFO, *Conservation and Management Measures*, Arts. 38(2)-38(4) and 43(1); NEAFC, *Scheme of Control and Enforcement*, Art. 41; CCAMLR, *Conservation Measure 10-07 (2009)*, para. 4; CCAMLR, *Conservation Measure 10-03 (2008),* para. 3; IOTC, *Resolution 01/03* Establishing a Scheme to Promote Compliance by Non-Contracting Party Vessels with Resolutions Established by IOTC, 2001, para. 3.

[164] ICCAT, *Recommendation on Ban on Landings and Transshipments of Non-Contracting Party Vessels*, para. 1.

[165] ICCAT, *Resolution by ICCAT Concerning More Effective Measures to Prevent, Deter, and Eliminate IUU Fishing by Tuna Longline Vessels*, 01-19 GEN, 22 February 2002, para. 8; ICCAT, *Port Inspection Scheme*, para. 3.

[166] NAFO, *Conservation and Enforcement Measure*, Annex XIII.

of fishing vessels in ports not only allows the collection of fisheries information necessary to determine the occurrence of IUU fishing but also the verification of fisheries information recorded in fishing logbooks.

Cooperation between port and flag States in implementing port State measures is a requirement in some RFMOs such as CCAMLR, IOTC, ICCAT, and NEAFC.[167] In particular, ICCAT requires its inspectors to draw up a standardised report which is sent to the flag State and ICCAT Secretariat within 10 days of the inspection.[168] The exchange of port inspection information with flag States is necessary to enable the latter to take punitive or corrective actions against the vessel involved in IUU fishing.

9.3.11.5 *Port Enforcement Actions*

If as a result of port inspections there are clear grounds for believing that a fishing vessel flying the flag of another member has been engaged in any activity contrary to regional conservation and management measures, a number of actions may be taken by the port State. These measures include, at the minimum, noting the infringement in the inspection report, ensuring the security of the evidence gathered during inspection, and communicating with the designated authority of the flag States of the inspected vessel.[169] Surveillance reports from boarding and inspections also serve as triggers for determining fisheries infringements or IUU fishing activities and for undertaking port State enforcement actions against vessels believed to have conducted IUU fishing.

RFMOs provide for specific enforcement actions against vessels believed to have engaged in or supported IUU fishing. Among various RFMOs, only CCAMLR provides for the denial of port access to fishing vessels which either declared that they have been involved in IUU fishing or failed to make a declaration, except in cases of emergency.[170] As discussed in Chapter 7, some CCAMLR members have denied port entry to vessels engaged in IUU fishing.

The prohibition of landing and transshipment of fish is the more common port State measure undertaken against vessels by members of RFMOs. A number of RFMOs prohibit the landing and transshipment of vessels of both members and non-members. Vessels which are prohibited from transshipping and landing fish

---

[167] CCAMLR, *Conservation Measures 10-06 (2008)*, para. 5; CCAMLR, *Conservation Measure 10-03 (2008)*, Art. 3; IOTC, *Resolution 01/03*, para. 5; IOTC, *Resolution 05/03*, para. 5; ICCAT, *Recommendation on Ban on Landing and Transshipments*, para. 4; NEAFC, *Scheme of Control and Enforcement*, Arts. 30; ICCAT, *Resolution by ICCAT Concerning Efforts to Improve the Completeness of Task I Reported Statistics*, 96-13 MISC, paras. 1-2.

[168] ICCAT, *Port Inspection Scheme*, para. 2.

[169] NAFO, *Conservation and Enforcement Measures*, Art. 36; NEAFC, *Scheme of Control and Enforcement*, Art. 28(1).

[170] CCAMLR, *Conservation Measure 10-03 (2008)*, Art. 2.

are those not included in the record of fishing vessels,[171] vessels appearing on the IUU list,[172] and vessels which have conducted fishing activities contrary to conservation and management measures, for example landing of undersized fish.[173] NEAFC also prohibits the supply of provisions, fuel or other services to vessels recorded in the IUU list.[174] For vessels of non-members, ICCAT, NAFO, NEAFC, IATTC, and IOTC provide that landing or transshipment of fish is prohibited if inspection reveals that the vessel has undermined conservation and management measures, unless the vessel establishes that the fish was caught outside the Convention area or in compliance with conservation and management measures.[175]

9.3.11.6 *Other Enforcement Measures Against IUU Fishing Vessels*

The requirement for the flag State to prove that its vessel has not conducted IUU fishing is supplemented by other requirements to ensure compliance with the management measures adopted by the RFMO. RFMO members are required to investigate fully any alleged violation by fishing vessels flying their flag and provide the progress of the investigation to the RFMO, as well as the details of

---

[171] ICCAT, *Recommendation by ICCAT Concerning the Establishment of an ICCAT Record of Vessels Over 24 Metres Authorised to Operation in the Convention Area*, 02-22 GEN, 04 June 2003; ICCAT, *Recommendation by ICCAT to Adopt Additional Measures Against Illegal, Unreported and Unregulated (IUU) Fishing*, 03-16 GEN, 19 June 2004; CCSBT, *Resolution on Amendment of the Resolution on "Illegal, Unregulated, and Unreported Fishing (IUU) and Establishment of a CCSBT Record of Vessels over 24 Metres Authorised to Fish for Southern Bluefin Tuna,"* Adopted at the Eleventh Annual Meeting, 19–22 October 2004, para. 8(a); IATTC, *Resolution C-03-07*, Resolution on the Establishment of a List of Longline Fishing Vessels Over 24 Metres (LSTLFVs) Authorised to Operate in the Eastern Pacific Ocean, 70th Meeting, Antigua, Guatemala, 24–27 June 2003, Art. 6(a); IOTC, *Resolution 07/02*, Art. 7(a).

[172] NEAFC, *Scheme of Control and Enforcement*, Art. 45(1); CCAMLR, *Conservation Measures 10-06 (2008)*, Art. 11(c).

[173] ICCAT, Recommendation on Additional Measures Against IUU Fishing; ICCAT, *Recommendation by ICCAT Concerning a Multi-year Conservation and Management Plan for Bluefin Tuna in the East Atlantic and Mediterranean*, 02-08 BFT, 04 June 2003, para. 9; ICCAT, *Recommendation by ICCAT Relating to the Rebuilding Program for North Atlantic Swordfish*, 02-02 SWO, para. 13; ICCAT, *Recommendation by ICCAT to Establish a Rebuilding Program for Western Atlantic Bluefin Tuna*, 98-07BFT, 21 June 1999, para. 8; ICCAT, *Recommendation by ICCAT on a Yellowfin Size Limit*, 72-01 YFT, 01 July 1973.

[174] NEAFC, *Scheme of Control and Enforcement*, Art. 45(1)(c); SEAFO, *Conservation Measure 09/07*, para. 8.

[175] ICCAT, *Recommendation on Ban on Landings and Transshipments*, para. 3; NAFO, Conservation and Management Measure, Arts. 41 and 43(2); NEAFC, *Scheme of Control and Enforcement*, Art. 45(3); CCAMLR, *Resolution 19/XXI*, Flags of Non-compliance, para. 4; CCAMLR, *Conservation Measure 10-06 (2008)*, para. 4; IATTC, *Resolution C-05-07*, para. 9(b); IOTC, *Resolution 09/03*, para. 12(b); IOTC, *Resolution 01/03*, para. 4; IOTC, *Resolution 99-02*, Calling for Actions Against Fishing Activities by Large-scale Flag of Convenience Longline Vessels, 1999, para. 2; IOTC, *Resolution 05/03*, para. 4.

the action taken or proposed to be taken in relation to the alleged violation.[176] Members of RFMOs are also required to cooperate with any member which has drawn the attention of the flag State to the violation of its fishing vessel in taking appropriate actions against the vessel.[177] If there is sufficient evidence that the vessel has committed a serious violation, the flag State member would need to institute proceedings against the vessel without delay and apply sanctions of sufficient severity.[178] If the vessel has committed a violation in the waters of a member coastal State, the flag State would also need to ensure that the vessel has submitted promptly to any sanctions which may be imposed by that coastal State in accordance with its national laws and regulations.[179] Similarly, if a vessel flying the flag of a member which is not on the record of fishing vessels has conducted fishing activities in the RFMO area, that member would need to take necessary measures to prevent the vessel from fishing in the area and report to the organisation the actions taken with respect to such vessel.[180]

In terms of applying sanctions against IUU fishing activities, ICCAT requires its members to impose adequate sanctions on their nationals and on their fishing vessels that conduct large scale pelagic driftnet fishing[181] and those that conduct other IUU fishing activities.[182] ICCAT also provides that illegal shipments of tuna shall be suspended into the territory of a member, or subject to administrative or other sanctions pending receipt of a properly completed document.[183] ICCAT further requests non-members whose vessels appear in the IUU list to take all

---

[176] *WCPF Convention*, Art. 25(2); IOTC, *Resolution 07/01*, To Promote Compliance By Nationals of Contracting Parties and Cooperating Non-contracting Parties with IOTC Conservation and Management Measures; ICCAT, *Recommendation by ICCAT Concerning the Duties of Contracting Parties and Cooperating Non-Contracting Parties, Entities, or Fishing Entities in Relation to their Vessels Fishing in the ICCAT Convention Area*, 03-12 GEN, 19 June 2004, para. 1(f); ICCAT, *Recommendation by ICCAT to Promote Compliance by Nationals of Contracting Parties, Cooperating Non-Contracting Parties, Entities or Fishing Entities with ICCAT Conservation and Management Measures*, 06-14 GEN, 13 June 2007, para. 1.

[177] *WCPF Convention*, Art. 25(6); ICCAT, *Resolution by ICCAT on Compliance with the ICCAT Conservation and Management Measures (Including Addendum)*, 94-09 GEN, para. 3; ICCAT, *Recommendation by ICCAT to Promote Compliance by Nationals of Contracting Parties, Cooperating Non-Contracting Parties, Entities or Fishing Entities*, para. 1(iii).

[178] *WCPF Convention*, Art. 25(3) and 25(7): NEAFC, *Scheme of Control and Enforcement*, Art. 31.

[179] *WCPF Convention*, Art. 25(4).

[180] WCPFC, *Conservation and Management Measures 2004-01*, para. 15.

[181] ICCAT, *Resolution by ICCAT on Large-scale Pelagic Driftnets*, 96-15 GEN, 03 February 1997.

[182] ICCAT, *Recommendation to Establish an IUU Vessels List*, para. 6(b).

[183] ICCAT, *Recommendation on Bigeye Tuna Statistical Document Program*, Annex 1, para. 4 and Annex 2, 5; ICCAT, *Recommendation Establishing a Swordfish Statistical Document Program*, Attachment 3, para. 5; IOTC, *Recommendation Concerning the IOTC Bigeye Tuna Statistical Document Programme*, Annex 1 para. 4 and Annex 2 para. 5

necessary measures to eliminate such IUU fishing activities, including if necessary, the withdrawal of the registration or of the fishing licences of the vessels in question.[184]

ICCAT has also established recommendations with respect to the identification of members and cooperating non-members which fail to discharge their duties under the ICCAT Convention in respect of conservation and management measures, or non-members which fail to discharge their duties to cooperate under international law by exercising effective control over their vessels.[185] The identification is based on the review of information such as catch data, trade information, statistical documentation, and catch documentation.[186] Such information becomes the basis for actions to be taken against the vessels.

Similar to ICCAT, NEAFC and NAFO require their members to apply measures that would effectively deprive fishing offenders of the economic benefit of their IUU activities or apply sanctions proportionate to the seriousness of the infringements.[187] Members are required to take immediate judicial or administrative action in the same manner as would have been the case when dealing with infringements of fisheries regulations in national waters.[188] In addition to the prohibition of landing, transshipment, and importation of fish, other RFMOs such as CCAMLR, IATTC, and IOTC also call on their members to take measures against vessels of non-members which have been confirmed to be involved in IUU fishing. The range of actions that may be taken include prohibiting the authorisation of such vessels to fish in waters under their national jurisdiction; prohibiting the chartering of such vessels; and refusal to grant their flag to such vessels.[189] Members of CCAMLR are also encouraged not to register or de-register vessels that have been placed on the provisional IUU list until such a time as the Commission has had the opportunity to examine the list.[190]

9.3.12 *Control of Fishing Capacity*

One measure adopted by RFMOs which has been linked to IUU fishing, although not directly provided for in the IPOA-IUU, is the control of excess fishing

---

[184] ICCAT, *Recommendation to Establish an IUU Vessels List*, para. 8.

[185] ICCAT, *Recommendation by ICCAT Concerning Trade Measures*, 06-13 GEN, 13 June 2007, para. 2(a)

[186] ICCAT, *Recommendation by ICCAT Concerning Trade Measures*, para. 2(b).

[187] NAFO, *Conservation and Enforcement Measures*, Art. 33(5); NEAFC, *Scheme of Control and Enforcement,* Art. 31.

[188] NAFO, *Conservation and Management Measures*, Art. 33(1); NEAFC, *Scheme of Control and Enforcement*, Art. 31.

[189] NEAFC, *Scheme of Control and Enforcement,* Art. 45(2); CCAMLR, *Conservation Measure 10-06 (2008)*, para. 18; CCAMLR, *Conservation Measure 10-07 (2008)*, para 11; IATTC, *Resolution C-05-07*, para. 9; IOTC, *Resolution 09-03*.

[190] CCAMLR, *Conservation Measure 10-06 (2008)*, para. 18.

capacity. ICCAT, for example, has adopted a resolution to urge Japan and Chinese Taipei to complete the scrapping of IUU vessels built in Japan and the re-registration of IUU vessels built in Chinese Taipei and owned by its residents to Chinese Taipei registry.[191] IOTC also provides that where there is a proposed transfer of capacity to their fleet, no vessels on the IUU vessels list of any RFMO may be transferred.[192] These measures ensure that fishing effort is limited and excess fishing capacity is discouraged by eliminating economic incentives conferred on IUU vessels and preventing the transfer of capacity of such vessels.

9.4 Measures Adopted by Other Regional Fisheries Organisations

Measures to effectively address IUU fishing have also been adopted, or at the very least discussed in other regional fisheries organisations. The Lake Victoria Fisheries Organization (LVFO), a regional organisation responsible for managing inland fisheries, has adopted a regional plan of action (RPOA) to combat IUU fishing. The LVFO RPOA-IUU aims to develop and implement coordinated, harmonised, unified and effective management measures to prevent, deter and eliminate IUU fishing in the Lake Victoria and its basin, and on all fish species to which the LVFO Convention applies.[193] Some of the elements of the LVFO RPOA-IUU are the development of national plans of actions by its members, the exercise of effective control over nationals, exchange of information on IUU fishing activities, registration and licensing of vessels, maintenance of a record of fishing vessels, conduct of regular patrols, and application of agreed conservation measures.

There have also been discussions in other regional fisheries organisations on IUU fishing related matters, particularly on the need to undertake studies and assessment on the relationship of fishing capacity and IUU fishing[194] and the

---

[191] ICCAT, *Supplemental Resolution by ICCAT to Enhance the Effectiveness of the ICCAT Measures to Eliminate Illegal, Unregulated and Unreported Fishing Activities by Large-Scale Tuna Longline Vessels in the Convention Area and Other Areas*, 00-19 GEN, 27 December 2000; ICCAT, *Resolution by ICCAT Concerning Cooperative Actions to Eliminate Illegal, Unreported and Unregulated Fishing Activities by Large-Scale Tuna Longline Vessels*, 02-26 GEN, 04 June 2003; *See also* Yann-huei Song, 'The Efforts of ICCAT to Combat IUU Fishing: The Roles of Japan and Taiwan in Conserving and Managing Tuna Resources', 24 *The International Journal of Marine and Coastal Law* (2009), at 101–139.

[192] IOTC, *Resolution 07/05*, Limitation of Fishing Capacity of IOTC Contracting Parties and Cooperating Non-Contracting Parties in Terms of Number of Longline Vessels Targeting Swordfish and Albacore, para. 6.

[193] LVFO, *Regional Plan of Action to Prevent, Deter and Eliminate Illegal, Unreported and Unregulated (IUU) Fishing on Lake Victoria and Its Basin*, Bagamoyo, Tanzania, 27 May 2004, Art. 1.

[194] FAO, Regional Office for Asia and the Pacific, APFIC Regional Consultative Workshop, Managing Fishing Capacity and Illegal, Unreported and Unregulated Fishing in Asia, 13–15 June 2007, Phuket, Thailand (Bangkok: FAO-RAP, 2007).

impact of subsidies on excess fishing capacity and IUU fishing.[195] A number of national and regional measures have also been identified by regional fisheries organisations as effective and more practical approaches to combating IUU fishing such as joint patrol and enforcement[196] and port inspections.[197] All of these initiatives illustrate the increasing cooperation among regional fisheries organisations to prevent, deter, and eliminate IUU fishing.

---

[195] WECAFC, Thirteenth Session, WECAFC Lesser Antilles Fisheries Committee, Tenth Session, Cartagena de Indias, Colombia, 21–24 October 2008, Summary Report of the Intersessional Activities and FAO projects in the WECAFC Region, WECAFC/XIII/08/5E, October 2008, paras. 27 and 29; Fishery Committee for the Eastern Central Atlantic, Nineteenth Session, Cotonou, Benin, 4–6 November 2008, Global Emerging Issues in Fisheries Development and Management Relevant to the Region, *CECAF/XIX/2008/8*, November 2008, para. 50.

[196] RECOFI, Fourth Session, Jeddah, Kingdom of Saudi Arabia, 07–09 May 2007, National and Regional Measures to Address Illegal, Unreported and Unregulated Fishing, RECOFI/IV/2007/4, May 2007; NPAFC, *Report of the Fifteenth Annual Meeting of the Commission*, Vladivostok, Russia, 8–12 October 2007.

[197] FAO Subregional Office for Southern Africa, Report of the Second Session of the South West Indian Ocean Fisheries Commission, Maputo, Mozambique, 22–25 August 2006, *FAO Fisheries Report No 823* (Rome: FAO, 2007), at para. 13.

Chapter Ten

Current State Practice to Combat IUU Fishing

As discussed in Chapter 4, there are two ways by which all States can give effect to the provisions of international instruments related to addressing IUU fishing: first, by adopting national plans of action and second, by incorporating provisions in national legislation to combat IUU fishing.[1] To implement the IPOA-IUU, a number of States and regional organisations have adopted plans of action specifically addressing IUU fishing. However, due to the lengthy process of enacting and amending legislation, not many States have directly provided for the prevention, deterrence and elimination of IUU fishing in national laws since the adoption of the IPOA-IUU in 2004. The two exceptions to this general practice amongst States is the United States of America (U.S.) and the European Union (EU) which have both adopted elaborate legal measures to address IUU fishing. These measures, as discussed in this chapter, not only provide strict controls against IUU fishing activities, but also measures directed to flag States of IUU vessels.

10.1 U.S. Measures to Combat IUU Fishing

In 2007, the U.S. adopted amendments to the *Magnuson-Stevens Fishery Conservation and Management Reauthorisation Act* on IUU fishing. Section 403 of the *Magnuson-Stevens Fishery Conservation and Management Reauthorisation Act* amends the *High Seas Driftnet Fishing Moratorium Protection Act* by adding section 609 on IUU fishing. The new section 609 has two main components: a definition of IUU fishing and measures for the identification and listing of foreign nations whose vessels have engaged in IUU fishing. Section 609(e)(3) defines IUU fishing as:

> (A) fishing activities that violate conservation and management measures required under an international fishery management agreement to which the United States is a party, including catch limits or quotas, capacity restrictions, and bycatch reduction requirements;
>
> (B) overfishing of fish stocks shared by the United States, for which there are no applicable international conservation or management measures or in areas with no applicable

[1] Food and Agriculture Organization (FAO), *International Plan of Action to Prevent, Deter, and Eliminate Illegal, Unreported, and Unregulated Fishing*, adopted on 23 June 2001 at the 120th Session of the FAO Council. *Hereinafter referred to as* IPOA-IUU. Paras. 16 and 25.

international fishery management organisation or agreement, that has adverse impacts on such stocks; and

(C) fishing activity that has an adverse impact on seamounts, hydrothermal vents, and cold water corals located beyond national jurisdiction, for which there are no applicable international conservation or management measures or in areas with no applicable international fishery management organisation or agreement.[2]

This definition extends the application of U.S. national legislation to areas beyond national jurisdiction and to fisheries which are currently unregulated. The definition further takes into account the impact of fishing activities on habitats and ecosystems, an area which is not discussed in the IPOA-IUU.

The *High Seas Driftnet Fishing Moratorium Protection Act* requires actions to be taken by the U.S. to address IUU fishing and by-catch of protected marine living resources by providing for the identification and listing of States whose vessels engage in IUU fishing. In January 2009, a rule to implement the identification and certification procedures to address IUU fishing activities was proposed. The proposed rule is accompanied by a draft environmental assessment, regulatory impact review, and regulatory flexibility act analysis.[3] An environmental assessment analyses the impacts of IUU fishing and by-catch of protected marine living resources to provide the public with a context for reviewing the proposed certification action.

Two separate procedures are being developed for the purpose of implementing the amended legislation. One procedure is for the certification of States whose fishing vessels engage in IUU fishing and the other is for States whose fishing vessels engage in activities resulting in the by-catch of protected living marine resources.[4] The objectives of the IUU certification procedure are to promote the sustainability of transboundary and shared fish stocks and enhance the conservation and recovery of protected living marine resources.[5] The proposed

[2] 16 USC 1826j HSDFMPA §609(e)(3).

[3] U.S. Department of Commerce, NOAA, National Marine Fisheries Service, Office of International Affairs, *Draft Environmental Assessment, Regulatory Impact Review, and Regulatory Flexibility Act Analysis for a Proposed Rule to Establish Identification and Certification Procedures for Nations Under the High Seas Driftnet Fishing Moratorium Protection Act*, January 2009.

[4] U.S., 50 CFR §300.202 and 203. Protected living marine resources are defined as non-target fish, sea turtles, or marine mammals that are protected under United States law or international agreement, including the Marine Mammal Protection Act, the Endangered Species Act, the Shark Finning Prohibition Act, and the Convention on International Trade in Endangered Species of Wild Flora and Fauna; but exclude species, except sharks, that are managed under the Magnuson-Stevens Fishery Conservation and Management Act, the Atlantic Tunas Convention Act or by any international fishery management organisation.

[5] U.S., Department of Commerce, National Oceanic and Atmospheric Administration, 'Magnuson-Stevens Fishery Conservation and Management Reauthorisation Act; Proposed Rule to Implement Identification and Certification Procedures to Address Illegal, Unreported, and Unregulated Fishing Activities and the By-catch of Protected Living Marine Resources (PLMRs)', *Federal Register* 74:9 (14 January 2009), at 2019.

rule is also envisioned to enhance the existing authority of the U.S. to enforce compliance with international fisheries management and conservation agreements. At the time of writing, the U.S. is soliciting comments on the IUU certification procedure and the draft environmental assessment, which are yet to be finalised.

10.1.1 *Procedures for the Identification of States Whose Fishing Vessels Have Engaged in IUU Fishing and By-catch of Protected Living Marine Resources*

Section 609(a) of the *High Seas Driftnet Fishing Moratorium Protection Act* provides for the identification of States, in a biennial report to Congress, whose vessels have engaged in IUU fishing in a number of circumstances: one, where the relevant fisheries management organisation has failed to implement effective measures to end the IUU activities of vessels of that State; two, if the State is not a party to, or does not maintain cooperating status with such organisation; and three, where there are no relevant fisheries organisations with a mandate to regulate the fishing activity in question. With respect to activities impacting negatively on protected marine living resources, the *High Seas Driftnet Fishing Moratorium Protection Act* provides for the identification and listing of a State if:

> (1) fishing vessels of that nation are engaged, or have been engaged during the preceding calendar year in fishing activities or practices;
>
> > (A) in waters beyond any national jurisdiction that result in bycatch of a protected living marine resource; or
> > (B) beyond the exclusive economic zone of the United States that result in bycatch of a protected living marine resource shared by the United States;
>
> (2) the relevant international organisation for the conservation and protection of such resources or the relevant international or regional fishery organisation has failed to implement effective measures to end or reduce such bycatch, or the nation is not a party to, or does not maintain cooperating status with, such organisation; and
>
> (3) the nation has not adopted a regulatory program governing such fishing practices designed to end or reduce such bycatch that is comparable to that of the United States, taking into account different conditions."[6]

The identification of States whose vessels engage in IUU fishing will be based on data gathered by the U.S. Government, as well as the data offered by other States, RFMOs, and international organisations, institutions, or arrangements that could support a determination that the vessels of that State have engaged in IUU fishing. The U.S. National Oceanographic and Atmospheric Administration (NOAA)

[6] 16 USC 1826k HSDFMPA §610(a).

National Marine Fisheries Service (NMFS) commits to exercise due diligence in reviewing and evaluating the information when identifying States whose vessels have engaged in IUU fishing.[7] The NMFS will be considering several criteria for determining whether information is appropriate for use in identifying and certifying States. These criteria include the corroboration of testimony and evidence; whether multiple sources have been able to provide information in support of an identification; the methodology used to collect information; specificity of the information provided; susceptibility of the information to falsification and alteration; and credibility of the individuals or organisations providing the information.[8]

Once it has been determined that the information is credible and supports the finding that the fishing vessels of identified States have been involved in IUU fishing, the NMFS, acting through or in cooperation with the U.S. State Department, will initiate bilateral discussions with identified States to seek corroboration of the alleged IUU activity or credible information that refutes such allegation. The NMFS will also communicate the requirements of the *High Seas Driftnet Fishing Moratorium Protection Act* to the State and encourage it to take corrective action to address the alleged IUU fishing or IUU fishing vessel in question. Only States whose vessels have proved to have conducted IUU fishing and have failed to take corrective actions against such vessels would have negative certification, and will be included in the list of States submitted to the U.S. Congress.

The other aspect of implementing Section 609 of the *High Seas Driftnet Fishing Moratorium Protection Act* is the establishment of a procedure for determining when a State has taken appropriate corrective action with respect to the offending activities or whether the relevant fisheries management organisation has implemented measures that are effective in ending IUU fishing by vessels of that State. For this purpose, the biennial report will include information on whether:

- The identified State has provided evidence documenting that it has taken appropriate enforcement or other responsive action to address IUU fishing or by-catch issues of its fishing vessels;
- The relevant RFMO has adopted effective measures and the identified member State has implemented effective measures to combat IUU fishing;
- The identified State has provided evidence documenting the adoption and enforcement of a regulatory programme to end or reduce by-catch of protected living marine resources that is comparable to that of the United States, accounting for different conditions; and

---

[7] U.S. *Federal Register* 74:9 of 2009, *supra* note 5, at 2021.

[8] *Ibid.*, at 2024.

- The identified State is enforcing a management plan containing requirements that will assist in gathering species-specific data to support international stock assessments and conservation enforcement efforts for protected living marine resources.[9]

According to the proposed rule, effective measures by States to address IUU fishing may include those which reflect the recommendations of international organisations to combat IUU fishing. These measures include:

- Data collection and catch reporting programmes, including observer programmes, catch documentation programmes, and trade-tracking schemes;
- Trade-related measures that seek to reduce or eliminate trade in fish and fish products derived from IUU fishing;
- At-sea or dockside boarding and inspection schemes;
- Programmes documenting whether fish were caught in a manner consistent with conservation and management measures;
- IUU vessel listing;
- Port state measures to prohibit landing and transshipment of unauthorised catch;
- Catch and effort monitoring, including licensing of fishing vessels, catch reporting, and vessel monitoring systems;
- By-catch reduction and mitigation strategies, techniques, and equipment;
- Programs for the identification and protection of vulnerable marine ecosystems;
- Efforts to improve and enhance fisheries enforcement and compliance, including through the development of effective sanctions and MCS systems; and
- Participation in voluntary international efforts to combat IUU fishing such as the International MCS Network.[10]

In considering whether States identified as having fishing vessels engaged in the by-catch of protected living marine resources have taken appropriate corrective action, the NMFS will consider programmes for data collection and sharing, including observer programmes, by-catch reduction and mitigation, and improving MCS activities.[11]

If the measures adopted by identified States are considered comparable to those adopted by the U.S., measures may be deemed to be sufficient for the purposes of implementing the IUU related provisions of the *High Seas Driftnet Fishing Moratorium Protection Act* and will be given positive certification. Part of the evaluation of whether or not a State has implemented measures that will

[9] *Ibid.*, at 2020.
[10] *Ibid.*, at 2023.
[11] *Ibid.*, at 2021 and 2022.

effectively address IUU fishing is the examination of the existence of adequate enforcement measures and capacity in the identified State, or the extent to which the State has taken actions to implement measures intended to address IUU fishing.[12] In order to assist States in implementing these provisions, the U.S. has committed to work with them on a bilateral and multilateral basis in the adoption of regulatory regimes designed to end or reduce by-catch which are comparable in effectiveness to those measures adopted by the U.S., taking into account relevant environmental and socio-economic conditions.[13]

10.1.2 *Alternative Procedures*

Where the Secretary of Commerce has not reached a certification determination of the identified State by the time of its next biennial report to the U.S. Congress, the *High Seas Driftnet Fishing Moratorium Protection Act* provides for the establishment of an alternative certification procedure based on a shipment-by-shipment, shipper-by-shipper, or other basis as long as specified conditions are met. The specified conditions include that the fishing vessel has not engaged in IUU fishing in violation of an international fishery management agreement to which the U.S. is a party, or the vessel has not been identified by an international fisheries management organisation as having participated in IUU fishing.[14] Importantly, the fish or fish products by vessels that qualify under the alternative procedure must be accompanied by a completed documentation of admissibility. Similarly, NMFS must determine that imports were harvested by practices that do not result in by-catch of protected living resources or were harvested by practices comparable to those required in the U.S. and that the vessel collects species specific by-catch data.[15] As an example, vessels engaged in pelagic longline fishing can qualify for alternative certification procedure only if they use circle hooks, careful handling and release equipment, and participate in training and observer programmes.[16]

10.1.3 *Effect of Certification Determination*

No remedial actions are required from a State identified as having vessels engaged in IUU fishing with positive certification. However, if the identified State fails to take sufficient action to address IUU fishing and does not receive positive certification, that State may face denial of port privileges, prohibitions on the import of

---

[12] *Ibid.*, at 2024.
[13] *Ibid.*, at 2021.
[14] 16 USC 1826k HSDFMPA §609(d)(2); 50 CFR §300.206.
[15] 16 USC 1826k HSDFMPA §610(c)(4); 50 CFR §300.207; U.S. *Federal Register* 74:9 of 2009, *supra* note 5, at 2025.
[16] U.S. *Federal Register* 74:9 of 2009, *ibid.*, at 2025.

certain fish and fish product into the U.S., as well as other appropriate measures consistent with international law, including WTO agreements.[17] In determining the appropriate course of action to be taken against the State, a number of factors are taken into account, such as the nature, circumstances, extent, duration and gravity of the IUU fishing activity, the degree of culpability, and history of IUU fishing. To ensure that the identified State has been given the opportunity to respond to the listing, the proposed rule provides that the Secretary of Commerce, in cooperation with the Secretary of State, may initiate further consultations prior to determining the course of action. To facilitate enforcement, the identified State may be required to submit documentation of admissibility along with fish or fish products not subject to import restrictions that may enter the U.S. territory.[18]

10.1.4 *Initial Implementation of the High Seas Driftnet Fishing Moratorium Protection Act*

In March 2008, the NMFS published a notice in its Federal Register requesting information on States whose vessels have engaged in IUU fishing, or have conducted activities that resulted in by-catch of protected living marine resources at any point during the two years preceding the biennial report.[19] Based on submitted reports, IUU vessel lists, peer reviewed literature and other information from individuals, non-governmental organisations and other States, the NMFS identified six States whose vessels have engaged in IUU fishing.[20] The activities of the relevant vessels of these States and information on other IUU activities which formed the bases for identification are included in the biennial report to Congress. However, it was recognised in the report that under the statutory language of the *High Seas Driftnet Fishing Moratorium Protection Act*, it is not clear whether the activities of these vessels could appropriately form the basis for the identification of these States.[21] Hence, no actions have been determined to be taken against the identified States or their vessels. The NMFS will still be consulting with these six States to adopt corrective measures, which will determine their positive or negative certification. In the case of States whose vessels have been engaged in the by-catch of protected living marine resources, no formal identification was done.

---

[17] 50 CFR §300.204 and 205.

[18] 50 CFR §300.202; U.S. *Federal Register* 74:9 of 2009, *supra* note 5, at 2024.

[19] U.S. Department of Commerce NOAA, 'Identification of Nations Whose Fishing Vessels are Engaged in Illegal, Unreported, or Unregulated Fishing and/or By-catch of Protected Living Marine Resources', *Federal Register* 73:56 (21 March 2008), at 15136.

[20] The identified nations are France, Italy, Libya, Panama, China, and Tunisia. *See* U.S. Department of Commerce, *Implementation of Title IV of the Magnuson-Stevens Fishery Conservation and Management Reauthorisation Act of 2006: Report to Congress Pursuant to Section 403(a) of the Magnuson-Stevens Fishery Conservation and Management Reauthorisation Act of 2006,* January 2009, at 89.

[21] *Ibid.*

The U.S. measures against IUU fishing are in their initial stages of implementation and the procedures for the identification and listing of States are yet to be fully developed and tested. The enforcement actions that would be undertaken by the U.S. against identified and listed States would determine how well other States will receive the new measures and whether or not any trade restrictive measures imposed on States would be challenged before the WTO.

10.2 EU Regulation to Combat IUU Fishing

The European Union has made the most far-reaching legislative efforts to specifically address IUU fishing through Council Regulation (EC) No 1005/2008 establishing a Community system to prevent, deter and eliminate IUU fishing.[22] This Regulation, adopted in 2008, puts in place measures to ensure that fish and fishery products derived from IUU fishing activities are not imported into the territories of EU member States. The measures established by the EU IUU Regulation are comprehensive in coverage and apply to all IUU fishing and associated activities carried out within Community waters, maritime zones under the jurisdiction or sovereignty of third States,[23] and on the high seas.[24] Fishing vessels subject to the EU IUU Regulation include any vessel of any size used for or intended for use for the purposes of commercial exploitation of fishery resources, including support ships, fish processing vessels, and vessels engaged in transshipment and carrier vessels equipped for the transportation of fishery products, except container vessels.[25]

In terms of the definition of IUU fishing, the EU IUU Regulation adopts both the scope and nature of IUU fishing under the IPOA-IUU and a list of activities which may be considered IUU fishing. Article 3 of the EU IUU Regulation not only provides for IUU fishing activities in violation of national and regional measures, but also other activities which are not directly related to fishing such as concealing, tampering or disposing of evidence relating to a fisheries investigation

---

[22] Council Regulation (EC) No 1005/2008 of 29 September 2008 establishing a Community system to prevent, deter and eliminate illegal, unreported and unregulated fishing, amending Regulations (EEC) No 2847/93, (EC) No 1936/2001 and (EC) No 601/2004 and repealing Regulations (EC) No 1093/94 and (EC) No 1447/1999, *OJ L 286/1 29.10.2008.* The regulation is scheduled to enter into force on 01 January 2010.

[23] *EC 1005/2008*, Art. 1(3). IUU fishing within maritime waters of overseas territories and countries included in Annex II of the EC Treaty are treated to be taking place within the maritime waters of third countries.

[24] *EC 1005/2008*, Art. 1(3).

[25] *EC 1005/2008*, Art. 2(5). This definition is similar to the definition in several international and regional fisheries instruments and national fisheries legislation. See for example, Article 1(c) of Convention on the Conservation of Highly Migratory Fish Stocks in the Western and Central Pacific Ocean, Honolulu, Hawaii, concluded 5 September 2000, entry into force 19 June 2004, [2004] ATS 15. *Hereinafter referred to as* WCPFC Convention.

and obstructing the work of inspection officers and fisheries observers.[26] The EU IUU Regulation also provides that a fishing vessel shall be presumed to be engaged in IUU fishing if it has transshipped or participated in joint fishing operations, supported or re-supplied other fishing vessels identified as having engaged in IUU fishing.[27]

In terms of product coverage, the EU IUU Regulation applies to any products which fall under Chapter 03[28] and Tariff headings 1604[29] and 1605[30] of the Combined Nomenclature established by EC No 2658/87 of 23 July 1987 on the tariff and statistical nomenclature and on the Common Customs tariff, with the exception of products listed in Annex 1 of the Regulation.[31]

There are six key elements of the EU IUU Regulation: port State control over third country fishing vessels, catch certification, EU IUU vessels listing, listing of non-cooperating States, control of nationals, and the Community Alert System.

10.2.1 *Port State Control over Third Country Fishing Vessels*

The EU IUU Regulation establishes a port State scheme for the inspection of third country fishing vessels calling at ports of EU member States. Transshipment of fish between third country fishing vessels or between vessels flying the flags of EU member States and third country vessels may only take place in designated ports and subject to specific conditions. One of the conditions is for masters of third country fishing vessels to notify, and submit specified information to the competent authorities of the EU member States at least three working days before the estimated time of arrival at the port.[32] The notification period may vary depending on the type of fishery product and the distance between fishing grounds and landing places. The information to be provided include: vessel identification; name of the designated port of destination; purposes of the call, landing, transshipment, onboard processing, and access to port services; fishing authorisation; trip information; estimated time of arrival at port; catches retained on board; and the zone or zones where the catch was taken.[33]

---

[26] *EC 1005/2008*, Art. 3(g) and (h).

[27] *EC 1005/2008*, Art. 3(j).

[28] Fish and crustaceans, molluscs and other aquatic invertebrates.

[29] Prepared or preserved fish; caviar and caviar substitutes prepared from fish eggs.

[30] Crustaceans, molluscs and other aquatic invertebrates, prepared or preserved.

[31] EC 1005/2008, Art. 2(8). The products listed in Annex I of the IUU Regulation upon its adoption are: freshwater fishery products; aquaculture products obtained from fry or larvae; ornamental fish; live oysters; scallops, including queen scallops, of the genera Pecten, Clamys or Placopecten, live fresh or chilled; Coquilles St Jacques (*Pecten maximus*), frozen; other scallops, fresh or chilled; mussels; snails, "others [sic] than those obtained from the sea"; and prepared and preserved molluscs.

[32] *EC 1005/2008*, Art. 6(1).

[33] *EC 1005/2008*, Art. 6(1).

If the vessel carries on board fishery products, the notice of entry into port is to be accompanied by a validated catch certificate in accordance with Chapter III of the EU IUU Regulation. It is only after the verification of such information and catch certificate that the vessel may be granted authorisation to enter into port.[34] The responsibility to carry out port inspections and verify the accuracy of the information transmitted by the third country fishing vessel in the prior notice of arrival and the catch certificate rests with the EU member State. The EU IUU Regulation requires EU member States to carry out inspections in their designated ports of at least 5 per cent of landing and transshipment operations by third country fishing vessels each year.[35] These vessels include those sighted and presumed to have conducted IUU fishing by the EU and RFMOs.

Port State inspectors have the duty to examine the fishing vessel, the fish catch, nets and other gears and equipment and any other relevant document deemed necessary by the inspector to verify compliance of the vessel with applicable laws, regulations, and international management and conservation measures.[36] If the information collected during the inspection provides evidence that a fishing vessel has engaged in IUU fishing, the vessel will not be authorised to land or transship its catch.[37]

Where the information provided by the fishing vessel is not complete or its verification is pending, an EU member port State may authorise port access or permit all or part of the landing in port, but would need to keep the fishery products concerned in storage under the control of the competent authorities, until the rest of the required information has been received or the verification process is completed.[38] If the verification process is not completed within 14 days of the landing, the EU member port State may confiscate and dispose of the fish in accordance with national laws.[39] Storage costs are required to be borne by the operators of the vessel.

A major issue with the port entry provisions of the EU IUU Regulation is that they do not seem to contain sufficient safeguards for third country fishing vessels against undue delay resulting from unfounded inspection or denial of port access. The only safeguards provided under the Regulation relate to cases of *force majeure*[40] and the vague requirement that EU member States shall undertake inspections and verifications "on the basis of risk management".[41] A requirement that inspections "cause minimum disturbance to the vessels activities and cause no deterioration in fish quality" was proposed by the EU but not included in the

[34] *EC 1005/2008*, Art. 7.
[35] *EC 1005/2008*, Art. 9(1).
[36] *EC 1005/2008*, Art. 10.
[37] *EC 1005/2008*, Art. 11(2).
[38] *EC 1005/2008*, Art. 7(3).
[39] *EC 1005/2008*, Art. 7(3).
[40] *EC 1005/2008*, Art. 4(2).
[41] *EC 1005/2008*, Arts. 9(1) and 17(3).

final text of the EU IUU Regulation.[42] A right to compensation and appeals concerning the actions of the EU member port State are also not included in the port State scheme under the Regulation.

While the effectiveness of port State measures to combat IUU fishing is universally acknowledged by international fisheries instruments, it is critical that the implementation of such measures achieve a balance between combating IUU fishing on the one hand and safety of fishing vessels and their crew and appropriate safeguards against abuse of port State powers on the other. These safeguards are recognised, for example, in the FAO Model Scheme for Port State Measures and the FAO Agreement on Port State Measures to Prevent, Deter and Eliminate Illegal, Unreported and Unregulated Fishing.[43]

10.2.2 *Catch Certification*

A key aspect of the EU IUU Regulation is the requirement for catch certification to ensure that fisheries products being imported into the EU are not derived from IUU fishing.[44] In general, the importation of fishery products into the EU is only allowed when accompanied by a catch certificate, completed by the master of the fishing vessel and validated by the flag State of the vessel. To be valid, the catch certificate must contain all information specified in the template documents shown in Annex II of the EU IUU Regulation,[45] namely:

- the name of the fishing vessel, home port and registration number, call sign, licence number, Inmarsat number and IMO number (if issued);
- the product (the type of species, catch areas and dates, estimated live weight and verified weight landed, as well as the applicable conservation and management measures and any transshipment at sea is also required); and
- declaration on export and import of the fishery product (including the vessel name and flag, flight number airway bill number, truck nationality and registration number, other transport documents and container number).

The indirect importation and exportation of fishery products are subject to the validation of a catch certificate by the competent authorities of the flag State of

[42] *See* Proposal for a Council Regulation Establishing a Community system to prevent, deter and eliminate illegal, unreported and unregulated fishing (presented by the Commission), Brussels, 17 October 2007, COM(2007) 602 final, 2007/0223(CNS), Art. 11(1).

[43] *See* FAO, Agreement on Port State Measures to Prevent, Deter and Eliminate Illegal, Unreported and Unregulated Fishing, Adopted in November 2009, Appendix V of the FAO Council, Hundred and Thirty-seventh Session, Rome, 28 September-02 October 2009, Report of the 88th Session of the Committee on Constitutional and Legal Matters (CCLM), 23–25 September 2009, *CL 137/5*, September 2009. *Hereinafter referred to as* the FAO Port State Measures Agreement. Arts. 18 and 19.

[44] *EC 1005/2008*, Art. 12(1).

[45] This catch certificate specimen has similar content to the standard *Dissostichus* catch document form used by CCAMLR and the statistical document forms used by IOTC, ICCAT, IATTC, and CCSBT.

the vessel.[46] Verifiable documentation or certification is required of products constituting one single consignment which are transported in the same form to the EU from a third country other than the flag State and products constituting one single consignment which have been processed in a third country other than the flag State.[47] Aside from validated catch certificates, proper documentation is required for every stage of transshipment or transit of the fishery product. Catch documents and any related documents that are validated in conformity with catch documentation schemes adopted by an RFMO and are recognised by the EU as complying with the requirements of the Regulation will be accepted as catch certificates in respect of the products from species to which such catch documentation schemes apply.[48]

The EU IUU Regulation empowers competent authorities of EU member States to carry out all of the controls they deem necessary for the verification of the catch certificate and other information provided.[49] In addition to the inspection of fishing vessels at port, the control measures may consist of examining the products, verifying declaration data and authenticity of documents, examining the accounts of operators and other records, inspecting means of transport, including containers and storage places of the products and carrying out official enquiries.[50] The competent authority of the EU member State may, for the purpose of verification, request the assistance of the competent authorities of the flag State or of another State other than the flag State from which the fishery products have been indirectly imported.[51]

Importers are required to submit validated catch certificates to the competent authorities of the EU member State in which the product is intended to be imported at least three working days before the estimated time of arrival into the territory of that State, or for another period provided according to the type of fishery product, distance to the place of entry, and the transport used.[52] However, an importer who has been granted the status of an approved economic operator has the option to merely advise the EU member State of the arrival of the products and keep the validated catch certificates for verification of the competent authority at a later stage when the fishery product enters the territory of the EU member State.[53] The status of an approved economic operator may be granted on the basis of a number of criteria including: the establishment of the importer

[46] *EC 1005/2008*, Art. 14 and 15.
[47] *EC 1005/2008*, Art. 14.
[48] *EC 1005/2008*, Art. 13.
[49] *EC 1005/2008*, Art. 17.
[50] *EC 1005/2008*, Art. 17(2).
[51] *EC 1005/2008*, Art. 17(6).
[52] *EC 1005/2008*, Art. 16. This requirement may be adapted according to the type of fishery product, distance to the place of entry, and the transport used.
[53] *EC 1005/2008*, Art. 16(2).

on the territory of that EU Member State; a sufficient number and volume of import operations; an appropriate record of compliance with the requirements of conservation and management measures; a satisfactory system of managing commercial and, where appropriate, transport and processing records; the existence of facilities with regard to the conduct of those checks and verifications; practical standards of competence or professional qualifications directly related to the activities carried out; and proven financial solvency.[54]

10.2.3 *IUU Vessel Listing*

The EU IUU Regulation provides for the creation of a Community IUU vessel list, applicable to both EU and third country vessels, which will contain information on vessels identified by the EU and its member States to have engaged in IUU fishing. The content of Community IUU vessel list will be based upon the compliance with the EU IUU Regulation, catch data, trade information obtained from national statistics and other reliable sources, vessel registers and databases, RFMO catch document or statistical document programmes, reports on sightings of presumed IUU vessels, information obtained by RFMOs, other relevant information obtained in ports or on fishing grounds and additional information provided by EU member States.[55] The Community IUU vessel list will also include IUU vessels listed by RFMOs on their respective IUU lists.[56]

A vessel may be removed from the Community IUU vessel list if the flag State demonstrates that the vessel did not engage in IUU fishing activities or if sanctions have been applied in response to the IUU fishing activity in question.[57] The EU may also consider removing a fishing vessel from the Community IUU vessel list if the owner or operator of the vessel provides evidence that the fishing vessel is no longer engaged in IUU fishing or if that vessel has sunk or has been scrapped.[58] In all other cases, the EU will only consider removing the fishing vessel from the IUU list if at least two years have lapsed since its listing during which time no further reports of alleged IUU fishing have been received by the EU. A fishing vessel may also be removed from the Community IUU list if the owner has submitted information of the vessel's full compliance with all applicable laws and regulations, and if the fishing vessel and its owner and operator have no links with other vessels, owners, and operators engaged in IUU fishing.[59]

---

[54] *EC 1005/2008*, Art. 16(3). These criteria are similar to those implemented to determine the list of authorised establishments complying with EU SPS Regulations.

[55] *EC 1005/2008,* Art. 25.

[56] *EC 1005/2008,* Art. 30.

[57] *EC 1005/2008,* Art. 28(1).

[58] *EC 1005/2008,* Art. 28(2).

[59] *EC 1005/2008,* Art. 28(2).

The range of actions that may be taken by EU member States against vessels on the Community IUU vessel list include: denial of fishing authorisations in the waters of EU member States; withdrawal of fishing authorisations or permits; prohibition on the chartering of such vessels; denial of port access and provisioning (except in cases of *force majeure*); confiscation of catches and fishing gear; and prohibition on the exportation and re-exportation of fishery products.[60] These measures are generally consistent with provisions under international fisheries instruments and conservation and management measures adopted by various RFMOs.

10.2.4 *Identification of Non-cooperating Third Countries*

In addition to the Community list of IUU vessels, the EU IUU Regulation provides for the establishment of a list of non-cooperating third countries. A State may be identified as a non-cooperating third country if it fails to discharge the duties incumbent upon it under international law as flag, port, coastal or market State and has failed to take action to prevent, deter and eliminate IUU fishing activities.[61] The listing of non-cooperating third countries is based on a number of factors, including:

- examination of measures taken by the State concerned in respect of recurrent IUU fishing activities carried out or supported by vessels flying its flag or by its nationals, or by vessels operating in its waters or using its ports, or access of fisheries products stemming from IUU fishing activities into its market;
- whether the State concerned effectively cooperates with the EU by providing responses to requests to investigate IUU fishing and associated activities of its vessels;
- whether the State concerned has taken effective enforcement measures in respect of operators responsible for IUU fishing, and whether sanctions of sufficient severity to deprive the offenders of the benefits accruing from these activities have been applied;
- the history, nature, circumstances, extent and gravity of the manifestations of IUU fishing activities considered;
- for developing States, the existing capacity of their competent authorities;
- the ratification of or accession to relevant international fisheries instruments;
- the status of the State concerned as a member to regional fisheries management organisations, or the State's agreement to apply the conservation and management measures established by such organisations;

---

[60] *EC 1005/2008,* Art. 37.
[61] *EC 1005/2008,* Art. 31(3).

- any acts or omissions by the State concerned that may have diminished the effectiveness of applicable laws, regulations or international conservation and management measures; and
- specific constraints of developing countries, particularly in MCS of fishing activities.[62]

The EU IUU Regulation requires the prohibition of the importation into the EU of fishery products caught by vessels flying the flag of non-cooperating third countries, and non-acceptance of catch certificates accompanying such products.[63] In cases where the identification of a non-cooperating State is justified by the lack of appropriate measures adopted by the State in relation to IUU fishing activities affecting a given stock or species, the prohibition of importation may only apply in respect of that stock or species.[64] The EU IUU Regulation also provides for the denunciation by the EU of any standing bilateral fisheries agreement or fisheries partnership agreements with such States, as well as refusal to enter into negotiations to conclude a bilateral fisheries agreement or fisheries partnership agreements with such States.[65] These measures, some of which are in the forms of economic sanctions, are more restrictive that those provided under the IPOA-IUU.

10.2.5 *Control of Nationals*

Another component of the EU IUU Regulation is the prohibition for nationals of EU member States from supporting or engaging in IUU fishing, including fishing activities onboard fishing vessels or as operators or beneficial owners of fishing vessels included in the Community IUU vessel list.[66] EU member States are required to cooperate amongst themselves and with third countries to take all appropriate measures in order to identify their nationals engaged in IUU fishing.[67] EU member States are further required to encourage their nationals to submit information pertaining to the legal, beneficial or financial interests or control of fishing vessels flagged to third countries and the names of the vessels concerned. EU member States have the responsibility not to grant any public aid to operators involved in IUU fishing.[68] The EU IUU Regulation also prohibits nationals of EU member States from selling or exporting any fishing vessel to operators involved in IUU fishing.[69]

---

[62] *EC 1005/2008,* Arts. 31(4), 31(5), 31(6), and 31(7).
[63] *EC 1005/2008,* Art. 38.
[64] *EC 1005/2008,* Art. 38(1).
[65] *EC 1005/2008,* Art. 38(8) and (9).
[66] *EC 1005/2008,* Art. 39(1).
[67] *EC 1005/2008,* Art. 39(2).
[68] *EC 1005/2008,* Art. 40(3).
[69] *EC 1005/2008,* Art. 40(2).

10.2.6 *The Community Alert System*

The EU IUU Regulation establishes a Community Alert System, where information obtained with respect to non-compliance with applicable laws, regulations or international conservation and management measures in respect of fishing vessels or fishery products from certain third countries will be published by the EU on its website and in the *Official Journal of the European Union* (C series).[70] This is intended to provide general notice to operators and to ensure that EU member States take appropriate measures in respect of third countries whose vessels may be engaging in IUU fishing activities. The Community Alert System will also include references to RFMOs whose conservation and management measures may have been violated.[71] Additionally, the EU IUU Regulation requires the Commission to take any measure necessary to ensure publicity of the list of non-cooperating States. The Commission will transmit the list of non-cooperating States to the FAO and RFMOs for the purposes of enhancing co-operation between the EU and these organisations to prevent, deter and eliminate IUU fishing.[72,73]

---

[70] *EC 1005/2008,* Art. 22.

[71] *EC 1005/2008,* Art. 24(5)(e).

[72] *EC 1005/2008,* Art. 35.

[73] For a detailed analysis of the EU IUU Regulation, *see* M. Tsamenyi, M.A. Palma, B. Milligan, and K. Mfodwo, 'Fairer Fishing?: The Impact on Developing Countries of the European Community Regulation on Illegal, Unreported and Unregulated Fisheries', *Economic Paper Series No. 86* (London: Commonwealth Secretariat, 2009); *See also* M. Tsamenyi, M.A. Palma, B. Milligan, and K. Mfodwo, 'The European Council Regulation on Illegal, Unreported and Unregulated Fishing: An International Fisheries Law Perspective', 25 *The International Journal of Marine and Coastal Law* (2010) 5–31.

Chapter Eleven

Conclusion

Illegal, unreported and unregulated fishing is considered a major threat to the sustainability of fisheries resources with far reaching ecological, environmental, social, and economic impacts. Not surprisingly, the IUU fishing concept has become the most widely embraced concept by international organisations, RFMOs, regional economic integration organisations and individual States. This raises the need to understand the scope and nature of IUU fishing and the range of measures adopted in international and regional instruments to address the problem. The book traced the development of the concept of IUU fishing and provided examples of activities violating national, regional and international fisheries management regulations and measures. As discussed in Chapter 2, determining the scope of IUU fishing in any jurisdiction depends largely on two elements: the first is what constitutes "fishing" in the national and regional context, and the second is with respect to the scope of fisheries violations or actions regarded as breaches of fisheries legislation at the national level and conservation and management measures of RFMOs at the regional and subregional levels. Analysis of these elements demonstrate that in reality, the scope and content of IUU fishing may differ from one State or RFMO to another.

The book examined at length the international legal and policy framework to combat IUU fishing within the context of sustainable fisheries, which comprise fisheries and non-fisheries specific instruments, both legally binding and voluntary in status. Chapter 3 demonstrates that while the IPOA-IUU has been adopted to deal specifically with IUU fishing, there are relevant environment, trade, and maritime safety instruments that not only provide the legal basis for, but also support and strengthen the implementation of the IPOA-IUU. Because international instruments promoting sustainable fisheries are known to have uneven effectiveness and their provisions sometimes have diverse application in States and RFMOs, it is important to understand the legal parameters within which States and regional organisations can implement these instruments to combat IUU fishing.

International fisheries related instruments provide the basis for the adoption and implementation of effective flag, coastal, port, market, and all State measures to prevent, deter and eliminate IUU fishing. As examined in Chapters 4 to 8, flag and coastal State measures addressing IUU fishing consist of fishing vessel registration, authorisation to fish, record of fishing vessels, and MCS related measures such as the licensing of foreign fishing vessels in the EEZ, VMS, observer programmes, boarding, inspection, arrest and judicial proceedings. Port and market measures include the designation of ports where foreign fishing vessels may be allowed entry, inspection of fishing vessels, traceability of fish, catch documentation, harmonised commodity description and coding systems and trade restrictive measures. Other more general measures that may be implemented by States include enforcement actions at sea and in port, application of sanctions of sufficient severity, and cooperation among each other. The book not only offers a thorough analysis of measures adopted under international instruments but also provides an examination analysis of some of the current practices of RFMOs, States, and non-governmental entities (i.e. private organisations) in addressing IUU fishing.

11.1 The Need for an Effective Domestic and Regional Framework

If the analysis is extended to existing national and regional frameworks to combat IUU fishing, one cannot challenge the fact that a major contributing factor to the proliferation of IUU fishing, despite all the international efforts to date, is the lack of adequate legal and policy framework for sustainable fisheries at the national and regional levels.

An effective national framework to address IUU fishing would need to comprise two aspects: one is a clear definition of IUU fishing and the second is a set of measures or strategies tailored to address specific IUU fishing concerns. The measures would need to take into account applicable international, regional, subregional, and bilateral agreements or arrangements on fisheries. The framework or approach would also need to be embedded in the legislation and policies of the State or States involved, and reviewed within a reasonable period of time in order to ensure that such framework remains current and adequate. It would also need to be a framework that is accepted by relevant stakeholders, primarily the fishing industry, in order to ensure successful implementation.

In light of existing State practice, a critical aspect in the framework to address IUU fishing is the adoption of adequate port and internationally agreed market related measures. The current trend in the implementation of domestic measures against IUU fishing demonstrates a rapid shift towards controlling IUU fishing at the points of market and trade entry, including the imposition of import restrictions against fish and fish products obtained through IUU fishing. More States are now beginning to implement or embrace trade related measures such as catch certification schemes. However, in order to ensure that no State is used as convenient entry for IUU caught fish, States would need to implement similar, if not harmonised, catch certification schemes within their respective domestic jurisdictions that would deter IUU related activities at every stage of the commodity chain. Catch certification would need effective tools for issuing and verifying licences and certificates from the point of harvesting to the processing, import, export, and re-export of fish.

Apart from a legal and policy framework, another critical element in effectively addressing IUU fishing is the establishment of a sound institutional arrangement. The application of relevant international instruments on fisheries suggests that a number of organisations with fisheries related functions would need to be engaged to ensure that IUU fishing does not occur. The IPOA-IUU in particular, calls on States, acting through RFMOs, to strengthen their institutional capacity to combat IUU fishing.[1] It further recognises the need for sufficient cooperation and information sharing between the agencies responsible for undertaking functions such as fishing vessel registration and issuing authorisations to fish.[2] The IPOA-IUU also requires the full participation of stakeholders in combating IUU fishing, including the industry, fishing communities, and non-governmental organisations.[3] These measures are consistent with the provisions of the FAO Code of Conduct, which emphasises the need to, among other things, establish an appropriate institutional framework to achieve the sustainable and integrated use of fisheries resources,[4] consult with relevant stakeholders,[5] conduct fisheries research,[6] and provide a mechanism to resolve fisheries conflicts.[7]

From the analysis, the components of an effective institutional framework for combating IUU fishing may be summarised into three. The first component is the involvement of all institutions which have functions related to addressing IUU fishing. These institutions include fisheries management authorities, policy formulation bodies, maritime administrations, enforcement agencies, justice departments and courts, foreign affairs departments, and port authorities. The second component of an effective institutional framework is the cooperation and coordination among government agencies; and the third component is the participation of relevant stakeholders.

At the regional level, the framework to tackle the IUU fishing probem should not only be limited to measures against vessels of non-cooperating members and the listing of IUU vessels, but also other measures that complement domestic actions such as authorisation to fish and transship fish, chartering of vessels, trade

---

[1] Food and Agriculture Organization (FAO), *International Plan of Action to Prevent, Deter, and Eliminate Illegal, Unreported, and Unregulated Fishing*, adopted on 23 June 2001 at the 120th Session of the FAO Council. *Hereinafter referred to as* IPOA-IUU. Para. 80.1.

[2] *IPOA-IUU*, para. 40.

[3] *IPOA-IUU*, para. 9.1.

[4] FAO, *Code of Conduct for Responsible Fisheries*, Adopted at the 28th Session of the FAO Conference, Rome, Italy, 31 October 1995. *Hereinafter referred to as* FAO Code of Conduct. Paras. 7.1.1, 10.1.1, and 10.1.3.

[5] *FAO Code of Conduct*, para. 10.1.2.

[6] *FAO Code of Conduct*, paras. 8.5.3 and 12.2.

[7] *FAO Code of Conduct*, para. 10.1.5.

documentation, and port measures. Among these measures, the listing of vessels engaged in IUU fishing may be considered as one the most significant steps taken by RFMOs to address the IUU fishing problem. However, as discussed in Chapter 9, the full scope of IUU listing measures is yet to be fully developed as there are a number of legal and policy considerations that have yet to be addressed by RFMOs. These considerations include the listing and deletion of vessels from IUU list and the mutual recognition of IUU lists across regional organisations. RFMOs would need to establish clear procedures and due processes in nominating and including vessels on their respective IUU vessel lits.

11.2 Emerging Issues in the Global Framework to Address IUU Fishing

As States strive to address the IUU fishing problem, an increasing adherence to the international legal and policy framework for sustainable fisheries may be observed. States manifest their commitment to combat IUU fishing and implement their obligations under international instruments through the adoption of national plans of action and legislative measures. States, through regional fisheries or economic bodies, have also intensified cooperative efforts through the development of regional plans of action and establishment of compliance measures to address the problem. As highlighted earlier, non-flag State compliance mechanisms such as trade or market related measures are beginning to supplant flag State implementation in the pursuit of eliminating IUU fishing. Examples of such measures include catch certification schemes by RFMOs; import prohibitions on Atlantic bigeye tuna and all its products imposed by ICCAT against non-members; eco-labelling standards applied by private companies to ensure that only legally-caught fish are traded in the market; and catch certification and verification requirements, together with the threat of trade and economic sanctions, under the EU IUU Regulation and amendments to the U.S. *Magnuson-Stevens Fishery Conservation and Management Reauthorisation Act and the High Seas Driftnet Fishing Moratorium Protection Act.* While questions have been raised about the fairness and equitability of some of these State and regional practices, the broadening application of stricter trade related measures against IUU fishing may suggest acceptance of their adoption to achieve sustainable fisheries.

The international community is also witnessing a progressive development of international fisheries law. A number of international initiatives to this end are evolving such as the impending implementation of the legally binding FAO agreement on port State measures, development of a draft WTO agreement on subsidies, and the possible development of a global record of fishing vessels. These developments are pivotal to the global efforts against IUU fishing and also have various national trade implications for States which need further investigation.

Apart from these major trends in implementing the international legal and policy framework to combat IUU fishing, there are three other emerging issues which are outside the scope of this book, but nevertheless merit the attention of States. These issues are: defining IUU fishing within the context of marine biodiversity conservation and environmental protection; categorising IUU fishing as an international environmental crime and a maritime security issue; and the concept of flag State responsibility and liability for IUU fishing.

11.2.1 *IUU Fishing and Marine Biodiversity Conservation*

Chapter 1 illustrates the impact of IUU fishing on the marine ecosystems and the marine environment. There is no argument that some of the IUU fishing activities, such as the use of illegal and destructive fishing methods have severe impact not only on target stocks but also on by-catch and other associated species, as well on fisheries habitats. However, as can be gleaned from the discussions in Chapter 2, biodiversity considerations are not directly covered in the scope of IUU fishing provided in paragraph 3 of the IPOA-IUU, unless such activities are directly prohibited under national laws (e.g. fishing in marine protected areas or for protected species), regional conservation and management measures (e.g. by-catch reduction), and international regulations (e.g. driftnet fishing).

The absence of clear regulations would make it difficult to show why a fishing activity with a negative impact on the ecosystem should automatically be considered as IUU fishing. For example, in the case of RFMOs, most of the by-catch related measures are in the form of non-binding resolutions or recommendations. Hence, any activity involving by-catch of non-target species such as sharks and sea turtles in an RFMO area may not be clearly considered as violations of obligations under regional conservation and management regulations. There is also uncertainty as to whether an activity may be considered IUU fishing merely because it has a known or potential negative effect on ecosystem health, raising the question as to how such impact may be treated in the legal sense. Furthermore, since the effects of IUU fishing on marine biodiversity and environment vary depending on the type of activity, regulations may need to qualify the level of adverse impact that would trigger IUU fishing in such circumstances. Although at its infancy, one example of an attempt to establish the legal link between IUU fishing and marine ecosystems is the amendments to the *U.S. Magnuson-Stevens Fishery Conservation and Management Reauthorisation Act, the High Seas Driftnet Fishing Moratorium Protection Act,* and ensuing regulations, which would determine whether vessels and States have failed to protect marine living resources, including dependent and associated species of particular interest to the U.S.

There are also other issues involved in ascertaining the relationship between IUU fishing and marine biodiversity. First, before any adverse impact is determined, an environmental assessment would need to be conducted which is a mechanism that may not be in place in every State. Second, the conservation of marine biodiversity and protection of the marine environment may not necessarily lie within the jurisdiction of fisheries departments, but rather environment departments. Establishing a fisheries violation based on marine biodiversity considerations may not be a straightforward process if potential jurisdictional overlap or conflict between agencies is not addressed. Third, it would be erroneous to assume that only fishing activities conducted by illegal fishers cause damage to the ecosystems and the environment because activities of legitimate fishers may also have similar impact. Sometimes it is difficult to distinguish between the negative impacts of fishing activities by legitimate fishers from those of illegal fishers. Fourth, adopting regulation to protect some parts of the ecosystem through gear modifications to minimise the by-catch of species, may result in unintended consequences for other components of the ecosystem such as seabirds, marine mammals and turtles. It is therefore important to take into account all of these factors before the link between IUU fishing and marine biodiversity conservation is be established in regulatory frameworks.

11.2.2 *IUU Fishing as an International Environmental Crime and a Maritime Security Concern*

The impact of IUU fishing on marine ecosystems and the environment has also given rise to an emerging debate on IUU fishing as an environmental crime and as a maritime security issue. IUU fishing has been identified by various international bodies as an international environmental crime, together with illegal trade in wildlife, illegal trade in zone-depleting substances, dumping and illegal transport of hazardous waste, and illegal logging and trade in timber.[8]

The identification of IUU fishing as an environmental crime is from the perspective of the link between the environment and sustainable development. Environmental crimes are broadly defined as illegal acts which directly harm the environment, are considered transboundary in nature and involve cross-border criminal syndicates.[9] The role of criminal law in the protection of the environment has been the priority of the United Nations through the Commission on Crime Prevention and Criminal Justice in the early 1990s.[10] The argument is that the environment must be protected in such a way that existing

[8] Environmental Investigation Agency, *Environmental Crime: A Threat to Our Future* (London: Emmerson Press, 2008), at 1; Royal Institute of International Affairs, Sustainable Development Programme, *International Environmental Crime: The Nature and Control of Environmental Black Markets*, Workshop Report, 2002, at 5. www.chathamhouse.org.uk. Accessed on 13 August 2009.

[9] Environmental Investigation Agency, *ibid.*, at 1.

[10] UN, Economic and Social Council, Commission on Crime Prevention and Criminal Justice, Report on the First Session, 21–30 April 1992, Economic and Social Council Official Records 1992 Supplement No. 10, *E/1992/30, E/CN.15/1992/7*, para. IV(1)(a). A number of institutions have

damage will be eliminated or at least reduced or restored, harm will be prevented, and risk will be minimised.[11] The goal of using criminal law as a supporting or independent tool to protect the environment is also considered to have a preventive effect, and possibly through moral stigma, may heighten environmental awareness.[12]

The relationship between IUU fishing and international environmental crime was raised at the ninth meeting of the United Nations Open-ended Informal Consultative Process on Oceans and the Law of the Sea (UNICPOLOS) in 2008. It was emphasised that some illegal fishing activities are conducted by transnational organised criminal networks.[13] Similarly, at the meeting of the Conference of Parties to the United Nations Convention Against Transnational Organised Crime, views were expressed that IUU fishing falls within the definition of the transnational organised crime provided under the Convention.[14] Organised criminal syndicates have been associated with illegal fishing for high value and low volume seafood products. There have been reports of the transport of illegally caught South African abalone combined with drug trafficking[15] or the use of

---

conducted studies on the protection of the environment by means of criminal law such as the United Nations Interregional Crime and Justice Research Institute (UNICRI), the United Nations Asia and Far East Institute of Crime Prevention and Treatment of Offenders (UNAFEI), Max Planck Institute for Foreign and International Criminal Law, Australian Institute of Criminology (AIC), and International Centre for Criminal Law Reform and Criminal Justice Policy. *See* Günter Henie, Mohan Prabhu, and Anna Alvazzi del Frate, eds, Environmental Protection-Potentials and Limits of Criminal Justice: Evaluation of Legal Structures, *United Nations Interregional Crime and Justice Research Institute Publication No. 56* (Freiburg im Breisgau: Ed. Iuscrim, Max Planck Institut für Ausländisches und Internationales Strafrecht, 1997), at 20.

[11] UN, Economic and Social Council, Substantive Session of 1993, 28 June to 30 July 1993, Report of the Commission on Crime Prevention and Criminal Justice on its Second Session, Vienna, 13–23 April 1993, *CE/1993/32, E/CN.15/1993/9*, 10 June 1993, Annex.

[12] *Ibid.*

[13] United Nations General Assembly (UNGA), Sixty-third Session, Item 73(a) of the provisional agenda, Oceans and the law of the sea, Report on the works of the United Nations and the Law of the Sea at its Ninth Meeting, *A/63/174*, 25 July 2008, para. 10(e); There have also been other reports on the use of fishing boats to transport illegal narcotics. See for example 'Cocaine traffickers develop new routes from Brazil', *Jane's Intelligence Review*, 01 January 2006, www.unodc.org. Accessed on 14 April 2009.

[14] United Nations (UN), Conference of Parties to the United Nations Convention Against Transnational Organised Crime, Report of the Conference of Parties to the United Nations Convention Against Transnational Organised Crime on its Fourth Session, Vienna, 8–17 October 2008, *CTOC/COP/2008/19*, 1 December 2008, para. 210; The United Nations Convention Against Transnational Organised Crime defines organised criminal group as "a structured group of three or more persons, existing for a period of time and acting in concert with the aim of committing one or more serious crimes or offences established in accordance with this Convention, in order to obtain, directly or indirectly, a financial or other material benefit. See *United Nations Convention Against Transnational Organised Crime*, concluded on 15 November 2000, in force 29 September 2009, 40 ILM 335 (2001); UN Doc. A/55/383 at 25 (2000); UN Doc. A/RES/55/25 at 4 (2001), Art. 2.

[15] Cand. jur. Gunnar Stølsvik, 'Transnational Organised Fisheries Crime as a Maritime Security Issue', Panel Presentation at the Ninth Meeting of the UN Open-ended Informal Consultative Process on Oceans and the Law of the Sea, 23–27 June 2008.

fishing vessels in the transport of cocaine to West Africa.[16] Criminal syndicates have also been associated with IUU fishing for Patagonian toothfish.[17] Studies have alluded to the possible involvement of outlaw gangs and international criminal networks in the illegal trade of Australian abalone.[18] Other high value, low volume fisheries such as shark fin, seahorse, eels, trepang, and sea urchin are also known to be vulnerable to the exploitation by criminal groups in Australia.[19]

However, there were divergent views at the ninth meeting of UNICPOLOS on the potential link between illegal fishing and transnational organised crime. While some delegations stressed the connection between illegal fishing and other illegal activities, some delegations were of the view that such link is not enough to designate illegal fishing as a transnational crime and that further studies are required to establish that link.[20] An area where the relationship between IUU fishing and transnational organised crime would need further clarification relates to identifying which types of IUU fishing may be considered as transnational "crime". To begin with, a basic characteristic of a crime is that it is an act in breach of the law, and is therefore illegal. However, illegal fishing is only one component of IUU fishing, and characterising IUU fishing as a "crime" misses the full scope and nature of the IUU fishing concept. If the scope and nature of IUU fishing under the IPOA-IUU are strictly followed, many types of unregulated fishing cannot be considered a crime. Similarly, unreported fishing resulting from lack of national or regional reporting regulations may not be considered illegal and therefore not a crime.

Another issue is the lack of a standardised penalty system for fisheries violations. Not all domestic legislation provide for criminal sanctions for fishing violations although the imposition of severe penalties on some illegal fishing activities have gained widespread acceptance. As discussed in Chapter 6, while some States treat fisheries violations as criminal, others rely on administration sanctions or use both.[21] The application of criminal sanctions to fisheries violations in the EEZ has some limitations under the LOSC. Nevertheless, the distinction between

---

[16] UN Office of Drugs and Crime, Cocaine Trafficking in Western Africa: Situation Report, October 2007. www.unodc.org. Accessed on 10 July 2009, at 8.

[17] Gavin Hayman and Duncan Brack, *International Environmental Crime: The Nature and Control of Environmental Black Markets* (London: Royal Royal Institute of International Affairs, 2002), at 7.

[18] See Rebecca Tailby and Frances Gant, 'The Illegal Market of Australian Abalone,' *Trends and Issues in Crime and Criminal Justice No 225* (Canberra: Institute of Criminology, 2002).

[19] Katherine M Anderson and Rob McCusker, 'Crime in the Australian Fishing Industry: Key Issues', *Trends and Issues in Crime and Criminal Justice No. 227* (Canberra: Institute of Criminology, 2005), at 3; *See also* Judy Putt and Katherine Anderson, 'A National Study of Crime in the Australian Fishing Industry', *Research and Public Policy Series No. 76* (Canberra: Australian Institute of Criminology, 2007).

[20] UN, *A/63/174*, *supra* note 13, paras. 71 and 73.

[21] P. Cacaud, M. Kuruc, and M. Spreij, Administrative Sanctions in Fisheries Law, *FAO Legislative Study No. 82* (Rome: FAO, 2003).

administrative and criminal sanctions applied by States against fisheries violations provide the basis for future discussions on the possible categorisation of IUU fishing as an international environmental crime.

Consideration would also need to be given to extending the debate to cover companies whose vessels engage in IUU fishing. If criminal law is going to be applied to IUU fishing related violations, it may be necessary to differentiate offences based on their seriousness (e.g. whether the IUU activity is intentional and reckless or a negligent act).[22] To address IUU fishing activities in different jurisdictions, the existence of regional extradition and mutual assistance in criminal matters may also need to be taken into account.[23]

There are other issues that would need to be considered in classifying IUU fishing as an international environmental crime. States would need to identify and agree on the components of IUU fishing which may be construed as environmentally harmful behaviour, and therefore criminal. Jurisdiction and liability would also need to be established through a multilateral convention.

11.2.3 *Flag State Responsibility and Liability*

Another emerging debate is the extent to which flag States should be held responsible for the IUU fishing activities of vessels flying their flags and the extent of their liability. To understand the complexities of this issue, it is necessary to briefly distinguish flag State duties from flag State responsibility.

As discussed in Chapter 5, at the core of the regulatory framework for shipping is the concept of flag State duties. Traditionally, Article 94(1) of the LOSC restates this customary international law rule: "Every State shall effectively exercise its jurisdiction and control in administrative, technical and social matters over ships flying its flag." In particular, the flag State is required, among other things, to "maintain a register of ships containing the names and particulars of ships flying its flag", and to "assume jurisdiction under its internal law over each ship flying its flag and its master, officers and crew in respect of administrative, technical and social matters concerning the ship."[24] These general flag State duties also apply to fishing vessels. This is, for example, confirmed by the preamble to the FAO Compliance Agreement.[25]

---

[22] UN, *E/CN.15/1993/9*, *supra* note 11, Annex.

[23] Henie, Prabhu, and del Frate, *supra* note 10, at 26.

[24] United Nations Convention on the Law of the Sea, Montego Bay, Jamaica, concluded on 10 December 1982, in force 16 November 1994, 1833 UNTS 3; 21 ILM 1261 (1982), Art. 94(2).

[25] The Preamble of the FAO Compliance Agreement states that "(C)onscious of the duties of every State to exercise effectively its jurisdiction and control over vessels flying its flag, including fishing vessels and vessels engaged in the transhipment of fish." FAO, Agreement to Promote Compliance with International Conservation and Management Measures by Fishing Vessels on the High Seas, Rome, Italy, concluded on 24 November 1993, entered into force 24 April 2003, 33 ILM 968. *Hereinafter referred to as* FAO Compliance Agreement.

A major gap in the international regulatory framework for shipping is the absence of an oversight body to verify and ensure the discharge by the flag State of its duties.[26] There is also no responsibility or liability attached to violations of the vessels flying the flag of a particular State as a result of its failure to discharge its duties under international law. Article 94 of the LOSC simply provides that: "A State which has clear grounds to believe that proper jurisdiction and control with respect to a ship have not been exercised may report the facts to the flag State." All that the flag State is required to do is to "investigate the matter and, if appropriate, take any action necessary to remedy the situation".

The term "responsibility" is used in the LOSC in the general context of State responsibility. For example, under the LOSC Article 31, the flag State bears "international responsibility for any loss or damage to the coastal State resulting from the non-compliance by a warship or other government ship operated for non-commercial purposes with the laws and regulations of the coastal State concerning passage through the territorial sea or with the provisions of this Convention or other rules of international law". General State responsibility also applies in relation to the wrongful activities by States Parties, State enterprises or natural or juridical persons which possess the nationality of States Parties or are effectively controlled by them or their nationals in the Area.[27] The failure by States to fulfil international obligations concerning the protection and preservation of the marine environment also gives rise to State responsibility and liability in accordance with international law.[28]

The concept of flag State responsibility in the fisheries context has its origins in bilateral access agreements between coastal States and distant water fishing nations following the extension of fisheries jurisdictions from the 1970s.[29] One of the conditions imposed by developing coastal States on foreign fishing access in their zones is the requirement that the flag States of the foreign fishing vessels assume responsibility to ensure that vessels flying their flags act in accordance with the terms of the fishing access agreements.[30] As an example, under a specific provision on "Flag State Responsibility", the Treaty on Fisheries Between the Governments of Certain Pacific Island States and the Government of the United States of America (U.S. Treaty) provides that the U.S. shall take the necessary

[26] See generally John N.K. Mansell, *Flag State Responsibility; Historical Development and Contemporary Issues* (Heiderlberg: Springer, 2009).

[27] *LOSC*, Art. 139(1).

[28] *LOSC*, Art. 235(1).

[29] Annick Van Houtte, 'Flag State Responsibility and the Contribution of Recent International Instruments in Preventing, Deterring, and Eliminating IUU Fishing', Appendix E in Report of the Expert Consultation on Fishing Vessels Operating Under Open Registries and their Impact on Illegal, Unreported and Unregulated Fishing, Miami, Florida, USA, 23–25 September 2003, *FAO Fisheries Report No. 722* (Rome; FAO, 2003), at 49.

[30] *Ibid.*

steps to ensure that its nationals and fishing vessels refrain from conducting unauthorised fishing in the Treaty area.[31] Such responsibility includes adopting all necessary measures to investigate an alleged breach of the U.S. Treaty, facilitating any claim arising out of the activities of U.S. vessels, and ensuring that offending vessels submit to the jurisdiction of the party concerned and is penalised by the U.S.[32]

Post LOSC international fisheries instruments have embraced the concept of "flag State responsibility". Direct reference to flag State responsibility can be found in the FAO Compliance Agreement and the UN Fish Stocks Agreement. In its Preamble, the FAO Compliance Agreement noted that the objective of the agreement can be achieved by specifying flag State responsibility over vessels on the high seas. The FAO Compliance Agreement then goes on to provide in detail the content of flag State responsibility for its fishing vessels.[33] Similarly, Article 18(2) of the UN Fish Stocks Agreement provides that a State shall only authorise a vessel to fish on the high seas if it is able to exercise effectively its responsibility in respect of such vessels under the LOSC and the UN Fish Stocks Agreement.

In the context of IUU fishing, the IPOA-IUU contains a number of provisions on flag State responsibility similar to the provisions in the FAO Compliance Agreement and the UN Fish Stocks Agreement. The IPOA-IUU not only recognises the primary responsibility of the flag State with respect to its vessels on the high seas,[34] but also clearly provides specific flag State measures such as fishing vessel registration, authorisation to fish, and record of fishing vessels. In particular, the IPOA-IUU provides that "(a) flag State should ensure, before it registers a fishing vessel that it can exercise its responsibility to ensure that the vessel does not engage in IUU fishing."[35]

Although these international fisheries instruments contain provisions relating to flag State responsibility, only the UN Fish Stocks Agreement relates such responsibility to the liability of flag States with respect to the activities of their vessels. Article 35 of the UN Fish Stocks Agreement provides that "States Parties to the Agreement are liable in accordance with international law for damage or loss attributable to them in regard to this Agreement"; however, there is no further elaboration of the scope and content of the liability of States Parties in this regard.

---

[31] Treaty on Fisheries Between the Governments of Certain Pacific Island States and the Government of the United States of America, Port Moresby, concluded on 2 April 1987, in force 15 June 1988. ATS 1988 No. 42. *Hereinafter referred to as the U.S. Treaty*, Art. 4.1.

[32] *U.S. Treaty*, Art. 4.1.

[33] *FAO Compliance Agreement*, Art. III.

[34] *IPOA-IUU*, para. 18.

[35] *IPOA-IUU*, para. 35.

Discussions on the concept of State responsibility in the International Law Commission (ILC) beginning the 1920s are relevant for the application of flag State responsibility and liability in the context of fisheries. The ILC deliberations resulted in the adoption of draft articles on the Responsibility of States for Internationally Wrongful Acts, which was brought to the attention of States for further action at the UN General Assembly in 2001.[36] Article 1 of the Articles on State Responsibility provides for the international responsibility of a State for every internationally wrongful act. The elements of an internationally wrongful act consist of an action or an omission that is attributable to the State under international law and constitutes a breach of an international obligation of the State.[37] Article 12 provides that "(t)here is a breach of an international obligation of a State when an act of that State is not in conformity with what is required of it by that obligation, regardless of its origin and character."

In the fisheries context, the question then becomes whether or not a flag State may be held responsible, and consequently, liable for the IUU fishing activities of its nationals, including vessels flying its flag. This issue must be distinguished from the discharge of the traditional flag State duties under Article 94 of the LOSC, which are not sufficient to hold States responsible for the IUU fishing activities of their vessels.

Arguably, emerging rules of international fisheries law as reflected in the UN Fish Stocks Agreement and the FAO Compliance Agreement, as well as non-binding instruments such as the IPOA-IUU, provide sufficient basis to make flag States responsible for the IUU fishing activities of their vessels under certain circumstances. This may be the case where a flag State has blatantly refused to take action against the activities of its vessels outside its jurisdiction, such as by failing to adopt appropriate measures required by international fisheries instruments to ensure compliance of the vessel with regional conservation and management measures, such as authorisations to fish, monitoring through VMS, and imposition of sanctions of adequate severity. State responsibility may also arguably arise where the flag State has refused to take action against its vessels involved in IUU fishing after such activities have been brought to its attention.

---

[36] The UN General Assembly commends the articles on responsibility of States for internationally wrongful acts to the attention of the Governments, without prejudice to the question of their future adoption or other appropriate action. *See* UNGA, Sixty-second session, Agenda item 78, Resolution adopted by the General Assembly [on the report of the Sixth Committee (A/62/446)], Responsibility of Sates for Internationally Wrongful Acts, *A/RES/62/61*, para. 1.

[37] Responsibility of States for Internationally Wrongful Acts, Text adopted by the Commission at its fifty-third session, in 2001, and submitted to the General Assembly as part of the Commission's report covering the work of that session. Text reproduced as it appears in the annex to General Assembly resolution 56/83 of 12 December 2001, and corrected by document A/56/49(Vol.I)Corr.4, Art. 2. Available at http://untreaty.un.org/ilc/research.htm. Accessed 20 August 2009.

Given the central role played by flag States in the regulation of fishing vessels, a major component of the global fight against IUU fishing must necessarily focus on the fundamental issue of responsibility and liability of flag States for the IUU fishing activities of their vessels. Recent measures adopted by the EU and the U.S. particularly on the listing of non-cooperating States and application of trade and economic sanctions against flag States whose vessels repeatedly engage in IUU fishing, may be seen as a gradual development in State practice to hold flag States liable for the IUU fishing activities of their vessels.

Appendix

International Plan of Action to Prevent, Deter and Eliminate Illegal, Unreported and Unregulated Fishing

I. Introduction

1. In the context of the Code of Conduct for Responsible Fisheries and its overall objective of sustainable fisheries, the issue of illegal, unreported and unregulated (IUU) fishing in world fisheries is of serious and increasing concern. IUU fishing undermines efforts to conserve and manage fish stocks in all capture fisheries. When confronted with IUU fishing, national and regional fisheries management organizations can fail to achieve management goals. This situation leads to the loss of both short and long-term social and economic opportunities and to negative effects on food security and environmental protection. IUU fishing can lead to the collapse of a fishery or seriously impair efforts to rebuild stocks that have already been depleted. Existing international instruments addressing IUU fishing have not been effective due to a lack of political will, priority, capacity and resources to ratify or accede to and implement them.
2. The Twenty-third Session of the FAO Committee on Fisheries (COFI) in February 1999 addressed the need to prevent, deter and eliminate IUU fishing. The Committee was concerned about information presented indicating increases in IUU fishing, including fishing vessels flying "flags of convenience". Shortly afterwards, an FAO Ministerial Meeting on Fisheries in March 1999 declared that, without prejudice to the rights and obligations of States under international law, FAO will develop a global plan of action to deal effectively with all forms of illegal, unregulated and unreported fishing including fishing vessels flying "flags of convenience" through coordinated efforts by States, FAO, relevant regional fisheries management bodies and other relevant international agencies such as the International Maritime Organization (IMO), as provided in Article IV of the Code of Conduct. The Government of Australia, in cooperation with FAO, organized an Expert Consultation on Illegal, Unreported and Unregulated Fishing in Sydney, Australia, from 15 to 19 May 2000. Subsequently, an FAO Technical Consultation on Illegal, Unreported and Unregulated Fishing was held in Rome from 2 to 6 October 2000 and a further Technical Consultation was held in Rome from 22 to 23 February 2001. The draft International Plan of

Action to Prevent, Deter and Eliminate Illegal, Unreported and Unregulated Fishing was adopted by the Consultation on 23 February 2001 with a request that the report be submitted to the Twenty-fourth Session of COFI for consideration and eventual adoption. COFI approved the International Plan of Action, by consensus, on 2 March 2001. In doing so, the Committee urged all Members to take the necessary steps to effectively implement the International Plan of Action.

II. Nature and Scope of IUU Fishing and the International Plan of Action

3. In this document:

3.1 Illegal fishing refers to activities:

3.1.1 conducted by national or foreign vessels in waters under the jurisdiction of a State, without the permission of that State, or in contravention of its laws and regulations;

3.1.2 conducted by vessels flying the flag of States that are parties to a relevant regional fisheries management organization but operate in contravention of the conservation and management measures adopted by that organization and by which the States are bound, or relevant provisions of the applicable international law; or

3.1.3 in violation of national laws or international obligations, including those undertaken by cooperating States to a relevant regional fisheries management organization.

3.2 Unreported fishing refers to fishing activities:

3.2.1 which have not been reported, or have been misreported, to the relevant national authority, in contravention of national laws and regulations; or

3.2.2 undertaken in the area of competence of a relevant regional fisheries management organization which have not been reported or have been misreported, in contravention of the reporting procedures of that organization.

3.3 Unregulated fishing refers to fishing activities:

3.3.1 in the area of application of a relevant regional fisheries management organization that are conducted by vessels without nationality, or by those flying the flag of a State not party to that organization, or by a fishing entity, in a manner that is not consistent with or contravenes the conservation and management measures of that organization; or

3.3.2 in areas or for fish stocks in relation to which there are no applicable conservation or management measures and where such fishing activities are conducted in a manner inconsistent with State responsibilities for the conservation of living marine resources under international law.

3.4 Notwithstanding paragraph 3.3, certain unregulated fishing may take place in a manner which is not in violation of applicable international law, and may not require the application of measures envisaged under the International Plan of Action (IPOA).

4. The IPOA is voluntary. It has been elaborated within the framework of the FAO Code of Conduct for Responsible Fisheries as envisaged by Article 2 (d).
5. The FAO Code of Conduct for Responsible Fisheries, in particular Articles 1.1, 1.2, 3.1, and 3.2 applies to the interpretation and application of this IPOA and its relationship with other international instruments. The IPOA is also directed as appropriate towards fishing entities as referred to in the Code of Conduct. The IPOA responds to fisheries specific issues and nothing in it prejudices the positions of States in other fora.
6. In this document:

(a) the reference to States includes regional economic integration organizations in matters within their competence;
(b) the term "regional" includes sub-regional, as appropriate;
(c) the term "regional fisheries management organization" means an intergovernmental fisheries organization or arrangement, as appropriate, that has the competence to establish fishery conservation and management measures;
(d) the term "conservation and management measures" means measures to conserve one or more species of living marine resources that are adopted and applied consistent with the relevant rules of international law;
(e) the term "1982 UN Convention" refers to the United Nations Convention on the Law of the Sea of 10 December 1982;
(f) the term "1993 FAO Compliance Agreement" refers to the Agreement to Promote Compliance with International Conservation and Management Measures by Fishing Vessels on the High Seas, approved by the FAO Conference on 24 November 1993.
(g) the term "1995 UN Fish Stocks Agreement" refers to the Agreement for the Implementation of the United Nations Convention on the Law of the

Sea of 10 December 1982 relating to the Conservation and Management of Straddling Fish Stocks and Highly Migratory Fish Stocks; and

(h) the term "Code of Conduct" refers to the FAO Code of Conduct for Responsible Fisheries.

7. This document is a further commitment by all States to implement the Code of Conduct.

III. Objective and Principles

8. The objective of the IPOA is to prevent, deter and eliminate IUU fishing by providing all States with comprehensive, effective and transparent measures by which to act, including through appropriate regional fisheries management organizations established in accordance with international law.
9. The IPOA to prevent, deter and eliminate IUU fishing incorporates the following principles and strategies. Due consideration should be given to the special requirements of developing countries in accordance with Article 5 of the Code of Conduct.

9.1 *Participation and coordination*: To be fully effective, the IPOA should be implemented by all States either directly, in cooperation with other States, or indirectly through relevant regional fisheries management organizations or through FAO and other appropriate international organizations. An important element in successful implementation will be close and effective coordination and consultation, and the sharing of information to reduce the incidence of IUU fishing, among States and relevant regional and global organizations. The full participation of stakeholders in combating IUU fishing, including industry, fishing communities, and non-governmental organizations, should be encouraged.

9.2 *Phased implementation*: Measures to prevent, deter and eliminate IUU fishing should be based on the earliest possible phased implementation of national plans of action, and regional and global action in accordance with the IPOA.

9.3 *Comprehensive and integrated approach*: Measures to prevent, deter and eliminate IUU fishing should address factors affecting all capture fisheries. In taking such an approach, States should embrace measures building on the primary responsibility of the flag State and using all available jurisdiction in accordance with international law, including port State measures, coastal State measures, market-related measures and measures to ensure that nationals do not support or engage in IUU fishing. States are encouraged to use all these measures, where appropriate, and to cooperate in order to ensure that measures are applied in an integrated manner. The action plan should address all economic, social and environmental impacts of IUU fishing.

9.4 *Conservation*: Measures to prevent, deter and eliminate IUU fishing should be consistent with the conservation and long-term sustainable use of fish stocks and the protection of the environment.

9.5 *Transparency*: The IPOA should be implemented in a transparent manner in accordance with Article 6.13 of the Code of Conduct.

9.6 *Non-discrimination*: The IPOA should be developed and applied without discrimination in form or in fact against any State or its fishing vessels.

IV. Implementation of Measures to Prevent, Deter and Eliminate IUU Fishing

All State Responsibilities

International Instruments

10. States should give full effect to relevant norms of international law, in particular as reflected in the 1982 UN Convention, in order to prevent, deter and eliminate IUU fishing.
11. States are encouraged, as a matter of priority, to ratify, accept or accede to, as appropriate, the 1982 UN Convention, the 1995 UN Fish Stocks Agreement and the 1993 FAO Compliance Agreement. Those States that have not ratified, accepted or acceded to these relevant international instruments should not act in a manner inconsistent with these instruments.
12. States should implement fully and effectively all relevant international fisheries instruments which they have ratified, accepted or acceded to.
13. Nothing in the IPOA affects, or should be interpreted as affecting, the rights and obligations of States under international law. Nothing in the IPOA affects, or should be interpreted as affecting, the rights and obligations contained in the 1995 UN Fish Stocks Agreement and the 1993 FAO Compliance Agreement, for States parties to those instruments.
14. States should fully and effectively implement the Code of Conduct and its associated International Plans of Action.
15. States whose nationals fish on the high seas in fisheries not regulated by a relevant regional fisheries management organization should fully implement their obligations under Part VII of the 1982 UN Convention to take measures with respect to their nationals as may be necessary for the conservation of the living resources of the high seas.

National Legislation

Legislation

16. National legislation should address in an effective manner all aspects of IUU fishing.

17. National legislation should address, inter alia, evidentiary standards and admissibility including, as appropriate, the use of electronic evidence and new technologies.

State Control over Nationals

18. In the light of relevant provisions of the 1982 UN Convention, and without prejudice to the primary responsibility of the flag State on the high seas, each State should, to the greatest extent possible, take measures or cooperate to ensure that nationals subject to their jurisdiction do not support or engage in IUU fishing. All States should cooperate to identify those nationals who are the operators or beneficial owners of vessels involved in IUU fishing.
19. States should discourage their nationals from flagging fishing vessels under the jurisdiction of a State that does not meet its flag State responsibilities.

Vessels without Nationality

20. States should take measures consistent with international law in relation to vessels without nationality on the high seas involved in IUU fishing.

Sanctions

21. States should ensure that sanctions for IUU fishing by vessels and, to the greatest extent possible, nationals under its jurisdiction are of sufficient severity to effectively prevent, deter and eliminate IUU fishing and to deprive offenders of the benefits accruing from such fishing. This may include the adoption of a civil sanction regime based on an administrative penalty scheme. States should ensure the consistent and transparent application of sanctions.

Non Cooperating States

22. All possible steps should be taken, consistent with international law, to prevent, deter and eliminate the activities of non-cooperating States to a relevant regional fisheries management organization which engage in IUU fishing.

Economic Incentives

23. States should, to the extent possible in their national law, avoid conferring economic support, including subsidies, to companies, vessels or persons that are involved in IUU fishing.

Monitoring, Control and Surveillance

24. States should undertake comprehensive and effective monitoring, control and surveillance (MCS) of fishing from its commencement, through the point of landing, to final destination, including by:

    24.1 developing and implementing schemes for access to waters and resources, including authorization schemes for vessels;
    24.2 maintaining records of all vessels and their current owners and operators authorized to undertake fishing subject to their jurisdiction;

24.3 implementing, where appropriate, a vessel monitoring system (VMS), in accordance with the relevant national, regional or international standards, including the requirement for vessels under their jurisdiction to carry VMS on board;

24.4 implementing, where appropriate, observer programmes in accordance with relevant national, regional or international standards, including the requirement for vessels under their jurisdiction to carry observers on board;

24.5 providing training and education to all persons involved in MCS operations;

24.6 planning, funding and undertaking MCS operations in a manner that will maximize their ability to prevent, deter and eliminate IUU fishing;

24.7 promoting industry knowledge and understanding of the need for, and their cooperative participation in, MCS activities to prevent, deter and eliminate IUU fishing;

24.8 promoting knowledge and understanding of MCS issues within national judicial systems;

24.9 establishing and maintaining systems for the acquisition, storage and dissemination of MCS data, taking into account applicable confidentiality requirements;

24.10 ensuring effective implementation of national and, where appropriate, internationally agreed boarding and inspection regimes consistent with international law, recognizing the rights and obligations of masters and of inspection officers, and noting that such regimes are provided for in certain international agreements, such as the 1995 UN Fish Stocks Agreement, and only apply to the parties to those agreements.

National Plans of Action

25. States should develop and implement, as soon as possible but not later than three years after the adoption of the IPOA, national plans of action to further achieve the objectives of the IPOA and give full effect to its provisions as an integral part of their fisheries management programmes and budgets. These plans should also include, as appropriate, actions to implement initiatives adopted by relevant regional fisheries management organizations to prevent, deter and eliminate IUU fishing. In doing so, States should encourage the full participation and engagement of all interested stakeholders, including industry, fishing communities and non-governmental organizations.

26. At least every four years after the adoption of their national plans of action, States should review the implementation of these plans for the purpose of

identifying cost-effective strategies to increase their effectiveness and to take into account their reporting obligations to FAO under Part VI of the IPOA.

27. States should ensure that national efforts to prevent, deter and eliminate IUU fishing are internally coordinated.

Cooperation between States

28. States should coordinate their activities and cooperate directly, and as appropriate through relevant regional fisheries management organizations, in preventing, deterring and eliminating IUU fishing. In particular, States should:

    28.1 exchange data or information, preferably in standardized format, from records of vessels authorized by them to fish, in a manner consistent with any applicable confidentiality requirements;
    28.2 cooperate in effective acquisition, management and verification of all relevant data and information from fishing;
    28.3 allow and enable their respective MCS practitioners or enforcement personnel to cooperate in the investigation of IUU fishing, and to this end States should collect and maintain data and information relating to such fishing;
    28.4 cooperate in transferring expertise and technology;
    28.5 cooperate to make policies and measures compatible;
    28.6 develop cooperative mechanisms that allow, inter alia, rapid responses to IUU fishing; and
    28.7 cooperate in monitoring, control and surveillance, including through international agreements.

29. In the light of Article VI of the 1993 FAO Compliance Agreement, flag States should make available to FAO and, as appropriate, to other States and relevant regional or international organizations, information about vessels deleted from their records or whose authorization to fish has been cancelled and to the extent possible, the reasons therefor.
30. In order to facilitate cooperation and exchange of information, each State and regional or international organization should nominate and publicize initial formal contact points.
31. Flag States should consider entering into agreements or arrangements with other States and otherwise cooperate for the enforcement of applicable laws and conservation and management measures or provisions adopted at a national, regional or global level.

Publicity

32. States should publicize widely, including through cooperation with other States, full details of IUU fishing and actions taken to eliminate it, in a manner consistent with any applicable confidentiality requirements.

Technical Capacity and Resources

33. States should endeavour to make available the technical capacity and resources which are needed to implement the IPOA. This should include, where appropriate, the establishment of special funds at the national, regional or global level. In this respect, international cooperation should play an important role.

Flag State Responsibilities

Fishing Vessel Registration

34. States should ensure that fishing vessels entitled to fly their flag do not engage in or support IUU fishing.
35. A flag State should ensure, before it registers a fishing vessel, that it can exercise its responsibility to ensure that the vessel does not engage in IUU fishing.
36. Flag States should avoid flagging vessels with a history of non-compliance except where:

    36.1 the ownership of the vessel has subsequently changed and the new owner has provided sufficient evidence demonstrating that the previous owner or operator has no further legal, beneficial or financial interest in, or control of, the vessel; or
    36.2 having taken into account all relevant facts, the flag State determines that flagging the vessel would not result in IUU fishing.

37. All States involved in a chartering arrangement, including flag States and other States that accept such an arrangement, should, within the limits of their respective jurisdictions, take measures to ensure that chartered vessels do not engage in IUU fishing.
38. Flag States should deter vessels from reflagging for the purposes of non-compliance with conservation and management measures or provisions adopted at a national, regional or global level. To the extent practicable, the actions and standards flag States adopt should be uniform to avoid creating incentives for vessel owners to reflag their vessels to other States.
39. States should take all practicable steps, including denial to a vessel of an authorization to fish and the entitlement to fly that State's flag, to prevent "flag hopping"; that is to say, the practice of repeated and rapid changes of a vessel's flag for the purposes of circumventing conservation and management measures or provisions adopted at a national, regional or global level or of facilitating non-compliance with such measures or provisions.
40. Although the functions of registration of a vessel and issuing of an authorization to fish are separate, flag States should consider conducting these functions in a manner which ensures each gives appropriate consideration to the other. Flag States should ensure appropriate links between the operation of

their vessel registers and the record those States keep of their fishing vessels. Where such functions are not undertaken by one agency, States should ensure sufficient cooperation and information sharing between the agencies responsible for those functions.

41. A Flag State should consider making its decision to register a fishing vessel conditional upon its being prepared to provide to the vessel an authorization to fish in waters under its jurisdiction, or on the high seas, or conditional upon an authorization to fish being issued by a coastal State to the vessel when it is under the control of that flag State.

Record of Fishing Vessels

42. Each flag State should maintain a record of fishing vessels entitled to fly its flag. Each flag State's record of fishing vessels should include, for vessels authorized to fish on the high seas, all the information set out in paragraphs 1 and 2 of Article VI of the 1993 FAO Compliance Agreement, and may also include, inter alia:

    42.1 the previous names, if any and if known;
    42.2 name, address and nationality of the natural or legal person in whose name the vessel is registered;
    42.3 name, street address, mailing address and nationality of the natural or legal persons responsible for managing the operations of the vessel;
    42.4 name, street address, mailing address and nationality of natural or legal persons with beneficial ownership of the vessel;
    42.5 name and ownership history of the vessel, and, where this is known, the history of non-compliance by that vessel, in accordance with national laws, with conservation and management measures or provisions adopted at a national, regional or global level; and
    42.6 vessel dimensions, and where appropriate, a photograph, taken at the time of registration or at the conclusion of any more recent structural alterations, showing a side profile view of the vessel.

43. Flag States may also require the inclusion of the information in paragraph 42 in their record of fishing vessels that are not authorized to fish on the high seas.

Authorization to Fish

44. States should adopt measures to ensure that no vessel be allowed to fish unless so authorized, in a manner consistent with international law for the high seas, in particular the rights and duties set out in articles 116 and 117 of the 1982 UN Convention, or in conformity with national legislation within areas of national jurisdiction.

45. A flag State should ensure that each of the vessels entitled to fly its flag fishing in waters outside its sovereignty or jurisdiction holds a valid authorization to fish issued by that flag State. Where a coastal State issues an authorization to fish to a vessel, that coastal State should ensure that no fishing in its waters occurs without an authorization to fish issued by the flag State of the vessel.
46. Vessels should have an authorization to fish and where required carry it on board. Each State's authorization should include, but need not be limited to:

    46.1 the name of the vessel, and, where appropriate, the natural or legal person authorized to fish;
    46.2 the areas, scope and duration of the authorization to fish; and
    46.3 the species, fishing gear authorized, and where appropriate, other applicable management measures.

47. Conditions under which an authorization is issued may also include, where required:

    47.1 vessel monitoring systems;
    47.2 catch reporting conditions, such as:

        47.2.1 time series of catch and effort statistics by vessel;
        47.2.2 total catch in number, nominal weight, or both, by species (both target and non-target) as is appropriate to each fishery period (nominal weight is defined as the live weight equivalent of the catch);
        47.2.3 discard statistics, including estimates where necessary, reported as number or nominal weight by species, as is appropriate to each fishery;
        47.2.4 effort statistics appropriate to each fishing method; and
        47.2.5 fishing location, date and time fished and other statistics on fishing operations.

    47.3 reporting and other conditions for transshipping, where transshipping is permitted;
    47.4 observer coverage;
    47.5 maintenance of fishing and related log books;
    47.6 navigational equipment to ensure compliance with boundaries and in relation to restricted areas;
    47.7 compliance with applicable international conventions and national laws and regulations in relation to maritime safety, protection of the marine environment, and conservation and management measures or provisions adopted at a national, regional or global level;

47.8 marking of its fishing vessels in accordance with internationally recognized standards, such as the FAO Standard Specification and Guidelines for the Marking and Identification of Fishing Vessels. Vessels' fishing gear should similarly be marked in accordance with internationally recognized standards;

47.9 where appropriate, compliance with other aspects of fisheries arrangements applicable to the flag State; and

47.10 the vessel having a unique, internationally recognized identification number, wherever possible, that enables it to be identified regardless of changes in registration or name over time.

48. Flag States should ensure that their fishing, transport and support vessels do not support or engage in IUU fishing. To this end, flag States should ensure that none of their vessels re-supply fishing vessels engaged in such activities or tranship fish to or from these vessels. This paragraph is without prejudice to the taking of appropriate action, as necessary, for humanitarian purposes, including the safety of crew members.

49. Flag States should ensure that, to the greatest extent possible, all of their fishing, transport and support vessels involved in transshipment at sea have a prior authorization to tranship issued by the flag State, and report to the national fisheries administration or other designated institution:

49.1 the date and location of all of their transshipments of fish at sea;

49.2 the weight by species and catch area of the catch transshipped;

49.3 the name, registration, flag and other information related to the identification of the vessels involved in the transshipment; and

49.4 the port of landing of the transshipped catch.

50. Flag States should make information from catch and transshipment reports available, aggregated according to areas and species, in a full, timely and regular manner and, as appropriate, to relevant national, regional and international organizations, including FAO, taking into account applicable confidentiality requirements.

Coastal State Measures

51. In the exercise of the sovereign rights of coastal States for exploring and exploiting, conserving and managing the living marine resources under their jurisdiction, in conformity with the 1982 UN Convention and international law, each coastal State should implement measures to prevent, deter and eliminate IUU fishing in the exclusive economic zone. Among the measures which the coastal State should consider, consistent with national legislation and international law, and to the extent practicable and appropriate, are:

51.1 effective monitoring, control and surveillance of fishing activities in the exclusive economic zone;
51.2 cooperation and exchange of information with other States, where appropriate, including neighbouring coastal States and with regional fisheries management organizations;
51.3 to ensure that no vessel undertakes fishing activities within its waters without a valid authorization to fish issued by that coastal State;
51.4 to ensure that an authorization to fish is issued only if the vessel concerned is entered on a record of vessels;
51.5 to ensure that each vessel fishing in its waters maintains a logbook recording its fishing activities where appropriate;
51.6 to ensure that at-sea transshipment and processing of fish and fish products in coastal State waters are authorized by that coastal State, or conducted in conformity with appropriate management regulations;
51.7 regulation of fishing access to its waters in a manner which will help to prevent, deter and eliminate IUU fishing; and
51.8 avoiding licensing a vessel to fish in its waters if that particular vessel has a history of IUU fishing, taking into account the provisions of paragraph 36.

Port State Measures

52. States should use measures, in accordance with international law, for port State control of fishing vessels in order to prevent, deter and eliminate IUU fishing. Such measures should be implemented in a fair, transparent and non-discriminatory manner.
53. When used in paragraphs 52 to 64, port access means admission for foreign fishing vessels to ports or offshore terminals for the purpose of, *inter alia*, refuelling, re-supplying, transshipping and landing, without prejudice to the sovereignty of a coastal State in accordance with its national law and article 25.2 of the 1982 UN Convention and other relevant international law.
54. Notwithstanding paragraphs 52, 53 and 55; a vessel should be provided port access, in accordance with international law, for reasons of *force majeure* or distress or for rendering assistance to persons, ships or aircraft in danger or distress.
55. Prior to allowing a vessel port access, States should require fishing vessels and vessels involved in fishing related activities seeking permission to enter their ports to provide reasonable advance notice of their entry into port, a copy of their authorization to fish, details of their fishing trip and quantities of fish on board, with due regard to confidentiality requirements, in order to ascertain whether the vessel may have engaged in, or supported, IUU fishing.

56. Where a port State has clear evidence that a vessel having been granted access to its ports has engaged in IUU fishing activity, the port State should not allow the vessel to land or transship fish in its ports, and should report the matter to the flag State of the vessel.
57. States should publicize ports to which foreign flagged vessels may be permitted admission and should ensure that these ports have the capacity to conduct inspections.
58. In the exercise of their right to inspect fishing vessels, port States should collect the following information and remit it to the flag State and, where appropriate, the relevant regional fisheries management organization:

    58.1 the flag State of the vessel and identification details;
    58.2 name, nationality, and qualifications of the master and the fishing master;
    58.3 fishing gear;
    58.4 catch on board, including origin, species, form, and quantity;
    58.5 where appropriate, other information required by relevant regional fisheries management organizations or other international agreements; and
    58.6 total landed and transshipped catch.

59. If, in the course of an inspection, it is found that there are reasonable grounds to suspect that the vessel has engaged in or supported IUU fishing in areas beyond the jurisdiction of the port State, the port State should, in addition to any other actions it may take consistent with international law, immediately report the matter to the flag State of the vessel and, where appropriate, the relevant coastal States and regional fisheries management organization. The port State may take other action with the consent of, or upon the request of, the flag State.
60. In applying paragraphs 58 and 59, States should safeguard the confidentiality of information collected, in accordance with their national laws.
61. States should establish and publicize a national strategy and procedures for port State control of vessels involved in fishing and related activities, including training, technical support, qualification requirements and general operating guidelines for port State control officers. States should also consider capacity-building needs in the development and implementation of this strategy.
62. States should cooperate, as appropriate, bilaterally, multilaterally and within relevant regional fisheries management organizations, to develop compatible measures for port State control of fishing vessels. Such measures should deal with the information to be collected by port States, procedures for information collection, and measures for dealing with suspected infringements by the vessel of measures adopted under these national, regional or international systems.

63. States should consider developing within relevant regional fisheries management organizations port State measures building on the presumption that fishing vessels entitled to fly the flag of States not parties to a regional fisheries management organization and which have not agreed to cooperate with that regional fisheries management organization, which are identified as being engaged in fishing activities in the area of that particular organization, may be engaging in IUU fishing. Such port State measures may prohibit landings and transshipment of catch unless the identified vessel can establish that the catch was taken in a manner consistent with those conservation and management measures. The identification of the vessels by the regional fisheries management organization should be made through agreed procedures in a fair, transparent and non-discriminatory manner.
64. States should enhance cooperation, including by the flow of relevant information, among and between relevant regional fisheries management organizations and States on port State controls.

Internationally Agreed Market–Related Measures

65. The measures in paragraphs 66 to 76 are to be implemented in a manner which recognizes the right of States to trade in fish and fishery products harvested in a sustainable manner and should be interpreted and applied in accordance with the principles, rights and obligations established in the World Trade Organisation, and implemented in a fair, transparent and non-discriminatory manner.
66. States should take all steps necessary, consistent with international law, to prevent fish caught by vessels identified by the relevant regional fisheries management organization to have been engaged in IUU fishing being traded or imported into their territories. The identification of the vessels by the regional fisheries management organization should be made through agreed procedures in a fair, transparent and non-discriminatory manner. Trade-related measures should be adopted and implemented in accordance with international law, including principles, rights and obligations established in WTO Agreements, and implemented in a fair, transparent and non-discriminatory manner. Trade-related measures should only be used in exceptional circumstances, where other measures have proven unsuccessful to prevent, deter and eliminate IUU fishing, and only after prior consultation with interested States. Unilateral trade-related measures should be avoided.
67. States should ensure that measures on international trade in fish and fishery products are transparent, based on scientific evidence, where applicable, and are in accordance with internationally agreed rules.
68. States should cooperate, including through relevant global and regional fisheries management organizations, to adopt appropriate multilaterally agreed

trade-related measures, consistent with the WTO, that may be necessary to prevent, deter and eliminate IUU fishing for specific fish stocks or species. Multilateral trade-related measures envisaged in regional fisheries management organizations may be used to support cooperative efforts to ensure that trade in specific fish and fish products does not in any way encourage IUU fishing or otherwise undermine the effectiveness of conservation and management measures which are consistent with the 1982 UN Convention.

69. Trade-related measures to reduce or eliminate trade in fish and fish products derived from IUU fishing could include the adoption of multilateral catch documentation and certification requirements, as well as other appropriate multilaterally-agreed measures such as import and export controls or prohibitions. Such measures should be adopted in a fair, transparent and non-discriminatory manner. When such measures are adopted, States should support their consistent and effective implementation.
70. Stock or species-specific trade-related measures may be necessary to reduce or eliminate the economic incentive for vessels to engage in IUU fishing.
71. States should take steps to improve the transparency of their markets to allow the traceability of fish or fish products.
72. States, when requested by an interested State, should assist any State in deterring trade in fish and fish products illegally harvested in its jurisdiction. Assistance should be given in accordance with terms agreed by both States and fully respecting the jurisdiction of the State requesting assistance.
73. States should take measures to ensure that their importers, transshippers, buyers, consumers, equipment suppliers, bankers, insurers, other services suppliers and the public are aware of the detrimental effects of doing business with vessels identified as engaged in IUU fishing, whether by the State under whose jurisdiction the vessel is operating or by the relevant regional fisheries management organizations in accordance with its agreed procedures, and should consider measures to deter such business. Such measures could include, to the extent possible under national law, legislation that makes it a violation to conduct such business or to trade in fish or fish products derived from IUU fishing. All identifications of vessels engaged in IUU fishing should be made in a fair, transparent and non-discriminatory manner.
74. States should take measures to ensure that their fishers are aware of the detrimental effects of doing business with importers, transshippers, buyers, consumers, equipment suppliers, bankers, insurers and other services suppliers identified as doing business with vessels identified as engaged in IUU fishing, whether by the State under whose jurisdiction the vessel is operating or by the relevant regional fisheries management organization in accordance with its agreed procedures, and should consider measures to deter such business. Such measures could include, to the extent possible under national law,

legislation that makes it a violation to conduct such business or to trade in fish or fish products derived from IUU fishing. All identifications of vessels engaged in IUU fishing should be made in a fair, transparent and non-discriminatory manner.

75. States should work towards using the Harmonized Commodity Description and Coding System for fish and fisheries products in order to help promote the implementation of the IPOA.
76. Certification and documentation requirements should be standardized to the extent feasible, and electronic schemes developed where possible, to ensure their effectiveness, reduce opportunities for fraud, and avoid unnecessary burdens on trade.

Research

77. States should encourage scientific research on methods of identifying fish species from samples of processed products. FAO should facilitate the establishment of a network of databases of genetic and other markers used to identify fish species from processed product, including the ability to identify the stock of origin where possible.

Regional Fisheries Management Organizations

78. States should ensure compliance with and enforcement of policies and measures having a bearing on IUU fishing which are adopted by any relevant regional fisheries management organization and by which they are bound. States should cooperate in the establishment of such organizations in regions where none currently exist.
79. As the cooperation of all relevant States is important for the success of measures taken by relevant regional fisheries management organizations to prevent, deter and eliminate IUU fishing, States which are not members of a relevant regional fisheries management organization are not discharged from their obligation to cooperate, in accordance with their international obligations, with that regional fisheries management organization. To that end, States should give effect to their duty to cooperate by agreeing to apply the conservation and management measures established by that regional fisheries management organization, or by adopting measures consistent with those conservation and management measures, and should ensure that vessels entitled to fly their flag do not undermine such measures.
80. States, acting through relevant regional fisheries management organizations, should take action to strengthen and develop innovative ways, in conformity with international law, to prevent, deter, and eliminate IUU fishing. Consideration should be given to including the following measures:

80.1 institutional strengthening, as appropriate, of relevant regional fisheries management organizations with a view to enhancing their capacity to prevent, deter and eliminate IUU fishing;

80.2 development of compliance measures in conformity with international law;

80.3 development and implementation of comprehensive arrangements for mandatory reporting;

80.4 establishment of and cooperation in the exchange of information on vessels engaged in or supporting IUU fishing;

80.5 development and maintenance of records of vessels fishing in the area of competence of a relevant regional fisheries management organization, including both those authorized to fish and those engaged in or supporting IUU fishing;

80.6 development of methods of compiling and using trade information to monitor IUU fishing;

80.7 development of MCS, including promoting for implementation by its members in their respective jurisdictions, unless otherwise provided for in an international agreement, real time catch and vessel monitoring systems, other new technologies, monitoring of landings, port control, and inspections and regulation of transshipment, as appropriate;

80.8 development within a regional fisheries management organization, where appropriate, of boarding and inspection regimes consistent with international law, recognizing the rights and obligations of masters and inspection officers;

80.9 development of observer programmes;

80.10 where appropriate, market-related measures in accordance with the IPOA;

80.11 definition of circumstances in which vessels will be presumed to have engaged in or to have supported IUU fishing;

80.12 development of education and public awareness programmes;

80.13 development of action plans; and

80.14 where agreed by their members, examination of chartering arrangements, if there is concern that these may result in IUU fishing.

81. States, acting through relevant regional fisheries management organizations, should compile and make available on a timely basis, and at least on an annual basis, to other regional fisheries management organizations and to FAO, information relevant to the prevention, deterrence and elimination of IUU fishing, including:

81.1 estimates of the extent, magnitude and character of IUU activities in the area of competence of the regional fisheries management organization;

81.2 details of measures taken to deter, prevent and eliminate IUU fishing;
81.3 records of vessels authorized to fish, as appropriate; and
81.4 records of vessels engaged in IUU fishing.

82. Objectives of institutional and policy strengthening in relevant regional fisheries management organizations in relation to IUU fishing should include enabling regional fisheries management organizations to:

82.1 determine policy objectives regarding IUU fishing, both for internal purposes and co-ordination with other regional fisheries management organizations;
82.2 strengthen institutional mechanisms as appropriate, including mandate, functions, finance, decision making, reporting or information requirements and enforcement schemes, for the optimum implementation of policies in relation to IUU fishing;
82.3 regularize coordination with institutional mechanisms of other regional fisheries management organizations as far as possible in relation to IUU fishing, in particular information, enforcement and trade aspects; and
82.4 ensure timely and effective implementation of policies and measures internally, and in cooperation with other regional fisheries management organizations and relevant regional and international organizations.

83. States, acting through relevant regional fisheries management organizations, should encourage non-contracting parties with a real interest in the fishery concerned to join those organizations and to participate fully in their work. Where this is not possible, the regional fisheries management organizations should encourage and facilitate the participation and cooperation of non-contracting parties, in accordance with applicable international agreements and international law, in the conservation and management of the relevant fisheries resources and in the implementation of measures adopted by the relevant organizations. Regional fisheries management organizations should address the issue of access to the resource in order to foster cooperation and enhance sustainability in the fishery, in accordance with international law. States, acting through relevant regional fisheries management organizations, should also assist, as necessary, non-contracting parties in the implementation of paragraphs 78 and 79 of the IPOA.
84. When a State fails to ensure that fishing vessels entitled to fly its flag, or, to the greatest extent possible, its nationals, do not engage in IUU fishing activities that affect the fish stocks covered by a relevant regional fisheries management organization, the member States, acting through the organization, should draw the problem to the attention of that State. If the problem is not

rectified, members of the organization may agree to adopt appropriate measures, through agreed procedures, in accordance with international law.

V. Special Requirements of Developing Countries

85. States, with the support of FAO and relevant international financial institutions and mechanisms, where appropriate, should cooperate to support training and capacity building and consider providing financial, technical and other assistance to developing countries, including in particular the least developed among them and small island developing States, so that they can more fully meet their commitments under the IPOA and obligations under international law, including their duties as flag States and port States. Such assistance should be directed in particular to help such States in the development and implementation of national plans of action in accordance with paragraph 25.
86. States, with the support of FAO and relevant international financial institutions and mechanisms, where appropriate, should cooperate to enable:
    - 86.1 review and revision of national legislation and regional regulatory frameworks;
    - 86.2 the improvement and harmonization of fisheries and related data collection;
    - 86.3 the strengthening of regional institutions; and
    - 86.4 the strengthening and enhancement of integrated MCS systems, including satellite monitoring systems.

VI. Reporting

87. States and regional fisheries management organizations should report to FAO on progress with the elaboration and implementation of their plans to prevent, deter and eliminate IUU fishing as part of their biennial reporting to FAO on the Code of Conduct. These reports should be published by FAO in a timely manner.

VII. Role of FAO

88. FAO will, as and to the extent directed by its Conference, collect all relevant information and data that might serve as a basis for further analysis aimed at identifying factors and causes contributing to IUU fishing such as, inter alia, a lack of input and output management controls, unsustainable fishery management methods and subsidies that contribute to IUU fishing.

89. FAO will, as and to the extent directed by its Conference, support development and implementation of national and regional plans to prevent, deter and eliminate IUU fishing through specific, in-country technical assistance projects with Regular Programme funds and through the use of extra-budgetary funds made available to the Organization for this purpose.
90. FAO should, in collaboration with other relevant international organizations, in particular IMO, further investigate the issue of IUU fishing.
91. FAO should convene an Expert Consultation on the implementation of paragraph 76 of the IPOA.
92. FAO should investigate the benefits of establishing and maintaining regional and global databases, including but not limited to, information as provided for in Article VI of the 1993 FAO Compliance Agreement.
93. The FAO Committee on Fisheries will, based on a detailed analysis by the Secretariat, biennially evaluate the progress towards the implementation of the IPOA.

Bibliography

1. Books, Book Chapters, and Monographs

Agnew, David J. and Barnes, Colin T. 'Economic Aspects and Drivers of IUU Fishing: Building a Framework'. In OECD. *Fish Piracy: Combating Illegal, Unreported and Unregulated Fishing* (Paris: OECD, 2004) 169–200.

Attard, David. *The Exclusive Economic Zone in International Law* (Oxford: Clarendon Press, 1987).

Baird, Rachel J. *Aspects of Illegal, Unreported and Unregulated Fishing in the Southern Ocean.* Review: Method and Technologies in Fish Biology and Fisheries Volume 5 (Dordrecht: Springer, 2006).

Balton, David. 'Dealing with the "Bad Actors" of Ocean Fisheries'. In OECD. *Fish Piracy: Combating Illegal, Unreported and Unregulated Fishing* (Paris: OECD, 2004) 57–63.

Balton, David. 'The Compliance Agreement'. In Hey, Ellen (ed.) *Developments in International Fisheries Law* (Hague: Kluwer Law International, 1999) 31–53.

Barrett, J.H., Locker, A.M. and Roberts, C.M. 'The Origins of Intensive Marine Fishing in Medieval Europe: the English Evidence'. 271 *Proceedings of the Royal Society B* (2004) 2417–2421. Cited in Roberts, Callum. *The Unnatural History of the Sea* (Washington: Island Press, 2007).

Barrett, J.H., Locker, A.M. and Roberts, C.M. 'Dark Age Economic Revisited: The English Fish Bone Evidence AD 600–1600'. 78 *Antiquity* (2004) 618–636. Cited in Roberts, Callum. *The Unnatural History of the Sea* (Washington: Island Press, 2007).

Barrett, J.H. 'Archaeo-ichthyological Evidence for Long-term Socio-economic Trends in Northern Scotland: 3500 BC to AD 1500'. 26 *Journal of Archaeological Science* (1999) 353–388. Cited in Roberts, Callum. *The Unnatural History of the Sea* (Washington: Island Press, 2007).

Benham, Awni. 'Ending Flag State Control'. In Kirchner, Andree (ed.) *International Marine Environment Law: Institutions, Implementation and Innovations* (Hague: Kluwer Law International, 2003) 123–135.

Berrill, Michael. *The Plundered Seas: Can the World's Fish Be Saved?* (San Francisco: Sierra Club Books, 1997).

Benitah, Marc. *The Law of Subsidies under the GATT/WTO System* (Hague: Kluwer Law International, 2001).

Birnie, Patricia and Boyle, Alan. *International Law and the Environment.* Third Edition (Oxford: Oxford University Press, 2002).

Boyle, Alan and Freestone, David (eds.). *International Law and Sustainable Development* (Oxford: Oxford University Press, 1999).

Brownlie, Ian. *Principles of Public International Law* (Oxford: Clarendon Press, 1998).

Burke, Lauretta; Kura, Yumiko; *et al. Pilot Analysis of Global Ecosystems: Coastal Ecosystems* (Washington, D.C.: World Resources Institute, 2001).

Burke, Laureta; Selig, Elizabeth; and Spalding, Mark. *Reefs at Risk in Southeast Asia* (U.S.: World Resource Institute, 2002).

Burnett, Mark; Dronova, Natalie; Esmark, Maren; Nelson, Steve; Rønning, Asle; and Spiridonov, Vassily. *Illegal Fishing in Arctic Waters: Catch of Today-Gone Tomorrow?* (Oslo: WWF International Arctic, 2008).

Churchill, R.R. and Lowe, A.V. *The Law of the Sea* (Manchester: Manchester University Press, 1999).

Churchill, R.R. *EEC Fisheries Law* (Dordrecht: Martinus Nijhoff Publishers, 1987).

Croxall, John P. and Trathan, Phil N. 'The Southern Ocean: A Model System for Conserving Resources'. In Glover, Linda K. and Earle, Sylvia A. (eds.) *Defying Ocean's End: An Agenda for Action* (Washington: Island Press, 2004).

Cushing, D.H. *The Provident Sea* (Cambridge: Cambridge University Press, 1988).

Dahmani, M. *The Fisheries Regime of the Exclusive Economic Zone* (Dordrecht: Martinus Nijhoff Publishers, 1987).

Douglas, N. (ed.) *Pacific Islands Yearbook*, 17th Edition (Suva: Fiji Times Ltd., 1994).

Edeson, William; Freestone, David; and Gudmundsdottir, Elly. *Legislating for Sustainable Fisheries: A Guide to Implementing the 1993 FAO Compliance Agreement and 1995 UN Fish Stocks Agreement* (Washington, D.C.: The World Bank, 2001).

Edeson, William. 'Towards Long-term Sustainable Use: Some Recent Developments in the Legal Regime of Fisheries'. In Boyle, Alan and Freestone, David (eds.) *International Law and Sustainable Development: Past Achievements and Future Challenges* (Oxford: Oxford University Press, 1999) 165–204.

Environmental Investigation Agency. *Environmental Crime: A Threat to Our Future* (London: Emmerson Press, 2008).

Gianni, Matthew. 'IUU Fishing and the Cost to Flag of Convenience Countries'. In OECD. *Fish Piracy: Combating Illegal, Unreported and Unregulated Fishing* (Paris: OECD, 2004) 285–288.

Goñi, Raquel. 'Fisheries Effects on Ecosystems'. In Sheppard, Charles R.C. (ed.) *Seas at the Millennium: An Environmental Evaluation. Vol. III: Global Issues and Processes* (Amsterdam: Pergamon Elsevier Science, 2000).

Hafner, Gerhard. 'General Principles of Sustainable Development: From Soft Law to Hard Law'. In Fitzmaurice, Malgosia and Szuniewicz, Milena (eds.) *Exploitation of Natural Resources in the 21st Century* (Hague: Kluwer Law International, 2003) 53–65.

Haward, Marcus and Bergin, Anthony. 'The Political Economy of Japanese Distant Water Tuna Fisheries'. 25 *Marine Policy* (2001) 91–101.

Hawthorne, Daniel and Minot, Francis. *The Inexhaustible Sea* (New York: Collier Books, 1961).

Henie, Günter; Prabhu, Mohan; and Alvazzi del Frate, Anna (eds.) Environmental Protection-Potentials and Limits of Criminal Justice: Evaluation of Legal Structures. *United Nations Interregional Crime and Justice Research Institute Publication No. 56* (Freiburg im Breisgau: Ed. Iuscrim, Max Planck Institut für Ausländisches und Internationales Strafrecht, 1997).

Henriksen, Tore; Hønneland, Geir; and Sydnes, Are. *Law and Politics in Ocean Governance: The UN Fish Stocks Agreement and Regional Fisheries Management Regimes* (Leiden: Martinus Nijhoff Publishers, 2006).

Hey, Ellen (ed.). *Developments in International Fisheries Law* (Hague: Kluwer Law International, 1999).

Hey, Ellen. 'The Fisheries Provisions of the LOS Convention'. In Hey, Ellen (ed.). *Developments in International Fisheries Law* (Hague: Kluwer Law International, 1999) 13–53.

Huxley, Thomax. 'The Fisheries Exhibition'. 28 *Nature* (1883). Cited in Ellis, Richard. *The Empty Ocean: Plundering the World's Marine Life* (Washington: Island Press, 2003).

Islam, M. Rafiquil. *International Trade Law of the WTO* (Oxford: Oxford University Press).

Johnston, Douglas M. *The International Law of Fisheries: A Policy-Oriented Inquiry* (New Haven: New Haven Press, 1987).

Charney, Jonathan and Alexander, Lewis (eds.). *International Maritime Boundaries* (Dordrecht: Martinus Nijhoff Publishers, 1993).

Deere, Carolyn. *Eco-labelling and Sustainable Fisheries* (FAO: Rome, 1999).

Detter, Ingrid. *The International Legal Order* (Hants: Dartmouth Publishing Company, 1994).

Grief, Nicholas. *Public International Law in the Airspace of the High Seas* (Dordrecht: Martinus Nijhoff Publishers, 1994).

Grotius, Hugo. *The Freedom of the Seas or The Right which Belongs to the Dutch to Take Part in the East Indian Trade* (New York: Oxford University Press, 1916).

Hall, Stephen J. *The Effects of Fishing on Marine Ecosystems and Communities* (Oxford: Blackwell Science Ltd., 1999).

Hatcher, Aaron. 'Incentives for Investments in IUU Fishing Capacity'. In OECD. *Fish Piracy: Combating Illegal, Unreported and Unregulated Fishing* (Paris: OECD, 2004) 239–254.

Hoydal, Kjartan. 'IUU Fishing in NEAFC: How Big is the Problem and What Have We Done?' In OECD. *Fish Piracy: Combating Illegal, Unreported and Unregulated Fishing* (Paris: OECD, 2004) 159–165.

Jenkins, Robert W.G. 'The Significant Trade Process: Making Appendix II Work'. In Hutton, Jon and Dickson, Barnabas (eds.) *Endangered Species Threatened Convention: The Past, Present and Future of CITES* (London: Earthscan Publications Ltd, 2000) 45–54.

Juda, Lawrence. *International Law and Ocean Use Management: the Evolution of Ocean Governance* (New York: Routledge, 1996).

Kasoulides, George C. *Port State Control and Jurisdiction: Evolution of the Port State Regime* (Netherlands: Martinus Nijhoff Publishers, 1993).

Kaye, Stuart. 'CCAMLR and Southern Ocean Fisheries'. In Bateman, Sam and Rothwell, Donald (eds.) *Southern Ocean Fishing: Policy Challenges for Australia* (Wollongong: Centre for Maritime Policy, 1998) 75–88.

Kaye, Stuart M. *International Fisheries Management* (Hague: Kluwer Law International, 2001).

Knight, Gary and Chiu, Hungdah. *The International Law of the Sea: Cases, Documents and Readings* (London: Elsevier Applied Science and UNIFO Publishers, Inc., 1991).

Kwiatkowska, Barbara. *The 200 Mile Exclusive Economic Zone in the New Law of the Sea* (Dordrecht: Martinus Nijhoff Publishers, 1989).

Lévy, Jean-Pierre and Schram, Gunnar G. *United Nations Conference on Straddling Fish Stocks and Highly Migratory Fish Stocks: Selected Documents* (Hague: Martinus Nijhoff Publishers, 1996).

Mansell, John N.K. *Flag State Responsibility: Historical Development and Contemporary Issues* (Heiderlberg: Springer, 2009).

Martin, Will; Lodge, Michael; Caddy, John; and Mfodwo, Kwame. *A Handbook for Negotiating Fishing Access Agreements* (Washington: WWF, 2001).

Matsushita, Mitsuo; Schoenbaum, Thomas J.; and Mavroidis, Petros C. *The World Trade Organization: Law, Practice and Policy* (Oxford: Oxford University Press, 2006).

Meere, Frank and Lack, Mary. *Assessment of Impacts of Illegal, Unreported and Unregulated (IUU) Fishing in the Asia Pacific* (Singapore: APEC Secretariat, 2008).

Miller, Denzil G.M. 'Patagonian Toothfish-The Storm Gathers'. In OECD. *Fish Piracy: Combating Illegal, Unreported and Unregulated Fishing* (Paris: OECD, 2004) 105–146.

Moore, Gerald. 'The Code of Conduct for Responsible Fisheries'. In Hey, Ellen (ed.) *Developments in International Fisheries Law* (Hague: Kluwer Law International, 1999) 85–105.

Mortimer, Jeanne A.; Donnelly, Marydele; and Plotkin, Pamela T. 'Sea Turtles'. In Sheppard, Charles R.C. (ed.) *Seas at the Millennium: An Environmental Evaluation. Vol. III: Global Issues and Processes* (Amsterdam: Pergamon Elsevier Science, 2000).

Nelson, Dolliver. 'The Development of the Legal Regime of High Seas Fisheries.' In Boyle, Alan and Freestone, David (eds.) *International Law and Sustainable Development* (Oxford: Oxford University Press, 1999) 113–134.

O'Connell, D.P. *The International Law of the Sea*, Vol. 1 (New York: Oxford University Press, 1983).

Oda, Shigeru. *International Control of Sea Resources* (Dordrecht: Martinus Nijhoff Publishers, 1989).

Organisation for Economic Co-operation and Development and Food and Agriculture Organisation (OECD/FAO). *Globalisation and Fisheries: Proceedings of an OECD-FAO Workshop* (OECD/FAO, 2007).

Palma, Mary Ann, and Tsamenyi, Martin. *Case Study on the Impacts of Illegal, Unreported and Unregulated Fishing in the Sulawesi Sea*. APEC Fisheries Working Group. FWG 02/2007. (Singapore: APEC Secretariat, April 2008).

Palma, Mary Ann. 'Chapter 3: Combating IUU Fishing: International Legal Developments'. In Hanich, Q. and Tsamenyi, M. (eds.) *Navigating Pacific Fisheries: Legal and Policy Trends in the Implementation of International Fisheries Instruments in the Western and Central Pacific Fisheries Region* (Wollongong: University of Wollongong Press, 2009) 71–103.

Pamborides, G.P. *International Shipping Law: Legislation and Enforcement* (Athens: Ant. N. Sakkoulas Publishers, 1999).

Poulantzas, Nicholas M. *The Right of Hot Pursuit in International Law* (Hague: Martinus Nijhoff Publishers, 2004).

Restrepo, Victor. 'Estimation of Unreported Catches by ICCAT'. In OECD. *Fish Piracy: Combating Illegal, Unreported and Unregulated Fishing* (Paris: OECD, 2004) 155–158.

Rigg, Kelly; Parmentier, Rémi; and Currie, Duncan. 'Halting IUU Fishing: Enforcing International Fisheries Agreements'. In OECD. *Fish Piracy: Combating Illegal, Unreported and Unregulated Fishing* (Paris: OECD, 2004) 369–399.

Roberts, Callum. *The Unnatural History of the Sea* (Washington: Island Press, 2007).

Robinson, Nicholas A. (ed). *Agenda 21 and the UNCED Proceedings*. Third Series: International Protection of the Environment (New York: Oceana Publications, Inc, 1992) 6 Volumes.

Roelofsen, C.G. 'The Sources of Mare Liberum; The Contested Origins of the Doctrine of the Freedom of the Seas.' In Heere, Wyobo P. (ed.) *International Law and its Sources: Liber Amirocum Maarten Bos* (Hague: T.M.C. Asser Institut, 1989) 93–123.

Rood, John and Schechter, Michael G. 'Globalisation and Fisheries: A Necessarily Interdisciplinary Inquiry'. In Taylor, William W., Schechter, Michael G. and Wolfson, Lois G. (eds.) *Globalisation: Effects on Fisheries Resources* (Cambridge: Cambridge University Press, 2007) 1–18.

Rozwadowski, Helen M. *The Sea Knows No Boundaries: A Century of Marine Science Under ICES* (Seattle: ICES and the University of Washington Press, 2002).

Sand, Peter H. *Transnational Environmental Law: Lessons in Global Change* (Hague: Kluwer Law International, 1999).

Sea Resources Management Sdn Bhd. *Case Study on Illegal, Unreported and Unregulated (IUU) Fishing off the East Coast of Peninsular Malaysia: Final Report* (Singapore: APEC Secretariat, 2008).

Sumaila, U.R; Alder, J.; Keith, H. 'The Cost of Being Apprehended Fishing Illegally: Empirical Evidences and Policy Implications'. In OECD. *Fish Piracy: Combating Illegal, Unreported and Unregulated Fishing* (Paris: OECD, 2004) 201–229.

Tsamenyi, Martin, Palma, M.A., Milligan, B. and Mfodwo, K. 'Fairer Fishing?: The Impact on Developing Countries of the European Community Regulation on Illegal, Unreported and Unregulated Fisheries'. *Economic Paper Series No. 86* (London: Commonwealth Secretariat, 2009).

United Nations Environment Programme, Division of Technology, Industry and Economics, *Certification and Sustainable Fisheries* (Switzerland: UNEP, 2009).

Vicuña, Francisco Orrego. *The Changing International Law on High Seas Fisheries* (Cambridge: University Press, 1999).

Wessells, Cathy. 'Barriers to International Trade in Fisheries'. Discussion Paper Prepared for the First FAO E-mail Conference on Fish Trade and Food Security, October-November 1998. Cited in Deere, Carolyn. *Eco-labelling and Sustainable Fisheries* (FAO: Rome, 1999).

Whitlow, Jon. 'The Social Dimension of IUU Fishing'. In OECD. *Fish Piracy: Combating Illegal, Unreported and Unregulated Fishing* (Paris: OECD, 2004) 231–238.

Williams, Meryl J. 'Enmeshed: Australia and Southeast Asia's Fisheries'. *Lowy Institute Paper 20* (Double Bay: Lowy Institute for International Policy, 2007).

Wilson, Douglas C. 'Fisheries Management as a Social Problem'. In Sheppard, Charles R.C. (ed.) *Seas at the Millennium: An Environmental Evaluation. Vol. III: Global Issues and Processes* (Amsterdam: Pergamon Elsevier Science, 2000).

Winters, L. Alan. 'EU's Preferential Trade Agreements: Objectives and Outcomes'. In Van Dijck, P. and Faber, G. (eds.) *The External Economic Dimension of the European Union* (the Hague: Kluwer Academic Publishers, 2000) 195–222.

Wolfrum, Rüdiger. 'The Protection and Management of Biological Diversity,' In Fred L Morrison and Rüdiger Wolfrum (eds.) *International, Regional and National Environmental Law* (Hague: Kluwer Law International, 2000).

World Wildlife Fund for Nature International Endangered Seas Campaign. *The Footprint of Distant Water Fleets on World Fisheries* (Surrey: WWF, 1998).

2. Journal Articles

Agnew, D.J. 'The Illegal and Unregulated Fishery for Toothfish in the Southern Ocean, and the CCAMLR Catch Documentation Scheme'. 24 *Marine Policy* (2000) 361–374.

Ainsworth, C.H. and Pitcher, T.J. 'Estimating Illegal, Unreported and Unregulated Catch in British Columbia's Marine Fisheries'. 75 *Fisheries Research* 75 (2005) 40–55.

Alcala, A.C. and Russ, G.R. 'Status of Philippine Coral Reef Fisheries'. 15 *Asian Fisheries Science* (2002) 177–192.

Allen, Craig. 'Doctrine of Hot Pursuit: A Functional Interpretation Adaptable to Emerging Maritime Law Enforcement Technologies and Practices'. 20 *Ocean Development and International Law* (1989) 309–342.

Aqorau, Transform. 'Illegal Fishing and Fisheries Law Enforcement in Small Island Developing States: The Pacific Islands Experience'. 15 *The International Journal of Marine and Coastal Law* (2000) 37–64.

Atta-Mills, John; Alder, Jackie; and Sumaila, Ussif Rashid. 'The Decline of a Regional Fishing Nation: The Case of Ghana and West Africa'. 28 *Natural Resources Forum* (2004) 13–21.

Baird, Rachel. 'CCAMLR Initiatives to Counter Flag State Non-enforcement in Southern Ocean Fisheries'. 36 *Victoria University of Wellington Law Review* (2005) 733–755.

Baird, Rachel. 'Illegal, Unreported and Unregulated Fishing: An Analysis of the Legal, Economic and Historical Factors Relevant to its Development and Persistence'. 5 *Melbourne Journal of International Law* (2004) 299–334.

Birnie, Patricia. 'New Approaches to Ensuring Compliance At Sea: The FAO Agreement to Promote Compliance with International Conservation and Management Measures by Fishing Vessels on the High Seas'. 8 *RECIEL* (1999) 48–55.

Brown, Chester. ''Reasonableness' in the Law of the Sea: The Prompt Release of the *Volga*'. 16 *Leiden Journal of International Law* (2003) 621–630.

Caddy, J.F. and Cochrane, K.L. 'A Review of Fisheries Management Past and Present and Some Future Perspectives for the Third Millennium'. 44 *Ocean and Coastal Management* (2001) 653–682.

Chaytor, Beatrice. 'An Examination of Trade and Environmental Policy Responses to the Challenge of Fisheries Subsidies'. 8 *Review of European Community & International Environmental Law* (1999) 275–283.

Edeson, William. 'Closing the Gap: The Role of "Soft" International Instruments to Control Fishing'. 20 *Australian Yearbook of International Law* (1999) 83–104.

Edeson, William. 'The International Plan of Action on Illegal, Unreported and Unregulated Fishing: The Legal Context of a Non-legally Binding Instrument'. 16 *International Journal of Marine and Coastal Law* (2001) 603–623.

Erceg, Diane. 'Deterring IUU Fishing Through State Control Over Nationals'. 30 *Marine Policy* (2006) 173–179.

Doulman, David and Terawasi, Peter. 'The South Pacific Regional Register of Foreign Fishing Vessels'. 14 *Marine Policy* (1990) 324–332.

Gullett, Warwick and Schofield, Clive. 'Pushing the Limits of the Law of the Sea Convention: Australian and French Cooperative Surveillance and Enforcement in the Southern Ocean'. 22 *The International Journal of Marine and Coastal Law* (2007) 545–583.

Gullett, Warwick. 'Taiwan's Engagement in the Developing International Legal Regime for Fishing: Implications for Taiwanese Officials and Fishers'. 3 *Taiwan Ocean Law Review* (2004) 163–216.

Haward, Marcus and Bergin, Anthony. 'Taiwan's Distant Water Tuna Fisheries'. 24 *Marine Policy* (2000) 33–43.

Henriksen, Tore. 'Revisiting the Freedom of Fishing and Legal Obligations on States Not Party to Regional Fisheries Management Organisations'. 40 *Ocean Development and International Law* (2009) 80–96.

'Illegal fishing is Destroying Stocks in Bering Sea'. 44 *Marine Pollution Bulletin* (2002) 92.

Johnson, Constance. 'Fisheries Enforcement in European Community Waters since 2002—Developments in Non-flag Enforcement'. 23 *The International Journal of Marine and Coastal Law* (2008) 249–270.

Kaczynski, Vlad M. and Fluharty, David L. 'European Policies in West Africa: Who Benefits from Fisheries Agreements?' 26 *Marine Policy* (2002) 75–93.

Kourti, N.; Shepherd, I.; Greidanus, H.; *et al.* 'Integrating Remote Sensing in Fisheries Control'. 12 *Fisheries Management and Ecology* (2005) 295–307.

Kuperan, K. and Sutinen, Jon G. 'Blue Water Crime: Deterrence, Legitimacy, and Compliance in Fisheries'. 32 *Law and Society Review* (1998) 309–337.

Lodge, Michael. 'Minimum Terms and Conditions of Access: Responsible Fisheries Management Measures in the South Pacific Region'. 16 *Marine Policy* (1992) 277–305.

Lugten, Gail L. 'Current Legal Developments: Food and Agriculture Organization'. 23 *The International Journal of Marine and Coastal Law* (2008) 761–767.

Lugten, Gail. 'Current Legal Developments: Food and Agriculture Organization, The FAO Global Record of Fishing Vessels, Refrigerated Vessels and Fishing Support Vessels'. 23 *International Journal of Marine and Coastal Law* (2008) 761–767.

Lugten, Gail L. 'Soft Law with Hidden Teeth: The Case for a FAO International Plan of Action on Sea Turtles'. 9 *Journal of International Wildlife Law and Policy* (2006) 155–173.

McDorman, Ted L. 'Stateless Fishing Vessels, International Law, and the UN High Seas Fisheries Conference'. 25 *Journal of Maritime Law and Commerce* (1994) 531–556.

Molenaar, Erik Jaap. 'The Concept of 'Real Interest' and Other Aspects of Co-operation through Regional Fisheries Management Mechanisms'. 15 *The International Journal of Marine and Coastal Law* (2000) 475–530.

Molenaar, Erik Jaap. 'Multilateral Hot Pursuit and Illegal Fishing in the Southern Ocean: The Pursuits of the Viarsa 1 and the South Tomi'. 19 *The International Journal of Marine and Coastal Law* (2004) 30–37.

Molenaar, Erik Jaap. 'Participation, Allocation and Unregulated Fishing: The Practice of Regional Fisheries Management Organisations'. 18 *The International Journal of Marine and Coastal Law* (2003) 457–480.

Molenaar, Erik Jaap. 'Port State Jurisdiction: Toward Comprehensive, Mandatory and Global Coverage'. 38 *Ocean Development and International Law* (2007) 225–257.

Pauly, Daniel, Christensen, Villy, Guénette, Sylvie, Pitcher, Tony J, Sumaila, U. Rashid, Walters, Carl J., Watson, R., and Zeller, Dirk. 'Towards Sustainability in World Fisheries'. 418 *Nature* (2002) 689–695.

Phillips, J.C. 'The Exclusive Economic Zone as a Concept in International Law'. 26 *International and Comparative Law Quarterly* (1977) 585–618.

Pitcher, Tony J; Watson, Reg; Forrest, Robyn; Þór ValtÝsson, Hreiðar; and Guénette, Sylvie. 'Estimating Illegal and Unreported Catches from Marine Ecosystems: A Basis for Change'. 3 *Fish and Fisheries* (2002) 317–339.

Potts, Tavis, and Haward, Marcus. 'International Trade, Eco-labelling, and Sustainable Fisheries—Recent Issues, Concepts, and Practices'. 9 *Environment, Development and Sustainability* (2007) 91–106.

Rayfuse, Rosemary. 'Countermeasures and High Seas Fisheries Enforcement'. 51 *Netherlands International Law Review* (2004) 41–76.

Riddle, Kevin W. 'Illegal, Unreported, and Unregulated Fishing'. 37 *Ocean Development and International Law* (2006) 282.

Roheim, Cathy and Sutinen, Jon G. 'Trade and Marketplace Measures to Promote Sustainable Fishing Practices'. 3 *Natural Resources, International Trade and Sustainable Development.* Issue (2006).

Rothwell, Donald R. and Stephens, Tim. 'Illegal Southern Ocean Fishing and Prompt Release: Balancing Coastal and Flag State Rights and Interests'. 53 *International and Comparative Law Quarterly* (2004) 171–187.

Tsamenyi, M.; Palma, M.A.; Milligan, B.; and Mfodwo, K. 'The European Council Regulation on Illegal, Unreported and Unregulated Fishing: An International Fisheries Law Perspective'. 25 *The International Journal of Marine and Coastal Law* (2010) 5–31.

Tsamenyi, Martin. 'The Legal Substance and Status of Fishing Entities in International Law: A Note'. 37 *Ocean Development and International Law* (2006) 123–131.

Song, Yann-huei. 'The Efforts of ICCAT to Combat IUU Fishing: The Roles of Japan and Taiwan in Conserving and Managing Tuna Resources'. 24 *The International Journal of Marine and Coastal Law* (2009) 101–139.

Sumaila, U.R.; Alder, J.; and Keith, H. 'Global Scope and Economics of Illegal Fishing'. 30 *Marine Policy* (2006) 696–703.

Swan, Judith. 'Port State Measures to Combat IUU Fishing: International and Regional Developments'. 7 *Sustainable Development Law and Policy* (2006) 38–43.

Sydnes, Are K. 'Regional Fisheries Organizations: How and Why Organizational Diversity Matters'. 32 *Ocean Development and International Law* (2001) 349–372.

Vorbach, Joseph E. 'The Vital Role of Non-Flag State Actors in the Pursuit of Safer Shipping'. 32 *Ocean Development and International Law* (2001) 27–42.

Warner-Kramer, Deirdre M. and Canty, Krista. 'Stateless Fishing Vessels: The Current International Regime and a New Approach'. 5 *Ocean and Coastal Law Journal* (2000) 227–243.

White, Alan T.; Vogt, Helge P.; and Arin, Tijen. 'The Philippine Coral Reefs Under Threat: The Economic Losses Caused by Reef Destruction'. 40 *Marine Pollution Bulletin* (2000) 598–605.

3. International Policy Instruments and Standards

The 2005 Rome Declaration on Illegal, Unreported and Unregulated Fishing. Adopted by the FAO Ministerial Meeting on Fisheries. Rome, 12 March 2005.

Convention on Biological Diversity (CBD). Decision VII/5. *Marine and Coastal Biological Diversity*. Adopted at the Seventh Meeting of the Conference of Parties. Kuala Lumpur, 9–20 February 2004.

CBD. Decision VIII/15. *Framework for monitoring implementation of the achievement of the 2010 target and integration of targets into the thematic programmes of work*. Adopted at the Conference of Parties at its Eighth Meeting. Curitiba, 20–31 March 2006.

Convention on International Trade in Endangered Species of Wild Fauna and Flora (CITES). Criteria for amendment of Appendices I and II. *Conf. 9.24 (Rev.CoP14)*. 2007.

CITES. Cooperation and Synergy with the Convention on Biological Diversity. *Conf. 10.4 (Rev. 14)*. Adopted at the Tenth Meeting of the Conference of the Parties. Harare, Zimbabwe, 9–20 June 1997.

CITES. Cooperation between CITES and the Commission for the Conservation of Antarctic Marine Living Resources Regarding Trade in Toothfish. *Conf.12.4*. Adopted at the Twelfth Meeting of the Conference of the Parties. Santiago, Chile, 3–15 November 2002.

CITES. Introduction from the Sea. *Conf.14.6*. Adopted at the Fourteenth Meeting of the Conference of the Parties. Hague, Netherlands, 3–15 June 2007.

Convention on Migratory Species. By-catch. UNEP/CMS/Resolution 8.14. Adopted at the Conference of the Parties at its Eighth Meeting. Nairobi, 20–25 November 2005.

Declaration of the International Conference on Responsible Fishing. Cancun, Mexico, 8 May 1992.

Food and Agriculture Organization (FAO). Ministerial Conference on Fisheries. *The Rome Consensus on World Fisheries*. Rome, 14–15 March 1995.

FAO. *The Rome Declaration on the Implementation of the Code of Conduct for Responsible Fisheries*. Adopted by the FAO Ministerial Meeting on Fisheries, Rome, Italy, 10–11 March 1999.

FAO. *The Standard Specifications for the Marking and Identification of Fishing Vessels* (Rome: FAO, 1989).

FAO/WHO Food Standards. Codex Alimentarius. *Recommended International Code of Practice for Crabs*. CAC/RCP 28–1983.

FAO/WHO Food Standards. Codex Alimentarius. *Code of Practice for Fish and Fishery Products*. CAC/RCP 52–2003.

FAO/WHO Food Standards. Codex Alimentarius. *Codex Standard for Quick Frozen Finfish, Uneviscerated and Eviscerated*. Codex Stan 36-1981. Rev. 1–1995.

FAO/WHO Food Standards. Codex Alimentarius. *Recommended International Code of Practice – General Principles of Food Hygiene*. CAC/RCP 1–1969.

FAO/WHO Food Standards. Codex Alimentarius. *Recommended International Code of Practice for Smoked Fish*. CAC/RCP 25–1979.

International Labour Organization (ILO). *Recommendation Concerning the Work in the Fishing Sector (R199)*. Geneva, Switzerland, adopted on 14 June 2007.

ILO. *Work in Fishing Recommendation 2007 (No. 199)*. Adopted at the Ninety-sixth Session, Geneva, 14 June 2007.

International Maritime Organization (IMO). *Code of Safety for Fishermen and Fishing Vessels* (London: IMO, 2005).

IMO. *FAO/ILO/IMCO Voluntary Guidelines for the Design. Construction and Equipment of Small Fishing Vessels* (London: IMO, 2005).

IMO. *FAO/ILO/IMO Document for Guidance on Training and Certification of Fishing Vessel Personnel* (London: IMO, 2001).

IMO. Resolution 3 of the 1995 STCW-F Convention. In *FAO/ILO/IMO Document for Guidance on Training and Certification of Fishing Vessel Personnel*. 1995.

Joint FAO/WHO Food Standard Programme. Codex Alimentarious Commission. Standard for Crackers from Marine and Freshwater Fish, Crustacean and Molluscan Shellfish. *Codex Stan 222–2001*.

Lymer, D., Funge-Smith, S., Clausen, J., and Miao, W. Status and Potential of Fisheries and Aquaculture in Asia and the Pacific 2008. *RAP Publication 2008/15* (Bangkok: FAO-RAP, 2008).

Ramsar Convention. *Criteria for Identifying Wetlands of International Importance*. Adopted by the 7th (1999) and 9th (2005) Meetings of the Conference of the Contracting Parties.

Ramsar Convention. Resolution IX.4. *The Ramsar Convention and conservation, production and sustainable use of fisheries resources*. Kampala, Uganda, 8–15 November 2005.

Ramsar Convention. Resolution IX.5. *Synergies with other international organisations dealing with biological diversity; including collaboration on, and harmonisation of, national reporting among biodiversity-related conventions and agreements*. Adopted at the Ninth Conference of the Contracting Parties. Kampala, Uganda, 8–15 November 2005.

Reykjavik Declaration on Responsible Fisheries in the Marine Ecosystem. Reykjavik, Iceland, 13 November 2001.

United Nations Conference on Environment and Development. *Agenda 21*. Chapter 17. Protection of the oceans, all kinds of seas, including enclosed and semi-enclosed seas, and coastal areas and the protection, rational use and development of their living resources. Rio de Janeiro, Brazil, 03–14 June 1992.

United Nations Environment Programme. Convention on Migratory Species. By-catch. Adopted by the Conference of the Parties at its Eighth Meeting, Nairobi, 20–25 November 2005. *UNEP/CMS/Resolution 8.14*. 2005.

World Customs Organization. *HS Nomenclature 2007 Edition*. 2007.

World Summit on Sustainable Development (WSSD). *Johannesburg Declaration on Sustainable Development*. Adopted at the 17th Plenary Meeting. 4 September 2002.

WSSD. *Plan of Implementation of the World Summit on Sustainable Development*. Johannesburg, South Africa, August 2002.

4. United Nations Documents

International Law Commission. Responsibility of States for Internationally Wrongful Acts, Text adopted by the Commission at its fifty-third session, in 2001, and submitted to the General Assembly as part of the Commission's report covering the work of that session. Text reproduced as it appears in the annex to General Assembly resolution 56/83 of 12 December 2001, and corrected by document A/56/49(Vol.I)Corr.4. Available at http://untreaty.un.org/ilc/research.htm. Accessed 20 August 2009.

United Nations (UN). Conference of Parties to the United Nations Convention Against Transnational Organised Crime. Report of the Conference of Parties to the United Nations Convention Against Transnational Organised Crime on its Fourth Session. Vienna, 8–17 October 2008. *CTOC/COP/2008/19*. 1 December 2008.

UN. Economic and Social Council. Commission on Crime Prevention and Criminal Justice. Report on the First Session, 21–30 April 1992. Economic and Social Council Official Records 1992 Supplement No. 10. *E/1992/30, E/CN.15/1992/7*.

UN. Economic and Social Council. Substantive Session of 1993, 28 June to 30 July 1993. Report of the Commission on Crime Prevention and Criminal Justice on its Second Session. Vienna, 13–23 April 1993. *CE/1993/32, E/CN.15/1993/9*. 10 June 1993.

United Nations General Assembly (UNGA). 71st Plenary Meeting. Large-scale pelagic driftnet fishing and its impact on the living marine resources of the world's oceans and seas. *A/RES/45/197*. 21 December 1990.

UNGA. 79th Plenary Meeting. Large-scale pelagic driftnet fishing and its impact on the living marine resources of the world's oceans and seas. *A/RES/46/215*. 20 December 1991.

UNGA. 85th Plenary Meeting. Large-scale pelagic driftnet fishing and its impact on the living marine resources of the world's oceans and seas. *A/RES/44/225*. 22 December 1989.

UNGA. 92nd Plenary Meeting. Unauthorised Fishing in Zones of National Jurisdiction and Its Impact on the Living Marine Resources of the World's Oceans and Seas. *A/RES/49/116*. 19 December 1994.

UNGA. Fiftieth Session. Agenda Item 96(c). Large-scale pelagic drift-net fishing and its impact on the living marine resources of the world's oceans and seas; unauthorised fishing in zones of national jurisdiction and its impact on the living marine resources of the world's oceans and seas; and fisheries by-catch and discards and their impact on the sustainable use of the world's living marine resources. *A/RES/50/25*. 5 December 1995.

UNGA. Fifty-eighth Session. Agenda Item 52(b). Resolution adopted by the General Assembly, Sustainable fisheries, including through the 1995 Agreement for the Implementation of the Provisions of the United Nations Convention on the Law of the Sea of 10 December 1982 relating to the Conservation and Management of Straddling Fish Stocks and Highly Migratory Fish Stocks, and related instruments. *A/RES/58/14*. 21 January 2004.

UNGA. Fifty-eighth Session. Item 53 of the Preliminary List. Oceans and Law of the Sea. Report of the Secretary General. *A/58/65*. 03 March 2003.

UNGA. Fifty-eighth Session. Item 53(b) of the Provisional Agenda, Oceans and the Law of the Sea: Sustainable Fisheries, including through the 1995 Agreement for the Implementation of the Provisions of the United Nations Convention on the Law of the Sea of 10 December 1982 Relating to the Conservation and Management of Straddling Fish Stocks and Highly Migratory Fish Stocks, and Related Instruments. Report of the Secretary-General. *A/58/215*. 05 August 2003.

UNGA. Fifty-first Session. Agenda Item 24(c). Large-scale pelagic driftnet fishing and its impact on the living marine resources of the world's oceans and seas; unauthorised fishing in zones of national jurisdiction and its impact on the living marine resources of the world's oceans and seas; and fisheries by-catch and discards and their impact on the sustainable use of the world's living marine resources. *A/RES/51/36*. 9 December 1996.

UNGA. Fifty-fourth Session. Agenda Items 40(a) and (c). Oceans and the Law of the Sea; Law of the Sea; Results of the Review by the Commission on Sustainable Development of the Sectoral Theme of "Oceans and Seas", Oceans and the Law of the Sea. Report of the Secretary-General. *A/54/429*. 30 September 1999.

UNGA. Forty-fourth Session. Large-scale pelagic drift-net fishing and its impact on the living marine resources of the world's oceans and seas. *A/44/445*. 21 December 1993.

UNGA. Fifty-fifth Session. Agenda Item 34(b). Large-scale pelagic drift-net fishing and its impact on the living marine resources of the world's oceans and seas; unauthorised fishing in zones of national jurisdiction and its impact on the living marine resources of the world's oceans and seas; and fisheries by-catch and discards and their impact on the sustainable use of the world's living marine resources. *A/RES/55/8*. 30 October 2000.

UNGA. Fifty-first Session. Agenda Item 24(c). Law of the sea: large-scale pelagic drift-net fishing and its impact on the living marine resources of the world's oceans and seas; unauthorized fishing in zones of national jurisdiction and its impact on the living marine resources of the world's oceans and seas; and fisheries by-catch and discards and their impact on the sustainable use of the world's living marine resources. Report of the Secretary-General. *A/51/404*. 25 September 1996.

UNGA. Fifty-ninth Session. Agenda Item 49(b). Resolution adopted by the General Assembly, Sustainable fisheries, including through the 1995 Agreement for the Implementation of the Provisions of the United Nations Convention on the Law of the Sea of 10 December 1982 relating to the Conservation and Management of Straddling Fish Stocks and Highly Migratory Fish Stocks, and related instruments. *A/RES/59/25*. 17 January 2005.

UNGA. Fifty-ninth Session. Item 50(b) of the Provisional Agenda. Oceans and the law of the sea. Sustainable fisheries, including through the Agreement for the Implementation of the Provisions of the United Nations Convention on the Law of the Sea of 10 December 1982 relating to the Conservation and Management of Straddling Fish Stocks and Highly Migratory Fish Stocks, and related instruments. Report of the Secretary-General. *A/59/62*. 26 August 2004.

UNGA. Fifty-ninth Session. Item 50(b) of the Provisional Agenda, Oceans and the Law of the Sea, Sustainable fisheries, including through the Agreement for the Implementation of the Provisions of the United Nations Convention on the Law of the Sea of 10 December 1982 Relating to the Conservation and Management of Straddling Fish Stocks and Highly Migratory Fish Stocks, and Related Instruments. Report of the Secretary-General. *A/59/298*. 26 August 2004.

UNGA. Fifty-ninth Session. Item 51 on the Preliminary List. Oceans and Law of the Sea, Consultative Group of Flag State Implementation. Report of the Secretary-General. *A/59/63*. 5 March 2004.

UNGA. Fifty-ninth Session. Item 51(a) of the Preliminary List. Oceans and the Law of the Sea. Report of the Secretary-General. *A/59/62*. 04 March 2004.

UNGA. Fifty-second Session. Agenda Item 39(c). Large-scale pelagic drift-net fishing and its impact on the living marine resources of the world's oceans and seas; unauthorised fishing in zones of national jurisdiction and its impact on the living marine resources of the world's oceans and seas; and fisheries by-catch and discards and their impact on the sustainable use of the world's living marine resources. A/RES/52/59. 26 November 1997.

UNGA. Fifty-seventh Session. Agenda Item 25(b). Large-scale pelagic drift-net fishing; unauthorised fishing in zones of national jurisdiction and on the high seas; fisheries by-catch and discards, and other developments. *A/RES/57/412*. 12 December 2002.

UNGA. Fifty-seventh Session. Agenda Item 25(b). Resolution adopted by the General Assembly (without reference to a Main Committee (A/57/L.49 and Add.1). Large-scale pelagic drift-net fishing, unauthorised fishing in zones of national jurisdiction and on the high seas/illegal, unreported and unregulated fishing, fisheries by-catch and discards, and other developments. *A/RES/57/142*. 26 February 2003.

UNGA. Fifty-second Session. Agenda Item 39(c). Oceans and the law of the sea: Large-scale pelagic drift-net fishing, unauthorized fishing in zones of national jurisdiction and fisheries by-catch and discards. Report of the Secretary-General. *A/52/557*. 31 October 1997.

UNGA. Fifty-second Session. Agenda Item 39(c). Resolution adopted by the General Assembly [without reference to a Main Committee (A/52/L.30 and Add.1). Large-scale pelagic drift-net fishing; unauthorised fishing in zones of national jurisdiction and on the high seas; fisheries by-catch and discards; and other developments. *A/RES/52/29*. 26 January 1998.

UNGA. Fifty-third Session. Agenda Item 38(b). Large-scale pelagic drift-net fishing; unauthorised fishing in zones of national jurisdiction and on the high seas; fisheries by-catch and discards, and other developments. *A/RES/53/33*. 24 November 1998.

UNGA. Fifty-third Session. Agenda Item 38(b). Oceans and the law of the sea: large-scale pelagic drift-net fishing, unauthorized fishing in zones of national jurisdiction and on the high seas, fisheries by-catch and discards, and other developments. Large-scale pelagic drift-net fishing, unauthorized fishing in zones of national jurisdiction and on the high seas, fisheries by-catch and discards, and other developments. Report of the Secretary-General. *A/53/473*. 8 October 1998.

UNGA. Forty-ninth Session. Large-scale pelagic drift-net fishing and its impact on the living marine resources of the world's oceans and seas. *A/49/436*. 19 December 1994.

UNGA. Forty-seventh Session. Large-scale pelagic drift-net fishing and its impact on the living marine resources of the world's oceans and seas. A/47/443. 22 December 1992.

UNGA. Report of the Review Conference on the Agreement for the Implementation of the provisions of the United Nations Convention on the Law of the Sea of 10 December 1982 Relating to the Conservation and Management of Straddling Fish Stocks and Highly Migratory Fish Stocks. New York, USA, 22–26 May 2006. Prepared by the President of the Conference with the Assistance of the Secretariat. *A/CONF.210/2006/15*. June 2006.

UNGA. Report of the Work of the United Nations Open-ended Informal Consultative Process on Oceans and the Law of the Sea at its Sixth Meeting. Sixtieth Session. Item 76(a) of the preliminary list. *A/60/99*. 07 July 2005.

UNGA. Sixtieth Session. Agenda Item 75(b). Resolution adopted by the General Assembly. Sustainable fisheries, including through the 1995 Agreement for the Implementation of the Provisions of the United Nations Convention on the Law of the Sea of 10 December 1982 relating to the Conservation and Management of Straddling Fish Stocks and Highly Migratory Fish Stocks, and related instruments. *A/RES/60/31*. 10 March 2006.

UNGA. Sixtieth Session. Item 76(a) of the preliminary list. Oceans and the law of the sea. Report on the work of the United Nations Open-ended Informal Consultative Process on Oceans and the Law of the Sea at its sixth meeting. Letter dated 7 July 2005 from the Co-Chairpersons of the Consultative Process addressed to the President of the General Assembly. *A/60/90*. 05 July 2005.

UNGA. Sixty-first Session. Agenda Item 71(b). Resolution adopted by the General Assembly. Sustainable fisheries, including through the 1995 Agreement for the Implementation of the Provisions of the United Nations Convention on the Law of the Sea of 10 December 1982 relating to the Conservation and Management of Straddling Fish Stocks and Highly Migratory Fish Stocks, and related instruments. *A/RES/61/105*. 6 March 2007.

UNGA. Sixty-second Session. Agenda Item 77(b). Resolution adopted by the General Assembly, Sustainable fisheries, including through the 1995 Agreement for the Implementation of the Provisions of the United Nations Convention on the Law of the Sea of 10 December 1982 relating to the Conservation and Management of Straddling Fish Stocks and Highly Migratory Fish Stocks, and related instruments. *A/RES/62/177*. 28 February 2008.

UNGA. Sixty-second Session. Agenda Item 77(b). Sustainable fisheries, including through the 1995 Agreement for the Implementation of the Provisions of the United Nations Convention on the Law of the Sea of 10 December 1982 relating to the Conservation and Management of Straddling Fish Stocks and Highly Migratory Fish Stocks, and related instruments. *A/66/177*. 28 February 2008.

UNGA. Sixty-second Session, Agenda Item 78. Resolution adopted by the General Assembly [on the report of the Sixth Committee (A/62/446)], Responsibility of States for Internationally Wrongful Acts, *A/RES/62/61*.

UNGA. Sixty-third Session. Item 73(a) of the provisional agenda, Oceans and the law of the sea, Report on the works of the United Nations and the Law of the Sea at its Ninth Meeting, *A/63/174*, 25 July 2008.

UNGA. Sixty-third Session. Item 73(a) of the provisional agenda, Oceans and the Law of the Sea, Available assistance to and measures that may be taken by developing States, in particular the least developed States and Small Island Developing States (SIDS), as well as coastal African States, to realize the benefits of sustainable and effective development of marine resources and uses of the oceans within the limits of national jurisdiction, Study Prepared by the Secretariat, *A/63/342,* 3 September 2008.

UNGA. United Nations Open-ended Informal Consultative Process on Ocean Affairs. First Meeting. 30 May-02 June 2000. Discussion Panel A Responsible Fisheries and Illegal, Unregulated and Unreported Fisheries. *A/AC.259/1*. 15 May 2001.

UN. Oceans and Law of the Sea. *Chronological Lists of Ratifications of, Accessions and Successions to the Convention and the Related Agreements as at 26 September 2008*. www.un.org. Accessed on 02 November 2008.

UN. Report of the Fifth Informal Consultations of the States Parties to the Agreement for the Implementation of the Provisions of the United Nations Convention on the Law of the Sea 10 December 1982 relating to the Conservation and Management of Straddling Fish Stocks and Highly Migratory Fish Stocks. New York, USA, 20–24 March 2006. *ICSP5/UNFSA/REP/INF.1*. April 2006.

UN. Report of the Fourth Informal Consultations of the States Parties to the Agreement for the Implementation of the Provisions of the United Nations Convention on the Law of the Sea 10 December 1982 relating to the Conservation and Management of Straddling Fish Stocks and Highly Migratory Fish Stocks. New York, USA, 31 May-3 June 2005. *ICSP4/UNFSA/REP/INF.1*. July 2005.

UN. Report of the Second Informal Consultations of the States Parties to the Agreement for the Implementation of the Provisions of the United Nations Convention on the Law of the Sea 10 December 1982 relating to the Conservation and Management of Straddling Fish Stocks and Highly Migratory Fish Stocks. New York, USA, 23–25 July 2003. *ICSP2/UNFSA/REP/INF.1*. August 2003.

UN. Report of the Third Informal Consultations of the States Parties to the Agreement for the Implementation of the Provisions of the United Nations Convention on the Law of the Sea 10 December 1982 relating to the Conservation and Management of Straddling Fish Stocks and Highly Migratory Fish Stocks. New York, USA, 8–9 July 2004. *ICSP3/UNFSA/REP/INF.1*. August 2004.

UN. *Resolution 2758 on Restoring the Rights of the PRC*. 25 October 1971.

5. Food and Agriculture Organization Publications

Bergh, Per Erik and Davies, Sandy. 'Fishery Monitoring, Control, and Surveillance'. In Cochrane, K.L. (ed.) A Fishery Manager's Guidebook: Management Measures and their Application. *FAO Fisheries Technical Paper No. 424* (Rome: FAO, 2002).

Bray, Kevin. 'A Global Review of Illegal, Unreported, and Unregulated Fishing'. FAO. Report of and Papers Presented at the Expert Consultation on Illegal, Unreported and Unregulated Fishing. Sydney, Australia, 15–19 May 2000. *FAO Fisheries Report No. 666* (Rome: FAO, 2001) 88–134.

Cacaud, P.; Kuruc, M.; and Spreij, M. Administrative Sanctions in Fisheries Law. *FAO Legislative Study No. 82* (Rome: FAO, 2003).

Chaves, Linda. 'Illegal, Unreported, and Unregulated Fishing: WTO-consistent Trade Related Measures to Address IUU Fishing.' FAO. Report of and Papers Presented at the Expert Consultation on Illegal, Unreported and Unregulated Fishing. Sydney, Australia, 15–19 May 2000. *FAO Fisheries Report No. 666* (Rome: FAO, 2001) 269–281.

Davies, Sandy I. and Reynolds, J. Eric (eds.). Guidelines for Developing an At-Sea Fishery Observer Programme. *FAO Fisheries Technical Paper No. 414* (Rome: FAO, 2002).

Doulman, David J. '2001 FAO International Plan of Action to Prevent, Deter and Eliminate Illegal, Unreported and Unregulated Fishing: Background and Progress Towards Implementation'. Presented at the *FAO Regional Workshop on the Elaboration of National Plans of Action to*

Prevent, Deter and Eliminate Illegal, Unreported and Unregulated Fishing: Southeast Asia Subregion. Penang, Malaysia, 10–14 October 2004.

Doulman, David. 'Illegal, Unreported and Unregulated Fishing: Mandate for an International Plan of Action'. FAO. Report of and Papers Presented at the Expert Consultation on Illegal, Unreported and Unregulated Fishing. Sydney, Australia, 15–19 May 2000. *FAO Fisheries Report No. 666* (Rome: FAO, 2001) 41–58.

Drammeh, Ousman K.L. 'Illegal, Unreported and Unregulated Fishing in Small-scale Marine and Inland Capture Fisheries'. FAO. Report of and Papers Presented at the Expert Consultation on Illegal, Unreported and Unregulated Fishing. Sydney, Australia, 15–19 May 2000. *FAO Fisheries Report No. 666* (Rome: FAO, 2001) 135–144.

Edeson, William. 'Tools to Address IUU Fishing: The Current Legal Situation'. FAO. Report of and Papers Presented at the Expert Consultation on Illegal, Unreported and Unregulated Fishing. Sydney, Australia, 15–19 May 2000. *FAO Fisheries Report No. 666* (Rome: FAO, 2001) 269–281.

Evans, David. 'The Consequences of Illegal, Unreported and Unregulated Fishing for Fishery Data and Management'. FAO. Report of and Papers Presented at the Expert Consultation on Illegal, Unreported and Unregulated Fishing. Sydney, Australia, 15–19 May 2000. *FAO Fisheries Report No. 666* (Rome: FAO, 2001) 222–233.

Food and Agriculture Organization (FAO). An Appraisal of the Suitability of the CITES Criteria for Listing Commercially-exploited Aquatic Species. *FAO Fisheries Circular No. 954* (FAO: Rome, 2000).

FAO. Appendix E. Major Issues Affecting the Performance of Regional Fishery Bodies. In Report of the Meeting of FAO and Non-FAO Regional Fishery Bodies or Arrangements. Rome, Italy, 11–12 February 1999. *FAO Fisheries Report No. 597* (Rome: FAO, 1999).

FAO. Coastal State Requirements for Foreign Fishing. *FAO Legislative Studies No. 57* (Rome: FAO, 1996).

FAO. Committee on Fisheries. Eleventh Session. Bremen, Germany, 2–6 June 2008. Trade Related Measures for Sustainability: Progress on a Binding Instrument on Port State Measures. *COFI:FT/XI/2008/6*.

FAO. Committee on Fisheries. Report of the Eighth Session of the Sub-committee on Fish Trade. Bremen, Germany, 12–16 February 2002. *FAO Fisheries Report No. 673* (Rome: FAO, 2002).

FAO. Committee on Fisheries. Report of the Tenth Session of the Sub-Committee on Fish Trade. Santiago de Compostela, Spain, 30 May-02 June 2006. *FAO Fisheries Report No. 807* (Rome: FAO, 2006).

FAO. Committee on Fisheries. Report of the Twenty-seventh Session of the FAO Committee on Fisheries. Rome, Italy, 5–9 March 2007. *FAO Fisheries Report No. 830* (Rome: FAO, 2007).

FAO. Committee on Fisheries. Subcommittee on Fish Trade. Eighth Session. Bremen, Germany, 12–16 February 2002. Item 8 of the Provisional Agenda. Traceability of Products from Fisheries and Aquaculture. *COFI:FT/VIII/2002/7* (Rome: FAO, 2002).

FAO. Committee on Fisheries. Sub-committee on Fish Trade. Tenth Session. Traceability and Labelling in Fish Trade. Santiago de Compostela, Spain, 30 May-02 June 2006. *COFI:FT/X/2006/6* (Rome: FAO, 2006).

FAO. Committee on Fisheries. Sub-Committee on Fish Trade. Eleventh Session. Bremen, Germany, 2–6 June 2008. 'Status and Important Recent Events Concerning International Trade in Fishery Products'. *COFI:FT/XI/2008/3*, (Rome: FAO, 2008).

FAO. Committee on Fisheries. Sub-committee on Fish Trade. Eleventh Session. Bremen, Germany, 2–6 June 2008. *CITES Issues with Respect to International Fish Trade*. COFI:FT/XI/2008/4. (Rome: FAO, 2008).

FAO. Committee on Fisheries. Twenty-fifth Session. Progress in the Implementation of the Code of Conduct for Responsible Fisheries and Related International Plans of Action. Rome, Italy, 24–28 February 2003. *COFI/2003/Rev.1*. Rome, FAO, 2003.

FAO Council. Hundred and Sixteenth Session. Rome, Italy, 14–19 June 1999. Current World Food Situation. *CL 116/2* (Rome, FAO, 1999).

FAO. Fisheries Department. Implementation of the International Plan of Action to Prevent, Deter and Eliminate Illegal, Unreported and Unregulated Fishing. *FAO Technical Guidelines for Responsible Fisheries No. 9* (Rome: FAO, 2002).

FAO. Fishery Resources Division and Fishery Policy and Planning Division. Fisheries Management. *FAO Technical Guidelines for Responsible Fisheries No. 4* (Rome: FAO, 1997).

FAO. Fisheries Department. The Ecosystem Approach to Fisheries. *FAO Technical Guidelines for Responsible Fisheries No. 4, Suppl. 2* (Rome: FAO, 2003).

FAO. Fisheries Department. Responsible Utilisation. *Technical Guidelines for Responsible Fisheries No. 7* (Rome: FAO, 1998).

FAO. Fishing Technology Service. Fishing Operations. *FAO Technical Guidelines for Responsible Fisheries No. 1* (Rome: FAO, 1996). 6 annexes.

FAO. Fishing Technology Service. Fishing Operations. 1. Vessel Monitoring Systems. *FAO Technical Guidelines for Responsible Fisheries No. 1 Suppl. 1* (Rome: FAO, 1998).

FAO. *Guidelines for Ecolabelling of Fish and Fish Products* (Rome: FAO, 2005).

FAO. Increasing the Contribution of Small-scale Fisheries to Poverty Alleviation and Food Security. *FAO Technical Guidelines for Responsible Fisheries No. 10* (Rome: FAO, 2005).

FAO. *Informal Open-ended Technical Meeting to Review the Annexes of the Draft Legally-binding Instrument on Port State Measures to Prevent, Deter and Eliminate IUU Fishing*. Rome, Italy, 25–27 November 2008.

FAO. *International Guidelines for the Management of Deep-sea Fisheries in the High Seas* (Rome: FAO, 2008).

FAO. *Model Scheme on Port State Measures to Combat Illegal, Unreported and Unregulated Fishing* (Rome: FAO, 2006).

FAO. Regional Office for Asia and the Pacific. APFIC Regional Consultative Workshop. Managing *Fishing Capacity and Illegal, Unreported and Unregulated Fishing in Asia*. Phuket, Thailand, 13–15 June 2007. (Bangkok: FAO-RAP, 2007).

FAO. Report of the Ad hoc Meeting on Management of Deepwater Fisheries Resources of the Southern Indian Ocean. Swakopmund, Namibia, 30 May-01 June 2001. *FAO Fisheries Report No. 652* (Rome: FAO, 2001).

FAO. 'Report of the Expert Consultation'. *Experts Consultation on Illegal, Unreported and Unregulated Fishing Organised by the Government of Australia in Cooperation with FAO.* Sydney, Australia, 15–19 May 2000. *FAO Fisheries Report No. 666* (Rome: FAO, 2001).

FAO. Report of the Expert Consultation of Regional Fisheries Management Bodies on Harmonization of Catch Certification. *FAO Fisheries Report No. 697*. La Jolla, United States of America, 09–11 January 2002 (Rome: FAO, 2002).

FAO. Report of the Expert Consultation on Data Formats and Procedures for Monitoring, Control and Surveillance. Bergen, Norway, 25–27 October 2004. *FAO Fisheries Report 761* (Rome: FAO, 2005).

FAO. Report of the Expert Consultation on the FAO Guidelines for Eco-labelling. Rome, Italy, 03–04 March 2008. *FAO Fisheries Report No. 864* (Rome: FAO, 2008).

FAO. Report of the Expert Consultation on the Development of a Comprehensive Global Record of Fishing Vessels. Rome, 25–28 February 2008. *FAO Fisheries Report No. 865* (Rome: FAO, 2008).

FAO. Report of the Expert Consultation on the FAO Guidelines for Eco-labelling. Rome, Italy, 03–04 March 2008. *FAO Fisheries Report No. 864* (Rome: FAO, 2008).

FAO. Report of the Expert Consultation on the Marking of Fishing Gear. Victoria, British Columbia, Canada, 14–19 July 1991. *FAO Fisheries Report No. 485* (Rome: FAO, 1993).

FAO. Report of the Expert Consultation to Draft a Legally-binding Instrument on Port State Measures. Washington, D.C., USA, 4–8 September 2007. *FAO Fisheries Report No. 846* (Rome, FAO, 2007).

FAO. Report of the Expert Consultation to Review Port State Measures to Combat Illegal, Unreported, and Unregulated Fishing. Rome, Italy, 04–06 November 2002. *FAO Fisheries Report No. 692* (Rome, FAO, 2002).

FAO. *Report of the FAO Council*. 102nd Session. Rome, Italy 9–20 November 1992 (Rome: FAO, 1992).

FAO. Report of the FAO/FFA Regional Workshop to Promote the Full and Effective Implementation of Port State Measures to Combat Illegal, Unreported and Unregulated Fishing. Nadi, Fiji, 28 August-01 September 2006. *FAO Fisheries Report No. 810* (Rome, FAO, 2007).

FAO, Report of the FAO Technical Working Group on the Management of Fishing Capacity, La Jolla, United States of America, 15–18 April 1998. *FAO Fisheries Report No. 586* (Rome: FAO, 1998).

FAO. Report of the Joint FAO/IMO Ad Hoc Working Group on Illegal, Unreported and Unregulated Fishing and Related Matters. Rome, 09–11 October 2000. *FAO Fisheries Report No. 637* (Rome, FAO, 2001).

FAO. Report of the Meeting of FAO and Non-FAO Regional Fishery Bodies or Arrangements. Rome, Italy, 11–12 February 1999. *FAO Fisheries Report No. 597* (Rome: FAO, 1999).

FAO. Report of the Second Technical Consultation on the Sustainability of the CITES Criteria for Listing Commercially-exploited Aquatic Species. Windhoek, Namibia, 22–25 October 2001. *FAO Fisheries Report No. 667* (Rome: FAO, 2001).

FAO. Report of the Twenty-sixth Session of the Committee on Fisheries. Rome, 7–11 March 2005. *FAO Fisheries Report No. 780* (Rome: FAO, 2005).

FAO. Report of the Sub-regional Fisheries Commission workshop on Vessel Monitoring Systems. Saly, Senegal, 14–17 October 2002. *FAO Fisheries Report No. 696* (Rome: FAO, 2002).

FAO. Report of the Technical Consultation on High Seas Fishing. *FAO Fisheries Report No. 484* (Rome: FAO, 1992).

FAO. Report of the Technical Consultation on Illegal, Unreported and Unregulated Fishing. Rome, Italy, 02–06 October 2000. *FAO Fisheries Report No. 634* (Rome, FAO, 2000).

FAO. Report of the Technical Consultation on Implementation Issues Associated with Listing Commercially-Exploited Aquatic Species on CITES Appendices. Rome, 25–28 May 2004. *FAO Fisheries Report No. 741* (Rome: FAO, 2004).

FAO. Report of the Technical Consultation on Technical Guidelines for Responsible Fish Trade. Rome, 5–7 November 2007. *FAO Fisheries Report No. 854* (Rome: FAO, 2007).

FAO. Report of the Technical Consultation on the Sustainability of the CITES Criteria for Listing Commercially-Exploited Aquatic Species. Rome, Italy, 28–30 June 2000. *FAO Fisheries Report No. 629* (Rome: FAO, 2000).

FAO. Report of the Technical Consultation to Review Port State Measures to Combat Illegal, Unreported and Unregulated Fishing. Rome, Italy, 31 August-02 September 2004. *FAO Fisheries Report No. 759* (Rome: FAO, 2004).

FAO. Report of the Third Ad Hoc Meeting of Intergovernmental Organisations on Work Programmes Related to Subsidies in Fisheries, Rome, Italy, 23–25 July 2003, *FAO Fisheries Report No. 719* (Rome: FAO).

FAO. *Report of the Twenty-fifth Session of COFI*. FAO, Rome, 24–28 February 2003.

FAO. *Report of the Twenty-Third Session of the Committee on Fisheries*. Rome, Italy, 15–19 February 1999.

FAO. 'Report on the Development of a Comprehensive Record of Fishing Vessels'. *COFI/2007/Inf.12*. In FAO. Committee on Fisheries. *Report of the Twenty-seventh Session of the FAO Committee on Fisheries*. Rome, Italy, 5–9 March 2007. *FAO Fisheries Report No. 830* (Rome: FAO, 2007).

FAO. *State of World Fisheries and Aquaculture 2004* (Rome: FAO, 2004).

FAO. *State of World Fisheries and Aquaculture 2006* (Rome: FAO, 2007).

FAO Subregional Office for Southern Africa. Report of the Second Session of the South West Indian Ocean Fisheries Commission. Maputo, Mozambique, 22–25 August 2006. *FAO Fisheries Report No. 823* (Rome: FAO, 2007).

FAO. 'Summary of the Report of the Joint FAO/IMO Ad Hoc Working Group on Illegal, Unreported and Unregulated Fishing and Related Matters'. Rome, 09–11 October 2000. *COFI/2001/Inf.10*. In FAO. Committee on Fisheries. Twenty-fourth Session. Rome, Italy, 26 February-02 March 2001.

FAO. *Technical Consultation to Draft a Legally-binding Instrument on Port State Measures to Prevent, Deter, and Eliminate Illegal, Unreported and Unregulated Fishing*. Rome, Italy, 23–27 June 2008.

FAO. *Technical Consultation to Draft a Legally-binding Instrument on Port State Measures to Prevent, Deter, and Eliminate Illegal, Unreported and Unregulated Fishing*. Rome, Italy, 26–29 January 2009.

FAO. 'Trade Related Measures for Sustainability: Progress on a Binding Instrument on Port State Measures,' *COFI:FT/XI/2008/6*. In FAO. Committee on Fisheries. Eleventh Session. Bremen, Germany, 2–6 June 2008.

FAO. 'FAO Workplan on CITES and Commercially-exploited Aquatic Species,' In FAO. Report of the Twenty-Fifth Session of the Committee on Fisheries. Rome, 24–28 February 2003. *FAO Fisheries Report No. 702* (Rome: FAO, 2003).

Fitzpatrick, John. 'Measures to Enhance the Capability of a Flag State to Exercise Effective Control Over a Fishing Vessel'. FAO. Report of and Papers Presented at the Expert Consultation on Illegal, Unreported and Unregulated Fishing. Sydney, Australia, 15–19 May 2000. *FAO Fisheries Report No. 666* (Rome: FAO, 2001) 205–221.

Flewwelling, Peter. An Introduction to Monitoring, Control and Surveillance for Capture Fisheries. *FAO Fisheries Technical Paper No. 338*. (Rome: FAO, 1995).

Flewwelling, Peter; Cullinan, Cormac, Balton, David, Sautter, Raymond P. and Reynolds, J. Eric. 'Guide to Monitoring, Control, and Surveillance Systems for Coastal and Offshore Capture Fisheries.' *FAO Fisheries Technical Paper No. 415* (Rome: Italy, 2003).

Franckx, Erik. 'Applications of the Term "Introduction from the Sea": Background Paper for the FAO'. Report of the Expert Consultation on Legal Issues Related to CITES and Commercially-exploited Aquatic Species. Rome, 22–25 June 2004. *FAO Fisheries Report No. 746*. (Rome: FAO, 2004).

Franckx, Erik. Fisheries Enforcement: Related Legal and Institutional Issues: National, Subregional or Regional Perspectives. *FAO Legislative Studies No. 71* (Rome: FAO, 2001).

Gabrielides, Gabriel P. 'Cyprus-A Case Study'. And Franco, Arnulfo. 'Panama-Case Study'. In Report of the Expert Consultation on Fishing Vessels Operating Under Open Registries and their Impact on Illegal, Unreported and Unregulated Fishing. *FAO Fisheries Report No. 722*. Miami, Florida, USA, 23–25 September 2003 (Rome: FAO, 2003).

Gréboval, Dominique. Report and Documentation of the International Workshop on Factors Contributing to Unsustainability and Overexploitation in Fisheries. Bangkok, Thailand, 04–8 February 2002. *FAO Fisheries Report No. 672* (Rome, FAO, 2002).

Gréboval, Dominique F. 'The International Plan of Action for the Management of Fishing Capacity and Selected Issues Pertaining Illegal, Unreported and Unregulated Fishing'. FAO, Report of and Papers Presented at the Expert Consultation on Illegal, Unreported and Unregulated Fishing, Sydney, Australia, 15–19 May 2000, *FAO Fisheries Report No. 666* (Rome: FAO, 2001) 234–243.

Huss, H.H., Ababouch, L., and Gram, L. Assessment and Management of Seafood Safety and Quality, *FAO Fisheries Technical Paper No. 444* (Rome: FAO, 2003).

Kelleher, Kieran. The Costs of Monitoring, Control and Surveillance of Fisheries in Developing Countries. *FAO Fisheries Circular No. 976* (Rome: FAO, 2002).

Kuemlangan, Blaise. 'National Legislative Options to Combat IUU Fishing'. FAO. Report of and Papers Presented at the Expert Consultation on Illegal, Unreported and Unregulated Fishing. Sydney, Australia, 15–19 May 2000. *FAO Fisheries Report No. 666* (Rome: FAO, 2001) 159–176.

Kurien, J. Responsible Fish Trade and Food Security. *FAO Fisheries Technical Paper No. 456* (Rome: FAO, 2005).

Lobach, Terje. Port State Control of Foreign Fishing Vessels. *FAO Fisheries Circular No. 987* (Rome: FAO, 2003).

Lobach, Terje. Port State Control of Foreign Fishing Vessels. *FAO Legal Papers Online No. 29* (Rome: FAO, 2002).

Marashi, S.H. The Role of FAO Regional Fishery Bodies in the Conservation and Management of Fisheries. *FAO Fisheries Circular No. 916* (Rome: FAO, 1996).

Martinez, Iciar; James, D.; Lungren, R.; Staples, D.; Funge-Smith, S.; and Clausen, J. Status and Potential of Fisheries and Aquaculture in Asia and the Pacific 2006. FAO Regional Office for Asia and the Pacific. *RAP Publication 2006/22*. Bangkok: FAO-RAP, 2006.

Martinez, Iciar; James, David; and Loréal, Henri. Application of Modern Analytical Techniques to Ensure Seafood Safety and Authenticity. *FAO Fisheries Technical Paper No. 455* (Rome: FAO, 2005).

Northridge, S.P. Driftnet Fisheries and their Impacts on Non-target Species: A Worldwide Review. *FAO Fisheries Technical Paper No. 320* (Rome: FAO, 1991).

Salzman, James. *Environmental Labelling in OECD Countries*, OECD: Paris and Karen West, 1991. Cited in Wessells, Cathy; Cochrane, Kevern; Derree, Carolyn; Wallis, Paul; and Willmann, Rolf. Product Certification and Ecolabelling for Fisheries Sustainability. *FAO Fisheries Technical Paper No. 422* (Rome: FAO, 2001).

Savini, Michel J. Summary Information on the Role of International Fishery Bodies with Regard to the Conservation and Management of Living Resources of the High Seas. *FAO Fisheries Circular No. 835. Rev. 1* (Rome: FAO, 1991).

Schrank, William E. 'Subsidies for Fisheries: A Review of Concepts'. Paper Presented at the Expert Consultation on Economic Incentives and Responsible Fisheries, Rome, Italy, 28 November–1 December 2000. *FAO Fisheries Report No. 638 Supplement* (Rome: FAO, 2000) 11–40.

Seijo, J.C. 'Estimating Fisheries Subsidies and Their Impact on Sustainability of Fish Resources: Dealing with Dynamics, Risk and Uncertainty.' Paper Presented at the Expert Consultation on Economic Incentives and Responsible Fisheries, Rome, Italy, 28 November–01 December 2000, *FAO Fisheries Report No. 638 Supplement* (Rome: FAO, 2000) 41–56.

Swan, Judith. Fishing Vessels Operating Under Open Registers and the Exercise of Flag State Responsibilities. *FAO Fisheries Circular No. 980* (Rome: FAO, 2002).

Swan, Judith. 'Regional Fishery Bodies and Governance: Issues, Action and Future Directions'. *FAO Fisheries Circular No. 959* (Rome: FAO, 2002).

Tokrisna, Ruangrai. 'WTO-Consistent Trade-Related Measures to Address IUU Fishing-Developing Countries Issues'. FAO. Report of and Papers Presented at the Expert Consultation on Illegal, Unreported and Unregulated Fishing. Sydney, Australia, 15–19 May 2000. *FAO Fisheries Report No. 666* (Rome: FAO, 2001) 282–293.

Tsamenyi, Martin and Molenaar, Erik Jaap. Satellite-based Vessel Monitoring Systems: International Legal Aspects and Developments in State Practice. *FAO Legal Papers Online No. 7* (Rome: FAO, 2000).

Van Houtte, Annick. 'Flag State Responsibility and the Contribution of Recent International Instruments in Preventing, Deterring, and Eliminating IUU Fishing'. Appendix E in Report of the Expert Consultation on Fishing Vessels Operating Under Open Registries and their Impact on Illegal, Unreported and Unregulated Fishing, Miami, Florida, USA, 23–25 September 2003. *FAO Fisheries Report No. 722*. (Rome: FAO, 2003).

Ward, J.M.; Kirkley, J.E.; Metzner, R.; and Pascoe, S. Measuring and Assessing Capacity in Fisheries. 1. Basic Concepts and Management Options. *FAO Fisheries Technical Paper No. 433/1* (Rome: FAO, 2004).

6. Other International Organisation Reports

Convention on International Trade in Endangered Species of Wild Fauna and Flora (CITES). Thirteenth meeting of the Conference of the Parties. *Cooperation and Synergy with the Convention on the Conservation of Migratory Species of Wild Animals (CMS)*. Conf.13.3. Bangkok, Thailand, 2–14 October 2004.

CITES. *Fourteenth Meeting of the Conference of the Parties*. The Hague, Netherlands, 3–15 June 2007.

CITES. Technical Workshop on the Conservation of Sea cucumbers in the Families *Holothuridae* and *Stichopodidae*, Kuala Lumpur, Malaysia, 1–3 March 2004.

International Law Commission. 1606th Meeting. Co-operation with other bodies. Item 10 of the Agenda. A/CN.4/SR.1606. 1 *Yearbook of the International Law Commission* (1980).

International Maritime Organization (IMO). Flag State Implementation. 9th Session. Agenda Item 15. 'Illegal, Unregulated, and Unreported Fishing and Related Matters'. Outcome of the First Meeting of the Joint FAO/IMO Ad Hoc Working Group on IUU Fishing and Related Matters. *FSI 9/15*. 08 November 2000.

IMO. Marine Environment Protection Committee. Forty-fourth Session. Agenda Item 20. Report of the Marine Environment Protection Committee on Its Forty-fourth Session. *MEPC 44/20*. 12 April 2000.

IMO. Maritime Safety Committee. 72nd Session. Agenda Item 23. Report of the Maritime Safety Committee on Its Seventy-Second Session. *MSC 72/23*. 31 May 2000.

IMO. Sub-committee on Flag State Implementation. Eighth Session. Agenda Item 19. Report to the Maritime Safety Committee and the Marine Environment Protection Committee. *FSI 8/19*. 14 February 1999.

Joint FAO/WHO Food Standard Programme. Codex Alimentarious Commission. Thirty-first Session. Geneva, Switzerland, 30 June-05 July 2008. Report of the Sixteenth Session of the Codex Committee on Food Import and Export Inspection and Certification Systems. Surfers Paradise, Queensland, Australia, 26–30 November 2007. *ALINORM 08/31/30*.

Joint FAO/WHO Food Standards Programme. Codex Alimentarius Commission. Twenty-eighth Session. Rome, Italy, 04–09 July 2005. Report of the Thirteenth Session of the Codex Committee on Food Import and Export Inspection and Certification Systems. Melbourne, Australia, 06–10 December 2004. *ALINFORM 05/28/30.*

Report of the World Commission on Environment and Development: Our Common Future. In UN Secretary General. Forty-second Session. Development and International Co-operation: Environment. Report of the World Commission on Environment and Development. *A/42/427.* 04 August 1987.

Secretariat of the Joint FAO/WHO Food Standards Programme. *Understanding the Codex Alimentarius* (Rome: WHO and FAO, 2006).

Toward the 2002 World Summit on Sustainable Development. Johannesburg, South Africa. Ensuring the Sustainable Development of Oceans and Coasts. Co-Chairs' Report from the Global Conference on Oceans and Coasts. UNESCO, Paris, 03–07 December 2001.

United Nations. Commission on Sustainable Development. Report on the Seventh Session. New York, 01 May and 27 July 1998, and 19–30 April 1999. Economic and Social Council Official Records. *1999 Supplement No. 9 (E/1999/29)* (New York: United Nations, 1999).

7. Regional Conventions, Policies, and Measures

African-Eurasian Waterbird Agreement (AEWA). Hague, the Netherlands, concluded on 16 June 1995, in force 1 November 1999, 2365 UNTS I-42632.

Agreement of the Establishment of the Indian Ocean Tuna Commission. Rome, Italy, concluded on 25 November 1993, in force 27 March 1996, ATS 1996 No. 20.

Agreement on the Conservation of Albatrosses and Petrels (ACAP). Cape Town, South Africa, concluded on 1 February 2001, in force 1 February 2004, 2258 UNTS 257.

Agreement for the Conservation of Cetaceans of the Black Sea, Mediterranean Sea and Contiguous Atlantic Area (ACCOBAMS). Monaco, concluded on 24 November 1996, in force 1 June 2001, 2183 UNTS 303.

Agreement on the Conservation of Small Cetaceans of the Baltic and North Seas (ASCOBANS). New York, United States, concluded on 13 September 1991, in force 29 March 1994, 1772 UNTS 217.

Agreement for the Conservation of Seals in the Wadden Sea. Bonn, Germany, concluded on 16 October 1990, in force 1 October 1991, www.cms.int.

Asia-Pacific Economic Cooperation (APEC). *Joint Ministerial Statement, Bali Plan of Action: Towards Healthy Oceans and Coasts for the Sustainable Growth and Prosperity of the Asia-Pacific Community.* Adopted at the 2nd APEC Ocean-related Ministerial Meeting, Second Ministerial Meeting, Bali, Indonesia, 16–17 September 2005.

APEC. *Seoul Oceans Declaration.* Adopted at the First APEC Ocean-related Ministerial Meeting. 25–26 April 2002, Seoul, Korea.

Association of Southeast Asian Nations (ASEAN). *ASEAN Political-Security Community Blueprint.* Adopted at the 14th ASEAN Summit. 26 February-01 March 2009, Cha-am, Phetchaburi, Thailand.

Castries Declaration. St Lucia, 20–24 November 1989.

Commission for the Conservation of Antarctic Marine Living Resources (CCAMLR). *Conservation Measure 10-02 (2008).* Licensing and Inspection Obligation of Contracting Parties with Regard to their Flag Vessels Operation in the Convention Area.

CCAMLR. *Conservation Measure 10-03 (2008).* Port Inspections of Vessels Carrying Toothfish.

CCAMLR. *Conservation Measure 10-04 (2007).* Automated Satellite-linked Vessel Monitoring Systems. Annex 10-04/B. Provisions on Secure and Confidential Treatment of Electronic Reports and Messages Transmitted Pursuant to Conservation Measure 10-04.

CCAMLR. *Conservation Measure 10-04 Annex 10-04/B.* Provisions on Secure and Confidential Treatment of Electronic Reports and Messages.

CCAMLR. *Conservation Measures 10-05 (2009).* Catch Documentation Scheme for *Dissostichus* spp.

CCAMLR. *Conservation Measure 10-06 (2008).* Scheme to Promote Compliance by Contracting Party Vessels with CCAMLR Conservation Measures.

CCAMLR. *Conservation Measure 10-07 (2009)*. Scheme to Promote Compliance by Non-Contracting Party Vessels with CCAMLR Conservation Measures.

CCAMLR. *Conservation Measure 10-08 (2008)*. Scheme to Promote Compliance by Contracting Party Nations with CCAMLR Conservation Measures.

CCAMLR. *Conservation Measure 10-09 (2009)*. Notification System for Transshipments within the Convention Area.

CCAMLR. *Conservation Measures 23-01 (2004)*. Five-day Catch and Effort Reporting System.

CCAMLR. *Conservation Measures 23-02 (1993)*. Ten-day Catch and Effort Reporting System.

CCAMLR. *Conservation Measures 23-03 (1991)* on Monthly Catch and Effort Reporting System.

CCAMLR. *Conservation Measures 23-06 (2004)* on Data Reporting System for Krill Fisheries.

CCAMLR. *Conservation Measure 32-01 (2001)* on the Prohibition of Directed Fishing for Finfish in Statistical Subarea 48.1.

CCAMLR. *Conservation Measure 32-09 (2004)* on the Prohibition of Directed Fishing for *Dissotichus spp*.

CCAMLR. *Scheme of International Scientific Observation*. Adopted at CCAMLR XI and amended at CCAMLR XVI.

Commission for the Conservation of Southern Bluefin Tuna (CCSBT). *Resolution on Establishing a Programme for Transshipment by Large-scale Fishing Vessels*. Adopted at the Fifteenth Annual Meeting. 14–17 October 2002.

CCSBT. *Resolution to Establish the Status of Co-operating Non-member of the Extended Commission and the Extended Scientific Committee*. 2003.

CCSBT. *Resolution on Illegal, Unregulated and Unreported Fishing (IUU) and Establishment of a CCSBT Record of Vessels over 24 Meters Authorised to Fish for Southern Bluefin Tuna*. 10 October 2003.

CCSBT. *Resolution on Amendment of the Resolution on "Illegal, Unregulated, and Unreported Fishing (IUU) and Establishment of a CCSBT Record of Vessels over 24 Metres Authorised to Fish for Southern Bluefin Tuna"*, Adopted at the Eleventh Annual Meeting. 19–22 October 2004.

CCSBT. *Scientific Observer Program Standards*. www.ccsbt.org. Accessed on 14 March 2004.

CCSBT. *Southern Bluefin Tuna Statistical Document Program*. October 2003.

Convention for the Conservation of Anadromous Stocks in the North Pacific Ocean. Moscow, concluded 11 February 1992, in force 16 February 1993, TIAS No 11465.

Convention on the Conservation of Highly Migratory Fish Stocks in the Western and Central Pacific Ocean, Honolulu, U.S., concluded 5 September 2000, entry into force 19 June 2004, [2004] ATS 15.

Convention for the Conservation of Southern Bluefin Tuna. Canberra, Australia. concluded on 10 May 1993, in force 20 May 1994, 1819 UNTS 359.

Convention for the Prohibition of Fishing with Long Driftnets in the South Pacific, Wellington, New Zealand, concluded on 24 November 1989, *and its Protocols,* Noumea, New Caledonia, concluded on 20 October 1990, in force 17 May 1991, 1899 UNTS 3; 29 ILM 1454 (1990).

Convention on Future Multilateral Cooperation in the Northwest Atlantic Fisheries. Ottawa, Canada, concluded 24 October 1978, in force 01 January 1979, 1135 UNTS 269; 34 ILM 1452 (1995).

Convention on the Conservation and Management of Fishery Resources in the Southeast Atlantic Ocean. Windhoek, Namibia, concluded on 20 April 2001, in force 13 April 2003; 41 ILM 257.

The Convention for the Conservation of Salmon in the North Atlantic Ocean. 01 October 1983.

European Union (EU). Community Action Plan for the Eradication of Illegal, Unreported and Unregulated Fishing. Brussels, Belgium. 28.5.2002 *COM(2002) 180 final*.

EUC. Council Regulation (EC) No. 104/2000 of 17 December 1999 on the common organisation of the markets in fishery and aquaculture products. *OJ L 17, 21.1.2000*.

EUC. Council Regulation (EC) 601/2004 of 22 March 2004 laying down certain control measures applicable to fishing activities in the area covered by the Convention on the conservation of Antarctic marine living resources and repealing regulations (EEC) No. 3943/90, EC 66/98 and (EC) 1721/1999. *OJ L 97/16 1.4.2004*.

EUC. Council Regulation (EC) No. 1005/2008 of 29 September 2008 establishing a Community system to prevent, deter and eliminate illegal, unreported and unregulated fishing, amending Regulations (EEC) No. 2847/93, (EC) No. 1936/2001 and (EC) No. 601/2004 and repealing Regulations (EC) No. 1093/94 and (EC) No. 1447/1999. *OJ L 286/1 29.10.2008*.

EUC. Commission Regulation (EC) No. 2090/98 as amended by EC No. 839/2002 of 21 May 2002 amending Regulation (EC) No. 2090/98 concerning the fishing vessel register of the Community. *OJ L 134*, 22/05/2002.

EC. Council Regulation (EC) No. 2371/2002 of 20 December 2002 on the conservation and sustainable exploitation of fisheries resources under the Common Fisheries Policy. *OJ L 358/59, 31.12.2002*.

Fishery Committee for the Eastern Central Atlantic. Nineteenth Session. Cotonou, Benin, 4–6 November 2008. Global Emerging Issues in Fisheries Development and Management Relevant to the Region. *CECAF/XIX/2008/8*. November 2008.

Forum Fisheries Agency (FFA). *Application for Registration*. 26 July 2005. www.ffa.int, Accessed on 15 February 2006.

FFA. *Arrangement Implementing the Nauru Agreement Setting Forth Minimum Terms and Conditions of Access to the Fisheries Zones of the Parties*. Nauru, 11 February 1982.

FFA. *Guidelines for Installation and Registration of Automatic Location Communicators*. 2004.

FFA. *Guide to Application and Renewal of Good Standing on the FFA Vessel Register*. www.ffa.int, Accessed on 15 February 2006.

FFA. *Nauru Agreement Concerning Cooperating in the Management of Fisheries of Common Interest*. Nauru, 11 February 1982.

FFA. *Members' Vessel Monitoring System*. www.ffa.int. Available at 12 January 2006.

FFA. *Procedure for the Operation of the VMS Register of Foreign Fishing Vessels*.

FFA. *Second Arrangement Implementing the Nauru Agreement Setting forth Additional Terms and Conditions of Access to the Fisheries of the Parties*, Koror, 19 September 1990.

General Fisheries Commission for the Mediterranean (GFCM). *Recommendation GFCM/2006/4*. Establishment of a List of Vessels Presumed to Have Carried Out Illegal, Unreported and Unregulated Fishing Activities in the GFCM Area.

GFCM. *Recommendation GFCM/2006/5*. Criteria for Obtaining the Status of Cooperating Non-Contracting Party in GFCM Area.

GFCM. *Recommendation GFCM/2008/1* on a Regional Scheme on Port State Measures to Combat Illegal, Unreported and Unregulated Fishing in the GFCM Area.

Indian Ocean Tuna Commission (IOTC). *Recommendation Concerning the IOTC Bigeye Tuna Statistical Document Programme*.

IOTC. *Resolution 01/02*. Relating to Control of Fishing Activities.

IOTC. *Resolution 01/03*. Establishing a Scheme to Promote Compliance by Non-Contracting Party Vessels with Resolutions Established by IOTC.

IOTC. *Resolution 01/05*. Mandatory Statistical Requirements for IOTC Members. 2001.

IOTC. *Resolution 01/06*. Recommendation by IOTC Concerning the IOTC Bigeye Tuna Statistical Document Programme. 2001.

IOTC. *Resolution 02/04*. Establishing a List of Vessels Presumed to Have Carried Out IUU Fishing in the IOTC Area.

IOTC. *Resolution 02/05*. Concerning the Establishment of an IOTC Record of Vessels Over 24 Meters Authorised to Operate in the IOTC Area.

IOTC. *Recommendation 02/07*. Concerning Measures to Prevent the Laundering of Catches by IUU Large-scale Tuna Longline Fishing Vessels.

IOTC. *Resolution 03/02*. Criteria for Attaining the Status of Cooperating Non-Contracting Party.

IOTC. *Resolution 05/03*. Relating to the Establishment of an IOTC Programme of Inspection in Port.

IOTC. *Resolution 07/01*. To Promote Compliance By Nationals of Contracting Parties and Cooperating Non-contracting Parties with IOTC Conservation and Management Measures.

IOTC. *Resolution 07/02*. Concerning the Establishment of an IOTC Record of Vessels Authorised to Operate in the IOTC Area.

IOTC. *Resolution 07/05*. Limitation of Fishing Capacity of IOTC Contracting Parties and Cooperating Non-Contracting Parties in Terms of Number of Longline Vessels Targeting Swordfish and Albacore.

IOTC. *Resolution 08/02*. Establishing a Programme for Transshipment by Large-scale Fishing Vessels.

IOTC. *Resolution 09/03*. Establishing a List of Vessels Presumed to Have Carried Out Illegal, Unregulated and Unreported Fishing in the IOTC Area.

IOTC. *Resolution 09/04*. Regional Observer Scheme.

IOTC. *Resolution 09/05.* To Prohibit the Use of Large-scale Driftnets on the High Seas in the IOTC Area of Competence.

IOTC. *Resolution 99/02.* Resolution Against Fishing Activities by Large-scale Flag of Convenience Longline Vessels.

Inter-American Tropical Tuna Commission (IATTC). Agreement on the International Dolphin Conservation Program. *Procedures for AIDCP Dolphin Safe Tuna Certification.* Amended 20 October 2005.

IATTC. *Resolution A-04-07* of the Agreement on the International Dolphin Conservation Program (AIDCP). 20 October 2004.

IATTC. *Resolution C-03-01.* Resolution on IATTC Bigeye Tuna Statistical Document Program. 24 June 2003.

IATTC. *Resolution C-03-04.* Resolution on At-Sea Reporting. 27 June 2003.

IATTC. *Resolution C-03-05.* Resolution on Data Provision. 27 June 2003.

IATTC. *Resolution C-03-07.* Resolution on the Establishment of a List of Longline Fishing Vessels Over 24 Meters (LSTLFVs) Authorised to Operate in the Eastern Pacific Ocean. 27 June 2003.

IATTC. *Resolution C-04-02.* Resolution on Criteria for Attaining the Status of Cooperating Non-Party or Fishing Entity in IATTC. Adopted at the 72nd Meeting, Lima, Peru, 14–18 June 2004.

IATTC. *Resolution C-04-03.* Resolution on a System of Notification of Sightings and Identification of Vessels Operating in the Convention Area.

IATTC. *Resolution C-04-04.* Resolution to Establish a List of Vessels Presumed to Have Carried Out Illegal, Unreported and Unregulated Fishing Activities in the Eastern Pacific Ocean. 72nd Meeting. Lima, Peru, 14–18 June 2004.

IATTC. *Resolution C-04-06.* Resolution on the Establishment of a Vessel Monitoring System.

IATTC. *Resolution C-05-07.* Resolution to Establish a List of Vessels Presumed to Have Carried Out Illegal, Unreported and Unregulated Fishing Activities in the Eastern Pacific Ocean.

IATTC. *Resolution C-07-02.* Criteria for Attaining the Status of Cooperating Non-party or Fishing Entity in IATTC.

IATTC. *Resolution C-08-02.* Establishing a Program for Transshipments by Large-Scale Fishing Vessels.

IATTC. *Resolution on the Participation of Taiwan in the Work of the Inter-American Tropical Tuna Commission.* Adopted During the 62nd Meeting of the IATTC, La Jolla, California, October 1998.

International Commission for the Conservation of Atlantic Tunas (ICCAT). Criteria for Attaining the Status of Cooperating Non-Contracting Party, Entity, or Fishing Entity in ICCAT. *03-20 MISC.*

ICCAT. General Outline of Integrated Monitoring Measures Adopted by ICCAT. *02-31 GEN.*

ICCAT. Recommendation Against Belize, Cambodia, Honduras, and St. Vincent and the Grenadines. 2000.

ICCAT. Recommendation by ICCAT Amending the Recommendation by ICCAT to Establish a List of Vessels Presumed to have carried out Illegal, Unreported and Unregulated Fishing Activities in the ICCAT Convention Area. *06-12 GEN.*

ICCAT. Recommendation by ICCAT Concerning the Ban on Landings and Transshipments of Vessels from Non-Contracting Parties Identified as Having Committed a Serious Infringement. *98-11 GEN.* 21 June 1999.

ICCAT. Recommendation by ICCAT Concerning Minimum Standards for the Establishment of a Vessel Monitoring System in the ICCAT Convention Area. *03-14 GEN.* 19 June 2004.

ICCAT. Recommendation by ICCAT Concerning the Continuance of Trade Measures Against Equatorial Guinea. *03-17 SANC.* 19 June 2004.

ICCAT. Recommendation by ICCAT Concerning the Duties of Contracting Parties and Cooperating Non-parties, Entities, or Fishing Entities in Relation to Their Vessels Fishing in the ICCAT Convention Area. *03-12 GEN.* 19 June 2004.

ICCAT. Recommendation by ICCAT Concerning the Establishment of an ICCAT Record Over 24 Meters Authorised to Operate in the Convention Area. *02-22 GEN.* 04 June 2003.

ICCAT. Recommendation by ICCAT Concerning the ICCAT Bigeye Tuna Statistical Document Program. *01-21 SDP.* 21 September 2002.

ICCAT. Recommendation by ICCAT Concerning the ICCAT Bluefin Tuna Statistical Document Program. *92-41 SDP.* 25 July 1993.

ICCAT. Recommendation by ICCAT Concerning the Implementation of the ICCAT Bluefin Tuna Statistical Document Program on Re-export. *97-04 SDP.* 12 December 1997.

ICCAT. Recommendation by ICCAT Concerning the Importation of Atlantic Bluefin Tuna, Atlantic Swordfish, and Atlantic Bigeye Tuna and their Products from Belize. *02-16 SANC*. 04 June 2003.

ICCAT. Recommendation by ICCAT Concerning the Importation of Bigeye Tuna and Its Products from Honduras. *02-18 SANC*. 04 June 2003.

ICCAT. Recommendation by ICCAT Concerning the Importation of Bigeye Tuna and Bigeye Tuna Products from St. Vincent and the Grenadines. *01-14 SANC*. 21 September 2002.

ICCAT. Recommendation by ICCAT Concerning Trade Measures. *06-13 GEN*. 13 June 2007.

ICCAT. Recommendation by ICCAT Concerning the Trade Sanction Against St. Vincent and the Grenadines. *02-20 SANC*. 04 June 2003.

ICCAT. Recommendation by ICCAT Concerning a Multi-year Conservation and Management Plan for Bluefin Tuna in the East Atlantic and Mediterranean. *02-08 BFT*. 04 June 2003.

ICCAT. Recommendation by ICCAT Concerning the Ban on Landing and Transshipments of Vessels from Non-Contracting Parties Identified as Having Committed a Serious Infringement. *98-11 GEN*. 21 June 1999.

ICCAT. Resolution by ICCAT Concerning the Composition of the Delegations of ICCAT Contracting Parties to CITES. *93-09 MISC*.

ICCAT. Recommendation of ICCAT Concerning the Establishment of an ICCAT Record of Vessels.

ICCAT. Recommendation by ICCAT Concerning Unreported Catches of Bluefin Tuna, Including Catches Classified as Not Elsewhere Included (NEI). *97-03 BFT*. 13 June 1998.

ICCAT. Recommendation by ICCAT Establishing a Programme for Transshipment. *06-11 GEN*. 13 June 2007.

ICCAT. Recommendation by ICCAT for a Revised ICCAT Port Inspection Scheme. *97-10 GEN*. 13 June 1998.

ICCAT. Recommendation by ICCAT for Bigeye Tuna Trade Restrictive Measures on Georgia. *03-18 SANC*. 19 June 2004.

ICCAT. Recommendation by ICCAT for Trade Restrictive Measures on Sierra Leone. *02-19 SANC*. 04 June 2003.

ICCAT. Recommendation by ICCAT on Vessel Chartering. *02-21 GEN*. 13 June 2003.

ICCAT. Recommendation by ICCAT on Establishing Statistical Document Programs for Swordfish, Bigeye Tuna, and Other Species Managed by ICCAT. *00-22 SDP*. 26 June 2001.

ICCAT. Recommendation by ICCAT Establishing a Swordfish Statistical Document Programme. *01-22 SDP*. 21 September 2002.

ICCAT. Recommendation by ICCAT on the Establishment of a Closed Area/Season for the Use of Fish Aggregation Devices (FADs). *99-01 TRO*. 15 June 2000.

ICCAT. Recommendation by ICCAT on Transshipments and Vessel Sightings. *97-11 GEN*. 13 June 1998.

ICCAT. Recommendation by ICCAT on the Validation of Bluefin Statistical Documents between ICCAT Contracting Parties Which are Members of the European Community. *96-10 SDP*. 04 August 1997.

ICCAT. Resolution by ICCAT Further Defining the Scope of IUU Fishing. *Resolution 01-18GEN*. 22 March 2002.

ICCAT. Recommendation by ICCAT on a Yellowfin Size Limit. *72-01 YFT*. 01 July 1973.

ICCAT. Recommendation by ICCAT on Transshipments and Vessel Sightings. *97-11 GEN*. 13 June 1998.

ICCAT. Recommendation by ICCAT Relating to the Rebuilding Program for North Atlantic Swordfish. *02-02 SWO*. 04 June 2003.

ICCAT. Recommendation by ICCAT to Establish a Rebuilding Program for Western Atlantic Bluefin Tuna. *98-07BFT*. 21 June 1999.

ICCAT. Recommendation by ICCAT Regarding Belize, Cambodia, Honduras, and St. Vincent and the Grenadines Pursuant to the 1998 Resolution Concerning the Unreported and Unregulated Catches of Tuna by Large-scale Longline Vessels in the Convention Area. *00-15 SANC*. 15 October 2001.

ICCAT. Recommendation by ICCAT Regarding Bolivia Pursuant to the 1998 Resolution Concerning the Unreported and Unregulated Catches of Tuna by Large-scale Longline Vessels in the Convention Area. *02-17 SANC*. 04 June 2003.

ICCAT. Recommendation by ICCAT to Adopt Additional Measures Against Illegal, Unreported and Unregulated (IUU) Fishing. *03-16 GEN.* 19 June 2004.

ICCAT. Recommendation by ICCAT to Establish a List of Vessels Presumed to Have Carried Out Illegal, Unreported and Unregulated Fishing Activities in the ICCAT Convention Area. *02-23 GEN.* 04 June 2003.

ICCAT. Recommendation by ICCAT to Establish a List of Vessels Presumed to have Carried out IUU Fishing Activities in the ICCAT Convention Area. *06-12 GEN.* 13 June 2007.

ICCAT. Recommendation by ICCAT to Establish a Plan to Rebuild Blue Marlin and White Marlin Populations. *00-13 BIL.* 26 June 2001.

ICCAT. Recommendation by ICCAT to Establish Multi-Annual Recovery Plan for Bluefin Tuna in the Eastern Atlantic and Mediterranean. *06-05 BFT.* 13 June 2007.

ICCAT. Recommendation by ICCAT to Promote Compliance by Nationals of Contracting Parties, Cooperating Non-Contracting Parties, Entities or Fishing Entities with ICCAT Conservation and Management Measures. *06-14 GEN.* 13 June 2007.

ICCAT. Recommendation Establishing a Swordfish Statistical Document Program. 2003.

ICCAT. Recommendation on Ban on Landings and Transshipments of Non-Contracting Party Vessels. 1998.

ICCAT. Recommendation on Bigeye Tuna Statistical Document Program. 2001.

ICCAT. Resolution by ICCAT Concerning a Management Standard for the Large-scale Tuna Longline Fishery. *01-20 GEN.* 22 February 2002.

ICCAT. Resolution by ICCAT Concerning Validation by a Government Official of the Bluefin Tuna Statistical Document. *93-02 SDP.* 30 November 1993.

ICCAT. Resolution by ICCAT on Cooperation with the Food and Agriculture Organization of the United Nations (FAO) with Regard to Study on the Status of Stocks and By-catches of Sharks Species. 21 December 1995.

ICCAT. Resolution by ICCAT Concerning a Management Standard for the Large-scale Tuna Longline Fishery. 22 February 2002.

ICCAT. Resolution by ICCAT Concerning Cooperative Actions to Eliminate Illegal, Unreported and Unregulated Fishing Activities by Large-Scale Tuna Longline Vessels. *02-26 GEN.* 04 June 2003.

ICCAT. Resolution by ICCAT Concerning More Effective Measures to Prevent, Deter, and Eliminate IUU Fishing by Tuna Longline Vessels, 01-19 GEN. 22 February 2002. ICCAT. Resolution by ICCAT Concerning the Effective Implementation of the ICCAT Bluefin Tuna Statistical Document Program. *94-05 SDP.* 23 January 1995.

ICCAT. Resolution by ICCAT Concerning the Measures to Prevent the Laundering of Catches by Illegal, Unreported and Unregulated (IUU) Large-scale Tuna Longline Fishing Vessels. *02-25 GEN.* 04 June 2003.

ICCAT. Resolution by ICCAT on Cooperation with the Convention on International Trade in Endangered Species of Wild Fauna and Flora (CITES). *93-08 MISC.*

ICCAT. Resolution by ICCAT on Compliance with the ICCAT Conservation and Management Measures (Including Addendum). *94-09 GEN.* 23 January 1995.

ICCAT. Resolution by ICCAT on Interpretation and Application of the ICCAT Bluefin Tuna Statistical Document Program. *94-04 SDP.* 23 January 1995.

ICCAT. Resolution by ICCAT on Large-scale Pelagic Driftnets. ICCAT. *96-15 GEN.* 03 February 1997.

ICCAT. Resolution by ICCAT Regarding the Agreement to Promote Compliance with International Conservation and Management Measures by Fishing Vessels on the High Seas. *94-08 GEN.* 23 January 1995.

ICCAT. Supplemental Resolution by ICCAT to Enhance the Effectiveness of the ICCAT Measures to Eliminate Illegal, Unregulated and Unreported Fishing Activities by Large-Scale Tuna Longline Vessels in the Convention Area and Other Areas. *00-19 GEN.* 27 December 2000.

International Baltic Sea Fishery Commission (IBSFC). *Fishery Rules of the International Baltic Sea Fishery Commission.* Revised Version Following the 29th Session Amendments. 2003.

International Pacific Halibut Commission (IPFC). *Pacific Halibut Fishery Regulations.* 2005.

IPHC. *Licensing Procedures for Area 2A.* No date.

Joint Ministerial Statement. *Regional Ministerial Meeting on Promoting Responsible Fishing Practices including Combating Illegal, Unreported, Unregulated (IUU) Fishing in the Region.* Bali, Indonesia, 04 May 2007.

Lake Victoria Fisheries Organisation. *Regional Plan of Action to Prevent, Deter, and Eliminate Illegal, Unreported and Unregulated (IUU) Fishing on Lake Victoria and Its Basin*. Bagamoyo, Tanzania, 27 May 2004.

North Atlantic Salmon Conservation Organization. Minimum Standard for Catch Statistics. Adopted by the Council at its Tenth Annual Meeting. Edinburgh, UK, 07–11 June 1993. *CNL(93)51*.

Northeast Atlantic Fisheries Commission (NEAFC). *Recommendation on a Scheme of Control and Enforcement in Respect of Fishing Vessels Fishing in Areas Beyond the Limits of National Fisheries Jurisdiction in the Convention Area*. 11 January 2005.

NEAFC. *Report of the 27th Annual Meeting of the North-east Atlantic Fisheries Commission*. 10–14 November 2008. NEAFC Headquarters, London. Volume I-Report.

NEAFC. *Scheme of Control and Enforcement*. London, February 2009.

NEAFC. *Scheme to Promote Compliance by Non-Contracting Party Vessels with Recommendations Established by NEAFC*.

Northwest Atlantic Fisheries Organisation (NAFO). 'Annex 19, Annual Compliance Review 2008'. In Report of the Fisheries Commission. 22–26 September 2008. *NAFO/FC Doc. 08/20*.

NAFO. *Conservation and Enforcement Measures*. NAFO/FC Doc. 09/1. 2009.

NAFO. *Conservation and Enforcement Measures*. Annex XIX. Rules on Confidentiality. Provisions on Secure and Confidential Treatment of Electronic Reports and Messages Transmitted Pursuant to Articles 26–27 of the Conservation and Enforcement Measures.

North Pacific Anadromous Fish Commission (NPAFC). *Report of the Fifteenth Annual Meeting of the Commission*. Vladivostok, Russia, 8–12 October 2007.

Regional Commission for Fisheries (RECOFI). Fourth Session. Jeddah, Kingdom of Saudi Arabia, 07–09 May 2007. National and Regional Measures to Address Illegal, Unreported and Unregulated Fishing. *RECOFI/IV/2007/4*. May 2007.

Regional Plan of Action (RPOA) to Promote Responsible Fishing Practices including Combating IUU Fishing in the Region. Bali, Indonesia, 04 May 2007.

Southeast Asian Fisheries Development Centre. *Regional Guidelines for Responsible Fishing Operations in Southeast Asia*. 1999.

Southeast Atlantic Fisheries Organisation (SEAFO). *Compliance Committee: IUU Vessels List*. www.seafo.org. Accessed on 12 February 2009.

SEAFO. *Conservation Measure 03/06* on an Interim Prohibition of Transshipments at Sea in the SEAFO Convention Area and to Regulate Transshipments in Port. Approved 04/10/2006.

SEAFO. *Conservation Measure 08/06* Establishing a List of Vessels Presumed to Have Carried Out Illegal, Unreported and Unregulated (IUU) Fishing Activities.

Southern African Development Community (SADC). *Protocol on Fisheries*. 26 August 2002.

SADC. Southern African Development Community. Statement of Commitment by SADC Ministers Responsible for Marine Fisheries on Illegal, Unreported and Unregulated Fishing. Windhoek, Namibia, 04 July 2008.

Southern Indian Ocean Fisheries Agreement. Rome, Italy, concluded on 29 December 2006, ATNIF 31.

Statutes of the South West Indian Ocean Fisheries Commission. *Resolution 1/127*. Adopted at the Hundred and Twenty-seventh Session of the FAO Council. Rome, 22–27 November 2004.

Western Central Atlantic Fishery Commission (WECAFC). Thirteenth Session. WECAFC Lesser Antilles Fisheries Committee. Tenth Session. Cartagena de Indias, Colombia, 21–24 October 2008. Summary Report of the Intersessional Activities and FAO projects in the WECAFC Region. WECAFC/XIII/08/5E. October 2008.

Western and Central Pacific Fisheries Commission (WCPFC). *Arrangement for the Participation of Fishing Entities*. Honolulu, USA, 05 September 2000.

WCPFC. Boarding and Inspection Procedures. *Conservation and Management Measure 2006-08*.

WCPFC. Charter Notification Scheme. *Conservation and Management Measure 2009-08*.

WCPFC. Conservation and Management Measure for Bigeye and Yellowfin Tuna in the Western and Central Pacific Ocean. *Conservation and Management Measure 2008-01*.

WCPFC. Conservation and Management Measure for the Regional Observer Programme. *Conservation and Management Measure 2007-01*.

WCPFC. Conservation and Management Measure on Regulation of Transshipment. *Conservation and Management Measure 2009-06*.

WCPFC. Conservation and Management Measure to Establish a List of Vessels Presumed to Have Carried Out Illegal, Unreported and Unregulated Fishing Activities in the WCPO. *Conservation and Management Measure 2007-03*. 07 December 2007.

WCPFC. Cooperating Non-Members. *Conservation and Management Measure 2009-11*.

WCPFC. Draft Conservation and Management Measure on Regulation of Transshipment (Revision from WCPFC-TCC4-2008/DP-06. Proposal Submitted by Republic of Marshall Islands. *WCPFC5-2008/DP02 Rev. 1*. 05 December 2008.

WCPFC. Draft IUU Vessel List and Current WCPFC IUU Vessel List. Prepared by the Secretariat. *WCPFC-TCC4-2008/11*. 15 September 2008.

WCPFC. Record of Fishing Vessels and Authorisation to Fish. *Conservation and Management Measure 2009-01*. 10 December 2004.

WCPFC. *Summary Report of the Fourth Regular Session*. Tumon, Guam, 3–7 December 2007.

WCPFC. Technical and Compliance Committee. *Conservation and Management Measure 2007-03: Outstanding Issues from WCPFC4*.

WCPFC. Technical and Compliance Committee. *Summary Report*. Fourth Regular Session. Pohnpei, Federated States of Micronesia. 2–7 October 2008.

WCPFC. Technical and Compliance Committee. *Summary Report*. Third Regular Session. 27 September-02 October 2007. Pohnpei, Federated State of Micronesia.

WCPFC. WCPFC Fishing Vessel Charter Arrangements Scheme. *WCPFC-TCC3-2007-23*. 11 September 2007.

8. Regional Organization Reports

Commission for the Conservation of Antarctic Living Marine Resources (CCAMLR). 'Annex 6: Communication Policy with Non-Contracting Parties Relating to Illegal, unreported and Unregulated Fishing in the CCAMLR Convention Area'. Report of the Sixteenth Meeting of the Scientific Committee. Hobart Australia, 27–31 October 1997. *SC-CCAMLR-XVI*. Tasmania, Australia, 1997.

CCAMLR. IUU Vessel List. www.ccamlr.org. Accessed on 15 January 2009.

CCAMLR. Report of the Fifteenth Meeting of the Commission. Hobart, Australia, 21 October-01 November 1996. *CCAMLR-XV*. Tasmania, Australia, 1996.

CCAMLR. Report of the Twenty-third Meeting of the Commission. Hobart, Australia, 25 October-05 November 2004. *CCAMLR-XXIII*. Tasmania, Australia, 2004.

CCAMLR. Report of the Twenty-fifth Meeting of the Commission. Hobart, Australia, 23 October-03 November 2004. *CCAMLR-XXIII*. Tasmania, Australia, 2006.

European Union (EUC). *Consultation on the Elaboration of a new Strategy against IUU fishing by the European Community, Response Document Resulting from a stakeholder consultation meeting*. Brussels, 20 February 2007.

EUC. European Economic and Social Committee and the Committee of the Regions. Communication from the Commission to the European Parliament on the new strategy for the Community to prevent, deter, and eliminate Illegal, Unreported and Unregulated Fishing. Brussels, 17.10.2007, *COM(2007) 601 final*.

EUC. Proposal for a Council Regulation for a Community system to prevent, deter and eliminate illegal, unreported and unregulated fishing (presented by the Commission). Brussels, 17 October 2007, *COM(2007) 602 final, 2007/0223(CNS)*.

Fifth Intergovernmental Consultation on the Southern Indian Ocean Fisheries Agreement. Mombassa, Kenya, 20–23 April 2005.

First Ad Hoc Meeting on Management of Deepwater Fisheries Resources of the Southern Indian Ocean. Swakopmund, Namibia, 30 May-01 June 2001.

Second Ad Hoc Meeting on Management of Deepwater Fisheries Resources of the Southern Indian Ocean. Fremantle, Western Australia, 20–22 May 2002.

Indian Ocean Tuna Commission (IOTC). IOTC IUU Vessels List. www.iotc.org. Accessed on 15 January 2009.

IOTC. *Report of the Seventh Session of the Scientific Committee*. Victoria, Seychelles, 08–12 November 2004.

Inter-American Tropical Tuna Commission (IATTC). IUU Vessels List. 13 January 2009. www.iattc .org. Accessed on 15 January 2009.
International Commission for the Conservation of Atlantic Tunas (ICCAT). List of Vessels Presumed to Have Carried Out IUU Fishing Activities in the ICCAT Convention Area. 11 December 2008. www.iccat.es. Accessed on 15 January 2009.
Lawson, Tim. 'Scientific Aspects of Observer Programmes for Tuna Fisheries in the Western and Central Pacific Ocean'. Paper Prepared for the WCPFC Scientific Committee, Second Regular Session. Manila, Philippines, 07–18 August 2006. *WCPFC-SC2-2006/ST WP-1.*
Moon, Dae-Yeon; Kim, Soon-Song; and Koh, Jeong-Rack. 'A Summary of the Korean Tuna Fishery Observer Programme for the Pacific Ocean (2004–2005)'. Prepared for the 1st Meeting of the WCPFC Scientific Committee Meeting. Noumea, New Caledonia, 08–19 August 2005. *WCPFC-SC1 ST IP-3.*
North Atlantic Salmon Conservation Organization. 'Unreported Catches-Returns by the Parties'. Report of the 21st Annual Meeting of the Council. Reykjavik, Iceland, 07–11 June 2004.
Northeast Atlantic Fisheries Commission. Illegal, Unregulated and Unreported Activity B List. 05 October 2007. www.neafc.org. Accessed on 15 January 2009.
Northwest Atlantic Fisheries Organization (NAFO). 'Annex 19, Annual Compliance Review 2008'. In Report of the Fisheries Commission. 22–26 September 2008.
NAFO. Illegal, Unregulated and Unreported Fishing. www.nafo.int. Accessed on 15 January 2009.
South Pacific Regional Fisheries Management Organisation (SPRFMO). First International Meeting on the Establishment of the South Pacific Regional Fisheries Management Organisation. Wellington, New Zealand, 14–17 February 2006. Information Paper Submitted by the Co-sponsors Australia, Chile, New Zealand, *SP/01/Inf3_rev1*, para. 3.
SPRFMO. 'Interim Measures Adopted by Participants in Negotiations to Establish South Pacific Regional Fisheries Management Organisation'. *Third International Meeting on the Establishment of the South Pacific Regional Fisheries Management Organisation*, Reñaca, Chile, 30 April- 04 May 2007.
Southwest Indian Ocean Fisheries Commission (SWIOFC). Reports of the First Intergovernmental Consultation on the Establishment of a Southwest Indian Ocean Fisheries Commission. Saint-Denis, Reunion, 06–09 February 2001.
SWIOFC. *Fourth Intergovernmental Consultation on the Establishment of a Southwest Indian Ocean Fisheries Commission.* Mahe, Seychelles, 13–16 July 2004.
SWIOFC. *Second Intergovernmental Consultation on the Establishment of a Southwest Indian Ocean Fisheries Commission.* Antananarivo, Madagascar, 25–28 September 2001.
SWIOFC. *Third Intergovernmental Consultation on the Establishment of a Southwest Indian Ocean Fisheries Commission.* Nairobi, Kenya, 27–30 January 2004.
Western and Central Pacific Fisheries Commission (WCPFC). Sixth Regular Session. Papeete, French Polynesia, 7–11 December 2009. *Summary Report.*
WCPFC. Technical and Compliance Committee. Fourth Regular Session. Pohnpei, Federated States of Micronesia, 2–7 October 2008. Conservation and Management Measure 2007-03: Outstanding Issues from WCPFC4. *WCPFC-TCC4-2008/12.* 29 August 2008.

9. Bilateral Agreements

1987 Treaty on Fisheries Between the Governments of Certain Pacific Island States and the Government of the United States of America.
Agreement between Sri Lanka and India on the Maritime Boundary between the Two Countries in the Gulf of Manaar and the Bay of Bengal and Related Matters. New Delhi, 23 March 1976.
Agreement between the Government of the Commonwealth of Australia and the Government of the Republic of Indonesia establishing Certain Seabed Boundaries, concluded on Canberra, Australia, concluded on 18 May 1971, in force 8 November 1973, ATS 1973 No. 31.
Agreement between the Government of the Commonwealth of Australia and the Government of the Republic of Indonesia establishing certain seabed boundaries in the area of the Timor and Arafura Seas, Supplementary to the Agreement of 18 May 1971, Jakarta, Indonesia, concluded on 9 October 1972, in force 8 November 1973, ATS 1973 No. 32.

Agreement between the European Community and the Republic of Kiribati on Fishing within the Kiribati Fishing Zone. *22.5.2003 EN OJ L 126/3.*

Agreement between the European Economic Community and the Government of the Revolutionary People's Republic of Guinea on fishing off the Guinean Coast, and its Current Protocol. 15 September 1980.

Agreement between the European Economic Community and the Government of Mauritius on Fishing in Mauritian Waters and its Protocol of 2004.

Agreement between the European Economic Community and the Republic of Seychelles on Fishing off Seychelles.

Agreement between the Government of Canada and the Government of the French Republic Concerning Their Mutual Fishing Relations Off the Atlantic Coast of Canada (1972).

Agreement between the Government of the Cook Islands and the Taiwan Deep Sea Tuna Boat Owners and Exporters Association Concerning the Licensing of Fishing Vessels of the Association to Fish Within the Exclusive Economic Zone of the Cook Islands. 30 January 1984.

Agreement between the Government of the French Republic and the Government of the Italian Republic on the Delimitation of the Maritime Boundaries in the Area of the Strait of Bonifacio. Paris, 28 November 1986.

Agreement between the Government of the Republic of Senegal and the European Economic Community on Fishing off the Coast of Senegal, and its Protocol of 2002.

Agreement between the Government of Tuvalu and the Deep-Sea Fisheries Association of the Republic of Korea Concerning the Licensing of Fishing Vessels of the Republic of Korea to Fish within the Fishery Limits of Tuvalu. 04 December 1982.

Agreement between the Kingdom of Sweden and the Union of Soviet Socialist Republics on the Principles for Delimitation of the Sea Areas in the Baltic Sea (1988).

Agreement between the United States of America and the Union of Soviet Socialist Republics on the Maritime Boundary (1990).

Agreement for the Implementation of the Provision of the United Nations Convention on the Law of the Sea of 10 December 1982 Relating to the Conservation and Management of Straddling Fish Stocks and Highly Migratory Fish Stocks. New York, 08 September 1995.

Agreement in the Form of an Exchange of Letters concerning the provisional application of the Protocol defining for the period of 1 January 2004 to 31 December 2008 the fishing opportunities and the financial contribution provided for in the Agreement between the European Economic Community and the Republic of Guinea on fishing off the Guinean Coast. *3.4.2004 EN OJ L 99/11.*

Agreement on Cooperative Enforcement of Fisheries Laws between the Government of Australia and the Government of the French Republic in the Maritime Areas Adjacent to the French Southern and Antarctic Territories, Heard Island and the McDonald Islands. Paris, 8 January 2007.

Agreement on Fisheries Between the Government of the Republic of Korea and the Government of the Cook Islands. Auckland, New Zealand, 25 August 1980.

Agreement on Fisheries Between the Government of the Republic of Korea and the Government of the Independent State of Papua New Guinea. Seoul, Korea, 25 January 1992.

Agreement on Fisheries Between the Government of Tuvalu and the Government of Japan, Suva, Tuvalu, 02 June 1986.

Agreement on Historic Waters of Vietnam and Kampuchea. 7 July 1982.

Agreement on the Delimitation of Maritime Boundaries Between Colombia and Haiti (1978).

Arrangement between the Department of Marine Affairs and Fisheries of the Republic of Indonesia and the Department of Agriculture of the Republic of the Philippines on the Utilisation of Part of the Total Allowable Catch in the Indonesian Exclusive Economic Zone, preceded by a Memorandum of Understanding between the two parties. Jakarta, 12 November 2001.

Arrangement between the Department of Marine Affairs and Fisheries of the Republic of Indonesia and the Ministry of Agriculture and Cooperatives of the Kingdom of Thailand on the Utilisation of Part of the Total Allowable Catch in the Indonesian Exclusive Economic Zone, preceded by the Memorandum of Understanding between the Indonesian and Thai Governments on Fisheries Cooperation. Jakarta, 17 January 2002.

Bilateral Arrangement between the Department of Marine Affairs and Fisheries of the Republic of Indonesia and the Ministry of Agriculture of the People's Republic of China on the Utilisation of

Part of the Total Allowable Catch in the Indonesian Exclusive Economic Zone, preceded by a Memorandum of Understanding between the two parties on fisheries cooperation. Beijing, 23 April 2001.

Conditions Governing Fishing Activities by EU Community Vessels Within the Kiribati Fishing Zone, *22.5.2003 EN OJ L 126/7.*

Convention on the Delimitation of Maritime and Submarine Areas between the Republics of Costa Rica and Ecuador (1985).

Exchange of Notes between the Government of the United Kingdom of Great Britain and Northern Ireland and the Government of the French Republic concerning the Activities of Fishermen in the Vicinity of the Channel Islands and the French coast of the Cotentin Peninsula (1992).

Exchange of Notes between the United Republic of Tanzania and Kenya Concerning the Delimitation of the Territorial Waters Boundary between the Two States (1976).

Fisheries Partnership Agreement between the Democratic Republic of São Tomé and Príncipe and the European Community. 01 June 2006.

Fisheries Partnership Agreement between the European Community and the Republic of Cape Verde, and its current Protocol. 01 September 2006.

Fisheries Partnership Agreement between the European Community and the Republic of Côte d'Ivoire. 05 April 2007.

Fisheries Partnership Agreement between the European Community and the Republic of Guinea-Bissau for the period 16 June 2007 to 15 June 2011.

Fisheries Partnership Agreement between the European Community and Islamic Republic of Mauritania and its Protocol of 2006.

Fisheries Partnership Agreement between the European Community and the Republic of Mozambique. 01 January 2007.

Fisheries Partnership Agreement between the European Community and the Republic of the Seychelles. 18 January 2005.

Fisheries Partnership Agreement between the European Community on the one hand, and the Republic of Kiribati, on the other. 16 September 2003.

Fisheries Partnership Agreement between the Gabonese Republic and the European Community, and its current Protocol. 03 December 1998.

Fisheries Partnership Agreement between the Republic of Madagascar and the European Community, 01 January 2007.

Maritime Boundary Agreement Between the United States of America and the Republic of Cuba (1977).

Maritime Delimitation Agreement between the Government of His Most Serene Highness the Prince of Monaco and the Government of the French Republic (1984).

Memorandum of Understanding between the Government of the Republic of Indonesia and the Government of Australia Concerning the Implementation of a Provisional Fisheries Surveillance and Enforcement Arrangement (1981).

Memorandum of Understanding between the Government of the Republic of Indonesia and the Government of Australia Concerning the Implementation of a Provisional Fisheries Surveillance and Enforcement Arrangement (1981).

Memorandum of Understanding between the Government of the Republic of Indonesia and the Government of Australia Concerning the Implementation of a Provisional Fisheries Surveillance and Enforcement Arrangement (1981).

Partnership Agreement between the European Community and Solomon Islands on Fishing off Solomon Islands. *13.4.2006 EN OJ L 105/34.*

Partnership Agreement between the European Community and the Federated States of Micronesia on fishing in the Federated States of Micronesia. 01 January 2005.

Partnership Agreement between the Members of the African, Caribbean and Pacific Group of States of the One Part, and the European Community and its Member States, of the Other Part. Cotonou, Benin, 23 June 2000.

Partnership Agreement in the fisheries sector between the European Community and the Union of the Comoros, and its Protocol. 20 July 1998.

Protocol defining for the period 1 January 2004 to 31 December 2006 the tuna fishing opportunities and the financial contribution provided for in the Agreement between the European Economic

Community and the Democratic Republic of Madagascar on fishing off Madagascar. www.intfish .net. Accessed on 14 February 2006.

Protocol defining for the period of 3 December 2003 to 2 December 2007, the fishing opportunities and the financial compensation provided for by the agreement between the European Economic Community and the government of Mauritius on fishing in the waters of Mauritius. www.intfish .net. Accessed on 14 February 2006.

Protocol setting out the fishing opportunities and financial contribution provided for in the Agreement between the European Community and the Republic of Guinea on fishing off the coast of Guinea for the period from 1 January 2009 to 31 December 2012, 19.6.2009 *EN L 156/40.*

Protocol setting out the fishing opportunities and financial contribution provided for the in the Fisheries Partnership Agreement between the European Community and Republic of Kiribati for the period from 16 September 2006 to 15 September 2012, *7.8.2007 EN OJ L 205/8.*

Protocol defining the fishing possibilities and the financial contribution provided for by the Agreement between the European Economic Community and the Republic of Seychelles on fishing off Seychelles; *22.5.2002 EN OJ L/134/40.*

Protocol establishing the fishing opportunities and the compensation provided for in the Agreement between the European Economic Community and the Government of the Republic of Guinea-Bissau on fishing off the coast of Guinea-Bissau for the period of 16 June 2001 to 15 June 2006. *22.1.2002 EN OJ L 19/35.*

Protocol setting out the fishing opportunities and financial contribution provided for in the Agreement between the European Economic Community and the Islamic Federal Republic of the Comoros on fishing off the Comoros for the period from 1 January 2005 to 31 December 2010.

Protocol setting out the fishing opportunities and financial contribution provided for in the Fisheries Partnership Agreement between the European Community and the Republic of Cape Verde on fishing off the Coast of Cape Verde for the period from 1 September 2006 to 31 August 2011, 30.12.2002 *EN OJ L 414/8.*

Protocol setting out the fishing opportunities and financial contribution provided for the in the Partnership Agreement between the European Community and the Federated States of Micronesia on fishing in the Federated States of Micronesia, *6.6.2006 EN OJ L 151/8.*

Protocol setting out the fishing opportunities and payments provided for in the Partnership Agreement between the European Community and Solomon Islands on fishing off Solomon Islands, 13.4.2006 *EN OJ L 105/39.*

Protocol setting out the fishing opportunities and payments provided for in the Partnership Agreement between the European Community and Solomon Islands on fishing off Solomon Islands. *13.4.2006 EN OJ L 105/40.*

Protocol setting out the fishing opportunities and the financial contribution provided for by the Agreement between the European Community and the Republic of Côte d'Ivoire on fishing off the coast of Côte d'Ivoire for the period from 1 July 2007 to 30 June 2013.

Protocol setting out the fishing opportunities and the financial contribution provided for in the Agreement between the European Economic Community and the Government of the Republic of Senegal on fishing off the coast of Senegal for the period from 1 July 2002 to 30 June 2006. *24.12.2002 EN OJ L 349/46.*

Protocol setting out the fishing possibilities and the financial contribution provided for in the Agreement between the European Community and the Republic of Kiribati on fishing within the Kiribati fishing zone. *22.5.2003 EN OJ L 126/5.*

Protocol setting out, for the period 1 July 2000 to 30 June 2003, the fishing opportunities and financial contribution provided for in the Agreement between the European Economic Community and the Republic of Côte d'Ivoire on fishing off the coast of Côte d'Ivoire, *12.4.2001 EN OJ L 102/3.*

Treaty between Australia and the Independent State of Papua New Guinea Concerning Sovereignty and Maritime Boundaries in the Area between the Two Countries, Including the Area known as Torres Strait, and Related Matters (1978).

Treaty between the Government of Australia and the Government of the French Republic on cooperation in the maritime areas adjacent to the French Southern and Antarctic Territories (TAAF), Heard Island and the McDonald Islands. Canberra, 24 November 2003.

Treaty on Fisheries Between the Governments of Certain Pacific Island States and the Government of the United States of America, Port Moresby, concluded on 2 April 1987, in force 15 June 1988. ATS 1988 No. 42.

Treaty between the Government of the United States of America and the Government of the United Kingdom of Great Britain and Northern Ireland on the Delimitation in the Caribbean of a Maritime Boundary Relating to Puerto Rico/U.S. Virgin Islands and the British Virgin Islands (1993).

Treaty between the Independent State of Papua New Guinea and Solomon Islands Concerning Sovereignty, Maritime and Seabed Boundaries between the Two Countries and Co-operation on Related Matters (1989).

Treaty between the Independent State of Papua New Guinea and Solomon Islands Concerning Sovereignty, Maritime and Seabed Boundaries between the Two Countries and Co-operation on Related Matters (1989).

Treaty Concerning Delimitation of Marine Areas and Maritime Cooperation Between the Republic of Costa Rica and the Republic of Panama (1980).

Treaty on Delimitation of Marine and Submarine Areas and Maritime Cooperation between the Republic of Colombia and the Republic of Costa Rica (1984).

Treaty on Maritime Boundaries between the United States of America and the United Mexican States (1978).

Treaty on the Delimitation of Marine and Submarine Areas and Associated Matters Between the Republic of Panama and the Republic of Colombia (1976).

Treaty to Resolve Pending Boundary Differences and Maintain the Rio Grande and Colorado River as the International Boundary Between the United States of America and the United Mexican States (1970).

10. National Legislation and Policies

Australia. Australian Seafood Industry Council. Code of Conduct for a Responsible Seafood Industry.

Australia. Department of Agriculture, Fisheries and Forestry. *National Plan of Action to Prevent, Deter, and Eliminate Illegal, Unreported and Unregulated Fishing*. July 2005.

Australia. *Fisheries Management Act 1991 (Cth)* as Amended.

Barbados. *Fisheries Act, Cap. 391 Fisheries (Management) Regulations*. 1998.

Benin. *Code des Douanes de la République du Bénin du 21 Novembre 1967.*

Canada. *Canada Marine Act* (1998, c. 10).

Canada. *Coastal Fisheries Protection Regulations*. 20 June 2001.

Canada. *Fisheries Act.*

Canada. Fisheries and Oceans Canada. *Canadian Code of Conduct for Responsible Fisheries Operations*. 1998.

Canada. *National Plan of Action to Prevent, Deter, and Eliminate Illegal, Unreported and Unregulated Fishing*. March 2005.

Chile. Ministry of Economy and Energy, National Fisheries Service, *National Plan of Action to Prevent, Deter and Eliminate Illegal, Unreported and Unregulated Fishing Activities*, 2004.

Côte d'Ivoire. *Code des Douanes Loi n° 64–291 du 1er Août 1964 telle que modifiée par l'ordonnance n° 88-225 du 2 Mars 1988.*

Federated States of Micronesia. *Chapter 3 of Title 18 of the Code of Federated States of Micronesia-Regulation of Foreign Vessels.*

Fiji. *Marine Spaces (Foreign Fishing Vessels) Regulations*. 1985.

Gabon. *Fisheries Regulations of 1995.*

Gabon. *Loi n° 015/2005 portant Code des pêches et de l'aquaculture en République Gabonaise.*

Gambia. *Fisheries Act 2007.*

Ghana. *Fisheries Act 2002* (Act No. 625 of 2002).

Ghana. *National Plan of Action to Prevent, Deter and Eliminate Illegal, Unreported and Unregulated Fishing,* June 2004.

Grenada. *Territorial Waters Act, No. 17 of 1978.*

Guinea. *Arrêté n° 0602/95/MPA/CAB portant réglementation de l'exercice de la pêche industrielle.*

Guinea. *Loi n° L/95/13/CTRN du 15 mai 1995 portant Code de la pêche maritime.*

Iceland. *Act Regarding Fisheries Within the Fisheries Jurisdiction of Iceland, 1976.*

India. Maritime Zones of India (Regulation of Fishing by Foreign Vessels) Rules, 1982.

Indonesia. *Law of the Republic of Indonesia No. 31 of 2004 Concerning Fisheries.*

Kingdom of Tonga. *Plan of Action to Prevent, Deter and Eliminate Illegal, Unreported and Unregulated Fishing,* September 2004.
Malawi. *Fisheries Conservation and Management Act, 1997 (No. 25 of 1997).*
Malaysia. *Fisheries Act 1985 (Act 317).*
Namibia. *Marine Resources Act 2000 (Act 27 of 2000).*
New Zealand. *Fisheries (Commercial Fishing) Regulations 2001*, Reg. 19(2).
New Zealand. *Fisheries (Commercial Fishing) Regulations 2001*, Reg. 17(10).
New Zealand. *Fisheries (Satellite Vessel Monitoring) Amendment Regulations 2001 (Nos. 1 and 2)*, 21 September 2001 and 12 November 2001.
New Zealand. *Fisheries (Satellite Vessel Monitoring) Amendment Regulations 2006*, 28 August 2006.
New Zealand. *Fisheries (Satellite Vessel Monitoring) Amendment Regulations 2007*, 15 October 2007.
New Zealand. *Fisheries (Satellite Vessel Monitoring) Regulations 1993.*
New Zealand. *Fisheries Act 1996 Amendment Act (No. 2) 1999.*
New Zealand. *Fisheries Act 1996* (No. 88) 1996.
New Zealand. Ministry of Fisheries, *New Zealand Plan of Action to Prevent, Deter and Eliminate Illegal, Unregulated and Unreported Fishing*, May 2004.
Nigeria. *Sea Fisheries Decree 1992* (No. 71) 1992.
Papua New Guinea. National Fisheries Authority Website. *Control Surveillance*. www.fisheries.gov .pg. Accessed on 17 August 2006.
Philippines. Department of Agriculture (DA), Fisheries Administrative Order (FAO) 199, *Guidelines on Fish Transshipment*, 04 April 2000.
Philippines. Department of Agriculture, FAO 200, *Guidelines and Procedures in Implementing Section 87 of the Philippine Fisheries Code of 1998*, 21 August 2000.
Philippines. Executive Order No. 236, *Organizing the National Committee on Illegal Entrants, Repealing EO 656 of 1981, to Ensure the Effective Coordination Among Member Agencies in the Investigation and Disposition of Cases Involving the Illegal Entry of Foreign Nationals and Vessels, and Providing Funds for its Operation*, 22 April 1995.
Philippines. Republic Act No. 8550. *The Philippine Fisheries Code of 1998,* 25 February 1998.
Republic of Korea. Ministry of Maritime Affairs and Fisheries, *National Plan of Action to Prevent, Deter and Eliminate Illegal, Unregulated and Unreported Fishing.*
Republic of Palau. *Vessel Monitoring System Regulations*, 12 October 2000.
Republic of Seychelles. *Fisheries Profile and National Plan of Action to Prevent, Deter, and Eliminate Illegal, Unreported and Unregulated Fishing*. 2004.
The Republic of Zambia. Ministry of Agriculture and Cooperatives, Fisheries Department. *National Plan of Action to Prevent, Deter, and Eliminate Illegal, Unreported and Unregulated Fishing.* Lusaka, Zambia, 09 October 2007.
Samoa. *Fisheries (Vessel Monitoring System) Regulations 1999*, 06 May 1999.
Slovenia. *Marine Fisheries Act (ZMR-2).*
Spain. *National Plan of Action to Prevent, Deter, and Eliminate Unreported and Unregulated Fishing*. November 2002.
Tanzania. *Territorial Sea and EEZ Act*, 1989.
Turkey. *Notification 1/1 Regulating Commercial Fishing (2007-43).*
United Kingdom, *The UK Action Plan on Implementing the Recommendations of the High Seas Task Force on Illegal, Unregulated and Unreported (IUU) Fishing*, May 2006.
United States of America (U.S.). 19 C.F.R. §101.4, Code of Federal Regulations Currentness, Title 19, *Customs Duties*, Chapter I, Bureau of Customs and Border Protection, Department of Homeland Security; Department of the Treasury (Refs & Annos), Part 101, General Provisions (Refs & Annos), §101.4.
U.S. Department of Commerce. *Implementation of Title IV of the Magnuson-Stevens Fishery Conservation and Management Reauthorisation Act of 2006: Report to Congress Pursuant to Section 403(a) of the Magnuson-Stevens Fishery Conservation and Management Reauthorisation Act of 2006*. January 2009.
U.S. Department of Commerce. National Oceanic and Atmospheric Administration. 'Identification of Nations Whose Fishing Vessels are Engaged in Illegal, Unreported, or Unregulated Fishing

and/or By-catch of Protected Living Marine Resources'. *Federal Register* 73: 56 (21 March 2008).

U.S. Department of Commerce. National Oceanic and Atmospheric Administration. 'Magnuson-Stevens Fishery Conservation and Management Reauthorisation Act; proposed Rule to Implement Identification and Certification Procedures to Address Illegal, Unreported, and Unregulated Fishing Activities and the By-catch of Protected Living Marine Resources (PLMRs)'. *Federal Register* 74: 9 (14 January 2009).

U.S. Department of Commerce. National Oceanic and Atmospheric Administration. National Marine Fisheries Service. Office of International Affairs. *Draft Environmental Assessment, Regulatory Impact Review, and Regulatory Flexibility Act Analysis for a Proposed Rule to Establish Identification and Certification Procedures for Nations Under the High Seas Driftnet Fishing Moratorium Protection Act.* January 2009.

U.S. *Food and Drug* Administration *Act.* Title 3, Section 306, Maintenance and Inspection of Records for Food. 9 December 2004.

U.S. *High Seas Driftnet Fishing Moratorium Protection Act 1995* (Title VI). 16 U.S.C. §§ 1826.

U.S. *Lacey Act* 16 U.S.C. §§ 3371-3378 (2002).

U.S. *Magnuson-Stevens Fishery Conservation and Management Act,* An Act to provide for the conservation and management of the fisheries and for other purposes. Public Law 94–265. As amended by the Magnuson-Stevens Fishery Conservation and Management Reauthorisation Act (Public Law 109–479).

U.S. *Marine Mammal Protection Act 1972, As Amended* 16 U.S.C. §§ 1361-1421h.

U.S. *National Plan of Action of the United States of America to Prevent, Deter and Eliminate Illegal, Unreported and Unregulated Fishing.* 20 February 2003.

U.S. *Fishermen's Protective Act of 1967.* 22 U.S.C.A (1964, suppl. 1973), as amended by Pub. L. 95-541, 42 Stat. 2057 (1978).

U.S. *Public Health Security and Bioterrorism Preparedness Response Act of 2002.*

11. Conference, Meeting, and Workshop Papers

Cacaud, Philippe. 'Legal Issues Relating to Vessel Monitoring Systems'. Technical Papers Presented at the Regional Workshop on Fisheries Monitoring, Control, and Surveillance. Kuala Lumpur and Kuala Terengganu, Malaysia, 29 June-3 July 1998. *GCP/INT/648/NOR—Field Report C-1/ Supp.2,* (Rome: FAO, 2000) 1–31.

Karp, William A. and McElderry, Howard. 'Catch Monitoring by Fisheries Observers in the United States and Canada'. In Nolan, C.P. (ed.) *Proceedings of the International Conference on Integrated Fisheries Monitoring. Sydney,* Australia, 01–05 February 1999 (Rome: FAO, 1999) 261–284.

Laubstein, Karl. 'Fisheries Observers on Fishing Vessels'. In Nolan, C.P. (ed.) *Proceedings of the International Conference on Integrated Fisheries Monitoring. Sydney,* Australia, 01–05 February 1999 (Rome: FAO, 1999) 161–173.

Marshall, Philip. 'Electronic Monitoring'. In Nolan, C.P. (ed.) *Proceedings of the International Conference on Integrated Fisheries Monitoring. Sydney,* Australia, 01–05 February 1999 (Rome: FAO, 1999) 303–315.

Mees, C. 'Points to Remember in Connection with Armed Boardings'. Technical Papers Presented at the Regional Workshop on Fisheries Monitoring, Control, and Surveillance. Kuala Lumpur and Kuala Terengganu, Malaysia, 29 June-3 July 1998. *GCP/INT/648/NOR1.* (Rome: FAO, 2000) 141–150.

National Marine Fisheries Service. Proceedings of the Third International Fisheries Observer Conference. New Orleans, Louisiana, USA, 18–21 November 2002. *U.S. Department of Commerce NOAA Technical Memorandum NMFS-SPO-64.* (Maryland: NMFS, 2004).

Navigs, S.A.R.L. 'Fishing Vessel Monitoring Systems: Past, Present and Future'. *Prepared for the High Seas Task Force* (Paris: OECD, 2005). www.high-seas.org. Accessed on 12 February 2006.

Nolan, C.P. (ed.) *Proceedings of the International Conference on Integrated Fisheries Monitoring. Sydney,* Australia, 01–05 February 1999 (Rome: FAO, 1999).

Meere, F.; Green, G. Hanich, Q.; Lack, M.; Sant, G. and Tsamenyi, M. 'CITES 'Introduction from the Sea': A Practical Way Forward'. Proceedings of a Joint TRAFFIC/ANCORS Workshop. 21 TRAFFIC Bulletin (2008).

Schmidt, Carl-Christian. 'Addressing Illegal, Unreported and Unregulated (IUU) Fishing'. *Paper Prepared for the International Fisheries Compliance Conference*. Brussels, Belgium, 29–30 September 2004.

Tsamenyi, Martin and Mfodwo, Kwame. 'Integrated Fisheries Monitoring: The Legal Framework'. In Nolan, CP (ed.) *Proceedings of the International Conference on Integrated Fisheries Monitoring*. Sydney, Australia, 01–05 February 1999 (Rome: FAO, 1999) 115–127.

12. Other References

Agencia de Informacao de Mocambique (Maputo). Mozambique: Country Seizes Namibian Pirate Fishing Ship "Antilles Reefer". 18 July 2008. http://allafrica.com/stories/200807180938.html. Accessed on 10 September 2008.

Aliño, Porfirio. 'Fisheries Resources of the Philippines'. Presentation during the Australian Consultation with the Philippines and Indonesia on the Identification of Researchable Options for the Development of Policy and Management Frameworks to Combat Illegal, unreported and Unregulated (IUU) Fishing Activities in Indonesian and Philippine Waters. In Centre for Maritime Policy and Oceans and Coastal Research Centre, Final Report on ACIAR Project No FIS/2000/163. (Wollongong, Australia, March 2002).

Anderson, Katherine M. and McCusker, Rob. 'Crime in the Australian Fishing Industry: Key Issues'. *Trends and Issues in Crime and Criminal Justice No. 227* (Canberra: Institute of Criminology, 2005).

Australian Fisheries Management Authority. *The Observer Program*. www.afma.gov.au. Accessed on 9 June 2008.

Australian Government. Attorney General's Department. *Customs and Fisheries—Armed Patrols to Protect Australia's Southern Ocean Interests*. 11 May 2004. www.ag.gov.au. Accessed on 01 June 2008.

Black, Richard. 'Action Urged on Illegal Fishing'. *BBC News*. www.bbc.co.uk. Accessed on 28 September 2007.

Benitah, Marc. 'Ongoing WTO Negotiations on Fisheries Subsidies'. *ASIL Insights*. The American Society of International Law Website, June 2004, www.asil.org. Accessed on 24 April 2008.

Ben-Yami, M. 'Column, How to Attain Decent Conditions of Work?' *World Fishing*. April 2008. www.benyami.org. Accessed on 10 September 2008.

Black, Richard. 'Action Urged on Illegal Fishing'. BBC News. www.bbc.co.uk. Accessed on 28 September 2007.

Bours, Hélène, Gianni, Matthew and Mather, Desley. 'Pirate Fishing Plundering the Oceans'. Greenpeace International Campaign Against Pirate Fishing. February 2001. www.greenpeace.org. Accessed on 31 March 2005.

'Canadian and U.S. Authorities break up seafood smuggling ring'. *CBC News*. 26 September 2007. www.cbc.ca. Accessed on 02 October 2007.

Centre for Maritime Policy and Oceans and Coastal Research Centre. The Identification of Researchable Options for the Development of Policy and Management Frameworks to Combat Illegal, Unreported and Unregulated (IUU) Fishing Activities in Indonesian and Philippine Waters. Final Report on ACIAR Project No. FIS/2000/163. (Wollongong, Australia, March 2002).

Churchill, Robin. 'The Meaning of the "Genuine Link" Requirement in Relation to the Nationality of Ships'. *A Study Prepared for the International Transport Workers' Federation*. October 2000. www.oceanlaw.net. Accessed on 15 January 2004.

Coffen-Smout, Scott. *Pirates, Warlords and Rogue Fishing Vessels in Somalia's Unruly Seas*. http://www.chebucto.ns.ca/~ar120/somalia.html. Accessed on 17 February 2006.

Collecte Localisation Satellites. 'Korea: A Distant Water Fishing Nation'. *ArgoNet Newsletter No. 7*, Winter 2002. www.cls.fr. Accessed on 11 April 2004.

Convention on International Trade in Endangered Species of Wild Fauna and Flora. *The CITES Species*. www.cites.org. Accessed on 22 April 2008.

'Crack down on foreign poachers, WWF urges government'. 13 September 2007. www.illegal-fishing.info. Accessed on 03 October 2007.

European Commission. Directorate General Fisheries. *European Distant Water Fishing Fleet*, April 2001. www.europa.eu.int. Accessed on 11 April 2004.

Earthtrust.org Website. *North Pacific Joint Observer Program 1989–1991* http://earthtrust.org/dnpaper/npac.html. Accessed on 22 April 2008.

Environmental Justice Foundation (EJF). *Illegal Driftnetting in the Mediterranean* (London: Environmental Justice Foundation, 2007).

EJF. *Las Palmas: Port of Convenience and Gateway to Europe*. www.ejf.org. Accessed on 20 November 2008.

EJF. *Pirates and Profiteers: How Pirate Fishing Fleets are Robbing People and Oceans*. London: UK, 2005. www.ejfoundation.org, Accessed on 15 February 2006.

EJF. *What's the Catch? Reducing By-catch in EU Distant Water Fisheries* (EJF: London, 2005).

European Commission. Community Research. Quality of Life and Management of Living Resources. *Traceability of Fish Products*. www.tracefish.org, Accessed on 12 February 2006.

European Commission. Community Research, Quality of Life and Management of Living Resources. *TraceCore-XML Standard Guidelines*. www.tracefish.org, Accessed on 12 February 2006.

'Fish Exports Dropped by 10% in 2006 due to Illegal Fishing'. *Eurofish Magazine*. April 2007. www.globefish.org, Accessed on 02 October 2007.

Food and Agriculture Organization (FAO) Website. *Convention on International Trade in Endangered Species of Wild Fauna and Flora (CITES) Implications to Fisheries*, www.fao.org. Accessed on 11 July 2008.

FAO. *Fact Sheet: The International Fish Trade and World Fisheries*. June 2008. www.fao.org. Accessed on 20 June 2008.

FAO Website. *Impacts of Globalisation*. www.fao.org. Accessed on 02 February 2009.

FAO Website. *IPOA-IUU National Plans of Action*. www.fao.org. Accessed on 04 June 2006.

FAO Website. *High Seas Vessels Authorization Record*. www.fao.org. Accessed on 13 May 2008.

FAO Website. *Regional Fishery Bodies and Arrangements*. www.fao.org. Accessed on 20 January 2009.

FAO Website. *Regional Fisheries Bodies Home Page*. www.fao.org. Accessed on 15 January 2009.

FAO Website. *World Trade Organization Agreements*. www.fao.org. Accessed on 11 July 2008.

Geddes, N.J.; Hollick, P.; and Trafford, K. 'Existing and Emerging Sensing Techniques for Fisheries Monitoring, Control and Surveillance'. *Prepared for the High Seas Task Force* (Paris: OECD, December 2005). www.high-seas.org. Accessed on 12 February 2006.

Gianni, Matthew. *High Seas Bottom Fisheries and Their Impact on the Biodiversity of Vulnerable Deep-Sea Ecosystems: Summary Findings*. IUCN, 2004, www.iucn.org. Accessed on 15 February 2006.

Greenpeace International. *Murky Waters: Hauling in the Net on Europe's High Seas Bottom Trawling Feet*. February 2006.

Greenpeace International. *Pirate Fishing-Stolen Fish, Stolen Futures*. www.greenpeace.org. Accessed on 14 May 2006.

Greenpeace International. *Tuna Melt-Down*. www.oceans.greenpeace.org. Accessed on 15 May 2006.

Halpern, Shanna L. *The United Nations Conference on Environment and Development: Process and Documentation*. Providence. http://www.ciesin.org/docs/008-585/unced-home.html. Accessed on 08 January 2008.

Hassan, M.G. 'IUU Fishing and Insecurity Impacts on Somali Fisheries and Marine Resources'. *Presented at the 4th Chatham House IUU Fishing Update and Stakeholder Consultation Meeting*. Chatham House, 31 March-01 April 2008. www.illegal-fishing.info. Accessed on 12 February 2009.

Hayman, Gavin and Brack, Duncan. *International Environmental Crime: The Nature and Control of Environmental Black Markets*. (London: Royal Institute of International Affairs, 2002).

High Seas Task Force. *Closing the Net: Stopping Illegal Fishing on the High Seas*, Governments of Australia, Canada, Chile, Namibia, New Zealand, and the United Kingdom, WWF, IUCN and the Earth Institute at Columbia University (London: DEFRA, 2006).

High Seas Task Force. 'High Seas Governance'. *Presented at the Meeting of the High Seas Task Force*. Paris, France, 09 March 2005. HSTF/09. February 2005. www.high-seas.org. Accessed on 12 January 2006.

High Seas Task Force. *Promoting Responsible Ports*. HSTF/06. (Paris: OECD, 2005). www.high-seas.org. Accessed on 12 February 2006.

'Illegal fishing off Somalia's coast has reached proportions of about USD 300 million annually'. 01 September 2005. www.mg.co.za. Accessed on 02 October 2007.

Institute for Security Studies Website. 10 April 2008: Corruption in Africa's Marine Fisheries. www.issafrica.org. Accessed on 11 April 2008.

'Interior ministry reveals huge seafood contraband on Sakhalin'. 03 November 2006, www.illegal-fishing.info. Accessed on 02 October 2007.

International Network for the Cooperation and Coordination of Fisheries Related Monitoring, Control and Surveillance Activities Website. www.imcsnet.org. Accessed on 17 February 2008.

Marine Resources Assessment Group Ltd (MRAG). *IUU Fishing on the High Seas: Impacts on Ecosystems and Future Science Needs Final Report*. London, August 2005. www.high-seas.org. Accessed on 10 February 2006.

MRAG. *Review of Impacts of Illegal, Unreported and Unregulated Fishing on Developing Countries: Final Report*. London: UK, July 2005. www.high-seas.org. Accessed on 10 February 2006.

MRAG and Fisheries Ecosystems Restoration Research, Fisheries Centre, University of British Columbia. *The Global Extent of IUU Fishing, Final Report*. www.illegal-fishing.info/uploads/MRAGExtentGlobalIllegalFishing.pdf. Accessed on 13 June 2008.

Marine Stewardship Council (MSC). *International Eco-labelling in Fisheries*. April 2004.

MSC Website. *Certified Fisheries*. www.msc.org. Accessed on 03 September 2008.

MSC Website. *Developing world fisheries embark on journey to MSC eco-label*. www.msc.org. Accessed on 17 March 2008.

McDiarmid, Bunny; Gotje, Martini; and Sack, Karen. *The Northwest Atlantic Fisheries Organisation: A Case Study in How RFMOs Regularly Fail to Manage Our Oceans*. 22 July 2005. www.greenpeace.org. Accessed on 15 March 2006.

Miho, David. "Captain Arrested Again for Shark Fins Aboard Boat." *Pacific Magazine*. 02 October 2007. www.pacificmagazine.net. Accessed on 04 October 2007.

Official Website of the Ramsar Convention. *What is the Ramsar Convention on Wetlands?* www.ramsar.org. Accessed on 25 April 2008.

Organisation for Economic Co-operation and Development (OECD). Maritime Transport Committee. *Ownership and Control of Ships*. March 2003. www.oecd.org. Accessed on 02 May 2008.

OECD. Steering Group on Corporate Governance. *Options for Obtaining Beneficial Ownership and Control Information*, September 2002. www.oecd.org. Accessed on 02 May 2008.

Pacnews. 'Taiwan tuna boat pays $500,000 pan'ga penalty for illegal fishing'. 18 December 2008. www.islandbusiness.com. Accessed on 01 February 2009.

Petersen, Arni and Green, David. *Seafood Traceability: A Practical Guide for the U.S. Industry*, Report prepared for the North Carolina Sea Grant, no date. www.ncseagrant.org. Accessed on 23 August 2008.

Putt, Judy and Anderson, Katherine. 'A National Study of Crime in the Australian Fishing Industry'. *Research and Public Policy Series No. 76* (Canberra: Australian Institute of Criminology, 2007).

Reuters Foundation. Somalia: Fishermen Appeal for Help Over Foreign Fishing Ships. www.alertnet.org. Accessed on 04 March 2006.

Royal Institute of International Affairs. Sustainable Development Programme, *International Environmental Crime: The Nature and Control of Environmental Black Markets* Workshop Report. 2002. www.chathamhouse.org.uk. Accessed on 13 August 2009.

Senator The Hon. Ian Macdonald. 'Record Arrests for Northern Illegal Fishing'. 03 January 2006. Ministry for Fisheries, Forestry, and Conservation Website. http://www.mffc.gov.au/releases/2006/06001m.html. Accessed on 16 February 2006.

Statement by the Coalition for Fair Fisheries Arrangements (CFFA) and Greenpeace. The 17th Session of the Fishery Committee for the East Central Atlantic (CECAF). Dakar, Senegal, 24–27 May 2000. www.globefish.org. Accessed on 14 February 2006.

Stølsvik, Cand. jur. Gunnar. 'Transnational Organised Fisheries Crime as a Maritime Security Issue'. Panel Presentation at the Ninth Meeting of the UN Open-ended Informal Consultative Process on Oceans and the Law of the Sea, 23–27 June 2008.

Stop Illegal Fishing Programme. Stop Illegal Fishing in Southern Africa. May 2008. www.illegal-fishing.info. Accessed on 10 February 2009.

Tailby, Rebecca and Gant, Frances. 'The Illegal Market of Australian Abalone'. *Trends and Issues in Crime and Criminal Justice No. 225* (Canberra: Institute of Criminology, 2002).

TRAFFIC News. 'Southern African States Move to Eradicate "Pirate" Fishing'. http://www.traffic.org/home/2008/7/11/southern-african-states-move-to-eradicate-pirate-fishing.html. 10 September 2008.

Treves, Tullio. 'Flags of Convenience Before the Law of the Tribunal'. *Paper Presented at a Law of the Sea Institute Conference entitled Multilateralism and International Ocean-Resources Law.* Boalt Hall School of Law, University of California, Berkeley, 21–22 February 2003. www.law.berkeley.edu. Accessed on 10 August 2004.

Unilever. *Fishing for the Future II: Unilever's Fish Sustainability Initiative (FSI).* FF/RLD/SP/5000/1003. www.unilever.com. Accessed on 17 March 2008.

United States. National Oceanic and Atmospheric Administration Fisheries. Office of Science and Technology. *National Observer Programme*, www.st.nmfs.gov. Accessed on 17 May 2008.

World Customs Organization (WCO). *HS Multi-Purposes Tool.* www.wco.org. Accessed on 16 January 2009.

WCO. *Position of Contracting Parties to the Harmonized System Convention and Non-Contracting Party Administrations.* 15 January 2009. www.wco.org. Accessed on 20 January 2009.

WCO. *Overview: What is the Harmonized System?* Accessed on www.wco.org. Accessed on 16 January 2009.

World Organisation for Animal Health Website. www.oie.int. Accessed on 13 July 2008.

Index

Legal Aspects of Sustainable Development

General Editor: David Freestone

1. K. Bastmeijer and T. Koivurova (eds), *Theory and Practice of Transboundary Environmental Impact Assessment*. 2008
   ISBN 978 90 04 16479 6

2. C. Voigt, *Sustainable Development as a Principle of International Law: Resolving Conflicts between Climate Measures and WTO Law*. 2008
   ISBN 978 90 04 16697 4

3. R. Warner, *Protecting the Oceans Beyond National Jurisdiction*. Strengthening the International Law Framework. 2009
   ISBN 978 90 04 17262 3

4. S.M.A. Salman, *The World Bank Policy for Projects on International Waterways*. An Historical and Legal Analysis. 2009
   ISBN 978 90 04 17837 3

5. T.G. Puthucherril, *From Shipbreaking to Sustainable Ship Recycling*. Evolution of a Legal Regime. 2010 ISBN 978 90 04 17491 7

6. M.A. Palma, M. Tsamenyi and W. Edeson, *Promoting Sustainable Fisheries*. The International Legal and Policy Framework to Combat Illegal, Unreported and Unregulated Fishing. 2010
   ISBN 978 90 04 17575 4

www.ingramcontent.com/pod-product-compliance
Lightning Source LLC
LaVergne TN
LVHW011109300926
849160LV00008B/320